Origins of the Human Brain

FONDATION FYSSEN

SYMPOSIA OF THE FYSSEN FOUNDATION

Social Relationships and Cognitive Development
*Edited by Robert A. Hinde, Anne-Nelly Perret-Clermont,
and Joan Stevenson-Hinde*

Thought Without Language
Edited by L. Weiskrantz

The Use of Tools by Human and Non-human Primates
Edited by Arlette Berthelet and Jean Chavaillon

Origins of the Human Brain
Edited by Jean-Pierre Changeux and Jean Chavaillon

Origins of the Human Brain

Edited by

Jean-Pierre Changeux and Jean Chavaillon

A Fyssen Foundation Symposium

CLARENDON PRESS · OXFORD
1996

Oxford University Press, Walton Street, Oxford OX2 6DP
Oxford   New York
Athens   Auckland   Bangkok   Bombay
Calcutta   Cape Town   Dar es Salaam   Delhi
Florence   Hong Kong   Istanbul   Karachi
Kuala Lumpur   Madras   Madrid   Melbourne
Mexico City   Nairobi   Paris   Singapore
Taipei   Tokyo   Toronto
and associated companies in
Berlin   Ibadan

Oxford is a trade mark of Oxford University Press

Published in the United States
by Oxford University Press Inc., New York

First published 1995
Reprinted 1995
First published in paperback 1996

A catalogue record for this book is available from the British Library

Library of Congress Cataloging in Publication Data
Origins of the human brain / edited by Jean-Pierre Changeux and Jean
Chavaillon.
(Symposia of the Fyssen Foundation.)
Includes bibliographical references and indexes.
1. Brain–Evolution–Congresses.   I. Changeux, Jean-Pierre.
II. Chavaillon, Jean.   III. Series: Fyssen Foundation symposium.
[DNLM: 1. Brain–anatomy & histology–congresses.   2. Cultural
Evolution–congresses.   3. Evolution–congresses.   4. Intelligence–
congresses.   5. Genetics, Population–congresses.   WL 300 069
1995]
QP376.075   1995   612.8'2–dc20   94–38955
ISBN 0 19 852390 4

Printed in Great Britain by
Bookcraft (Bath) Ltd, Midsomer Norton, Avon

Foreword

The resulting work of the fifth interdisciplinary Fyssen symposium is the basis of this publication. This symposium took place between the 14 and 17 December 1990 in the Pavillon Henri IV at St Germain en Laye.

It was officially opened with the speech to the participants by Mme Fyssen, the President of the Foundation. She expressed the wish that 'the issues dealt with in these discussions should be considered at a major international level, thus focusing on the scientific activity of the Foundation and the success of the work it fosters'.

Professors Jean-Pierre Changeux and Jean Chavaillon chaired the symposium. They also actively participated in its organization, ably assisted by the Fyssen Foundation secretarial department.

There is an obvious direct link between the central theme of the symposium, 'The Origins of the Human Brain: Palaeontogy, Molecular Biology and Developmental Genetics' and the aims of the Foundation itself. These aims are to 'encourage all forms of scientific enquiry into cognitive mechanisms that underly animal and human behaviour and into their ontogenetic and phylogenetic development' and are particularly concerned with 'animal reasoning ability, and on a wider level, animal and human cognitive processes, and their biological and cultural bases'.

Summaries of the papers were sent to all symposium participants beforehand. The papers were presented both in English and in French. Simultaneous translation was carried out thanks to the competence, particularly in the more problematic fields, of Mme Candelier, Mme Sartin, and M Pinhas whose contributions were invaluable.

Subsequent to the symposium, the participants revised all texts so that points put forward during the discussions could be taken into account.

All the debates were recorded. The difficult task of preparing a condensed overview was undertaken by Catherine Vidal and Laurent Cohen for the section on neurobiology and by Pascal Picq for the section on palaeontology.

Julie Pattinson carried out the French–English translation of the papers and their successive updated versions, with unfailing patience.

The paper by Professor Roger Saban was translated by Anthony Saul.

Lastly, we should like to acknowledge our grateful thanks to Mme Fyssen for her constant attendance at meetings, her active and gracious participation in the symposium itself, and her continual encouragement of our work.

Contents

Part III: Culture

Part IV: Intelligence

Participants

Robert Boyd

Department of Anthropology, University of California, 405 Hilgard Avenue, Los Angeles, CA 900-1553, USA

Rebecca L. Cann

Department of Genetics and Molecular Biology, J. A. Burns School of Medicine. University of Hawaii at Manoa, 1960 West Road, Honolulu, Hawaii, 96822, USA

Jean-Pierre Changeux (co-editor)

Laboratoire de Neurobiologie Moléculaire, Institut Pasteur, 25/28 rue du Docteur Roux, 75724 Paris Cedex 15, France

Jean Chavaillon (co-editor)

Laboratoire de Recherche sur l'Afrique Orientale, UPR 311-CNRS, 1 Place Aristide Briand, 92195 Meudon, France

Yves Coppens

Chaire de Paléoanthropologie et Préhistoire, Collège de France, Laboratoire d'Anthropologie Biologique, Musée de l'Homme, 17 Place du Trocadéro, 75116 Paris, France

Henri Delporte

Le Perron, Lezigneux, 42600 Montbrisson, France

Robert A. Hinde

St John's College, Cambridge CB2 1TP, UK

Ralph L. Holloway

Department of Anthropology, Columbia University, New York, NY 10027, USA

Bernardo A. Huberman

Dynamics of Computation Group, Xerox Palo Alto Research Center, 3333 Coyote Hill Road, Palo Alto, CA 94304, USA

Claudia Kappen

Department of Biochemistry and Molecular Biology, Mayo Clinic Scottsdale, 13400 East Shea Boulevard, Scottsdale, A285259, USA

Janusz K. Kozłowski

Instytut Archeology, Jagiellonski Uniwersytet, ul. Golebia 11, 31-007 W. Krakow, Poland

Jean-Louis Mandel

Laboratoire de Génétique Moléculaire des Eucaryotes, Institut de Chimie Biologique, Faculté de Médecine, 11 rue Human, 67085 Strasbourg Cedex, France

Steven Pinker

Department of Brain and Cognitive Sciences, Massachusetts Institute of Technology, Cambridge, MA 02139, USA

David Premack

Laboratoire de Psycho-Biologie du Développment–CNRS–EPHE, 41 rue Gay-Lussac, 7005 Paris, France

Pasko Rakic

Section of Anatomy, Yale University School of Medicine, P.O. Box 3333, New Haven, CT 06510-8001, USA

André Roch Lecours

Laboratoire Théophile-Alajouanine, Centre de recherche du Centre hospitalier Côte-des-Neiges, Université de Montréal, 4565 Chemin de la Reine Marie, Montréal, Québec H3W 1W5, Canada

Peter J. Richerson

Centre for Population Biology, University of California, Davis, CA 95616, USA

Frank H. Ruddle

Department of Biology and Department of Human Genetics, Kline Biology Tower, Yale University, P.O. Box 6666, New Haven, CT 06511-8112, USA

Roger Saban

111 rue de Cambronne, 75015 Paris, France

Phillip V. Tobias

Department of Anatomy and Human Biology, Palaeo-Anthropology Research Unit, University of the Witwatersrand, 7 York Road, Parktown, 2193 Johannesburg, South Africa

Bernard Vandermeersch

Laboratoire d'Anthropologie, Université de Bordeaux, Avenue des Facultés, 33405 Talence Cedex, France

Lawrence Weiskrantz

Department of Experimental Psychology, University of Oxford, South Parks Road, Oxford OX1 3UD, UK

Introduction

Throughout his history, from the beginning of time, man has been concerned with his origins and, more exactly, with the origins of the underlying thinking process essential to him. Only a gathering of scientists from varied fields of research could provide the starting-point for a debate based on the most recent scientific data on the evolutive origins of the human brain. This issue linked up directly with the objectives of the Fyssen Foundation and was the decisive factor for its scientific committee, as from April 1989, in setting up its fifth colloquium on that theme.

The Foundation gave a privileged place to discussion between research scientists focusing on the past, the palaeontologists and archaeologists and those focusing on the present and even the future, the molecular biologists, neurobiologists, and geneticians of populations. They were concerned with a common theme: the human brain's evolution, its mechanism, and its behavioural and social functions. The obligatory disciplines of human palaeontology and cultural evolution were deliberately backed up not only by the most modern theories on population genetics and molecular genetics, but also by the extraordinary leap forward in research in the fields of neurobiology of the learning process and of the higher functions of the brain. All this constitutes an extremely complex and difficult problem.

Within a few million years the anatomical organization of the central nervous system and its associated functions has evolved at a spectacular rate. The importance of the conscious realm of the brain has become predominant in behaviour control. Furthermore, memory capacity has increased and diverse cultures have formed and developed. These developments in structural complexity and brain performance only correspond to slight genomic changes. A genetic evolution dictates the development of the overall 'envelope' which provides access to its functions.

First of all, what particular process works on the genes in question, undoubtedly few in number, which nevertheless causes a major impact on central nervous system development? The most plausible hypothesis is that these genes have an effect on the course of development itself. The wide extent of the effect produced on the morphogenesis would be directly linked to the fact that they come to bear on the process at an early stage. This would explain the progressive parcellation of the brain cortex into areas specialized for particular operations. It would also explain the differential expansion of some of these zones, such as the pre-frontal cortex, referred to by Alexander Luria as the 'organ of civilization'.

The issue remains puzzling over the conditions in which this type of biological evolution could have occurred in the pre-human and then the human populations. To what extent did behaviour and modes of social life have a retroactive effect on genetic evolution? What consequences did language appearance and its development have on the brain's morphogenesis? This is open to discussion. However, we may not limit ourselves to genetic determinants since one of the most striking characteristics in the development of the human central nervous system is the 'awareness' of the physical and cultural environment. The latter leaves within the extremely complex network of developing synaptic links, an 'imprint' which will mark the individual throughout their life and will ensure its permanence of evolution through varied cultures.

If we compare the different anatomical and biological stages, which are linked with cultural levels, we are inevitably led, through ontogenetic and phylogenetic developments, from the stage of the first Hominids, that is *Australopithecus* and *Homo habilis*, to that of the social and cultural conflict between *Homo sapiens neanderthalensis* and *Homo sapiens sapiens*.

Bipedal locomotion was the starting-point. The body became erect and the upper limbs were freed up and evolved physically and technically to carry out new functions. Eight million years ago in East Africa and Equatorial Africa the Gregory Rift Valley clift modified the relief, the climate, and the vegetation. In the West, the tropical forest remained and was still populated by the Panids (chimpanzees and gorillas), whilst in the East the savannah grew up and was more favourable towards the physical, cerebral, and social development of our ancestors. The brain of our earliest ancestors, dated between 5 and 5 million years, was still small (400 cm^3). It presented a high coefficient of encephalization and a modern organization. The relative size of the brain to the body would be modified rapidly towards growth in brain size. Two groups of Hominids would evolve at a parallel rate up until approximately 1 million years. These were the first Men and the Australopithecines and in particular the subspecies known as robust. In the Hominid the beginnings of asymmetrical occipital and frontal lobes may be observed. At the same time, we may note the development of the frontal, parietal, and temporal lobes. It may be supposed that these new neurobiological foundations enabled *Homo habilis* to use spoken communication, that is to say, language. There has been a continual interplay between the quantitative and qualitative transformations of the human brain and the changes in Hominid behaviour. The meat diet called for and justified new strategies: exploitation of carrion and hunting. These activities were accompanied by a more elaborate social life and also by more frequent and more specific communication. Eyesight also adapted itself to a different kind of pursuit, no longer oriented to gathering but to animal capture.

A transition was made from the collecting of inanimate objects, such as rocks, vegetal objects, and carcasses, to the exploring of a space where live beings are at large.

The notion of culture is particular to the Primates. There are signs of it amongst the first men and then its various forms are developed in direct correlation with the development of the cortical areas. Manual skill, technology, tools, domestic fire, structures, religion, and art may all be observed to have variants in the techniques and behaviours involved. This was evident in the past, but is even more the case nowadays. We may perhaps distinguish the Hominids and in particular *Homo sapiens neanderthalensis* from *Homo sapiens sapiens* by their parallel evolution of body and brain morphology and culture. It is, however, true to say that they existed side by side in Western Europe and the Near East. In France, the Neanderthals were already if not instigators, at least users, of the new technologies and ways of life which characterized Upper Palaeolithic man.

It is true that the Cro-Magnon took the place of the Neanderthal, but this may be due more to cultural issues. We may point to an upsurge of traditions of former technologies and also of new ones related to the varied occupations and activities of modern man. The manual, practical tool may from that point onwards become a household object, imbued with aesthetic or symbolic power.

The advantage that man had over the chimpanzee, his Primate brother, was skilled know-how and technology. He transmitted the methods of this technology to his descendants and gradually was able to perfect it, thanks to his intelligence. In the Upper Palaeolithic period, man moved within a new dimension, which was at one and the same time ecological, social, and philosophical. There was a rich blending of symbolism with his activities and way of life. His new fields of action were art, ornament, and religion. Metaphysical thinking may be perceived: art was simultaneously a representation of reality, aesthetics, language, and writing.

The first men were able to exchange information. However, this was factual and as brief and fleeting as words themselves. By using schemas, codes, signs, ideograms, and then alphabetic writing, man was able to leave messages allowing the past to be evoked and the future to be organized. Man's thoughts were set down in words, then signs, and finally, writing, in an increasingly permanent fashion.

The cultural evolution of the group is subject to the laws of natural selection. Society, which is imbued with culture in its wide sense, may acquire new behaviour patterns, take advantage of the exchange of views, and challenge the ideas of its members, before decision-making takes place once the choice has been made. This constitutes the collective intelligence in which thinking is organized and comes to maturity for better or worse, in the form of word power.

J.-P.C.

J.C.

Part I

Anatomy of the brain

1

The first modern men

BERNARD VANDERMEERSCH

In the last few years there have been two equally important research results which have vastly modified our conception of the origins of modern man: new fossil evidence and the absolute dating of both recent and former discoveries. I use the term modern man to mean all the fossils which are assigned to the subspecies of *Homo sapiens sapiens* together with all the present day populations. As such they have a general brain structure identical to that of present day man, a reduction in superstructures (for example, the subdivision into two parts of the supra-orbitalis contour) and a refined quality. The latter, although it does not necessarily affect all the details of the anatomy, renders the similarities with present day populations extremely close. Nearly all of their metrical and morphological characteristics fall into the range of present day human variation. Even if occasional archaic traits may persist, they may generally be encountered in the form of individual variants, in one or another living population. For example, the *torus occipitalis* and the *torus supra-orbitalis* are not unknown in modern individuals (Larnach and MacIntosh 1970; Hublin 1978).

The classical evolutionary outline established up to the fifties held that modern man appeared in his present day morphology at the beginning of the Upper Palaeolithic. This was at the same time as certain characteristics of the human psyche, considered as the most elevated, made their appearance. Thus it was that art was considered as one of the prerogatives of modern man, as was the case for articulated language, according to certain anthropologists and prehistorians. There was a double advantage to this theory: on the one hand, a form of parallelism between biological and cultural evolution could be set up, since each biological transformation would have corresponded to a newly acquired cultural characteristic. On the other hand, the theory showed that if man had indeed followed the same type of evolution as the other Primate groups, the last stage of this evolution had been marked by transformations at the level of the psyche. These transformations constituted, to a certain extent, a break with what came before. The discovery of fossils with modern morphology at levels well below the Upper Palaeolithic definitively crushed this notion and challenged our ideas about the link between biology and culture in the Upper Pleistocene. Excavations in the last 20 years in Africa, Asia, and the Levant have played an essential role in setting up new notions about the origin of modern man, his relations with cultural phenomena and, thus, on his psychic development.

There is no question that all the discoveries that have taken place cannot be presented here. However, I should like to summarize briefly the data at our disposal presently. This will then serve as a basis for further discussion.

Africa

South Africa

Our attention is drawn to two deposits in South Africa, Border Cave and Klasies River Mouth.

At Border Cave, the remains of two individuals were discovered between 1941 and 1974. The first two were found in 1941, out of their stratigraphic context, during work extracting sediment to be used as fertilizer. Their stratigraphic position is, therefore, uncertain. The third one found in a ditch in 1942, is that of a child. Thirty years later, during new excavations, an adult mandible was discovered. The stratigraphic position of the first three is subject to debate. Nevertheless, as the 1942 excavations showed that the 4WA layer was intact, the first two individuals are dated later than this layer. Furthermore, a likely correlation with the level located just above was able to be set up, owing to the presence of sediment still adherent to the skull (Cook *et al.* 1945). The child, BD3, is likely to have issued from the 1GBS layer. Lastly, in 1974 a mandible was found *in situ* in layer 3WA and is, therefore the only item with an unquestionable stratigraphic position.

All these fossils undoubtedly have modern morphology and the stone industries accompanying them are related to the MSA facies. In 1980, P.B. Beaumont had considered an age of between 90 000 and 110 000 years for these fossils. More recently, following dating by ESR, he was obliged to date them as younger. The first two individuals would certainly be less than 90 000 years, the child was likely to be between 70 000 and 80 000 years, and the mandible discovered in 1974, between 50 000 and 65 000 years (Grun *et al.* 1990).

In 1968, J.J. Wywer came across items in another deposit, the Klasies River Mouth. There were two mandibles still with teeth and other very fragmented remains, all in a clear stratigraphic position. The mandibles have an astoundingly modern morphology and are accompanied by MSA stone tools. The most recent datings have attributed an age in the order of 100 000 for these items.

In at least two of the South African deposits, we have data which agree, tending to show the presence of populations of the modern type between 50 000 and 100 000 years.

East Africa

The most important fossils from this region are undoubtedly the two skulls discovered in 1967 in the Kibish formation of the Omo valley. The first one,

Omo 1, was found on the spot in limb 1 of the formation. It was accompanied by fauna and a few flake tools, though in insufficient numbers to define an industry.

The second one, Omo 2, was lying on the ground surface over 1 km away and had no accompanying evidence of fauna or industry while the sedimentary sequence is the same in the two sites and the proportions of nitrogen and uranium would tend to show that the two items were contemporaneous, this is nevertheless not an accepted fact.

I shall not give here a detailed description of these fossils (Day 1973; Bräuer 1984). Omo 1 is completely modern, whereas Omo 2 still has some archaic characteristics. Amongst the latter we may cite an upper mastoid bone in continuity with the torus occipitalis, the inion and the opisthocranion that coincide, and a thick temporal bone. According to some authors, these differences would seem to be sufficient to justify relating Omo 1 to a more modern group (Day and Stringer 1982; Bräuer 1984). If indeed they are of the same age, it would have to be conceded that two different populations lived simultaneously in the same region of East Africa. Whatever the outcome of this debate may be, it is nevertheless true that Omo 1 is modern and that its dating is a marker of the age of *Homo sapiens sapiens* in East Africa.

The data that we have at our disposal at the present time show that this skull issues from the basis of limb 1. The dating that C14 has established from a level clearly situated higher is over 40 000 years. Furthermore according to Butzer, as quoted by Bräuer (1989), limb 1 would appear to be linked to the isotopic stage 5 and, thus, would be aged over 75 000 years. Finally, an age of 130,000 years was also obtained by uranium/thorium. We may thus accept an age in the order of 100 000 years for this skull. According to Bräuer (1989), it represents a modern, robust man who could belong to the ancestral line of present day humans, including those outside Africa.

There are other pieces of documented evidence. However, these are either extremely badly dated, such as the skull from Singa in the Sudan or both badly dated and too fragmentary for their chronological and anthropological position to be specified, as for example, the mandibular fragment from Dire Dawa.

These results may seem scanty, but the point is to note that all the data that we possess on the Upper Pleistocene of East Africa and South Africa, even if it is frequently imprecise, tends to point to the same conclusion: modern man (*Homo sapiens sapiens* was present in sub-Saharan Africa approximately 100 000 years ago.

Asia

The Asiatic fossils of modern man are, by contrast, more recent. In South East Asia, the Niach skull from Borneo would seem to issue from a layer dated at 40 000 years. However, it has been suggested that this skull is not in its original context but is an intrusive burial, from a burial place that had been dug, starting from a more recent level. Thus, this date is generally put into question. Additionally, at Tabon, in the Philippines, a skull has been dated at 24 000 years. The latter two fossils are thus relatively recent and their morphology enables us to set up links with the present day Melanesians and Australians. It is important to note here the fact that Australia was certainly populated 40 000 years ago and archaeological evidence enables us to consider a likely first population between 50 000 and 60 000 years (Boudler 1992). There is hardly any doubt that the first settlers, who were the direct ancestors of the present day aborigines, were already *Homo sapiens sapiens*.

All the modern fossils of continental Asia have characteristics that relate them to the present day Mongoloids and most of them are of a late date, that is, less than 20 000 years. Nevertheless, recently the Liujiang skull, which was discovered in 1958 in the province of Guangxi, was dated through the uranium process at 67 000 years (Wu Xinzhi 1992). If these dates are further confirmed, this would turn out to be the oldest fossil likely to be directly related to present day populations.

The Near East

Two deposits in the Levant draw our primary attention, those of Skhul and Qafzeh.

The Skhul cave is situated on Mount Carmel, in the Wadi-el-Mughara, approximately 20 km south of Haifa. In 1931 and 1932, D. Garrod and Th. McCown excavated from a Mousterian layer the remains of ten skeletons, including three children and several isolated bones. The same team was to extract a skeleton and some isolated bones from the nearby cave of Tabun in archaeological layers of the same type.

Two years later, R. Neuville, who was the French consul in Jerusalem at the time, undertook the excavation of the Qafzeh cave situated in Galilee, 2.5 km from Nazareth. There he discovered five new subjects, once again from the Mousterian levels.

The Mount Carmel fossils were studied by McCown and Keith (1939). At first they considered the possibility of the presence of two populations: a more modern one at Skhul and another Neanderthal one at Tabun. They subsequently linked the two together to form a single group close in nature to the European Neanderthal man in accordance with the classical viewpoint on our continent. In other words, Neanderthal man was related to the Mousterian industries and, later on, modern man with the Upper Palaeolithic ones.

However, Coon (1939) also considered the likelihood that the Mount Carmel men (or at least some of them) could have issued from interbreeding with Neanderthal men. A representative of this interbreeding would have been the Tabun fossil and other as yet undiscovered *Homo sapiens sapiens* in the area. This interpretation had the enormous merit of introducing for the first time the idea of a likely contemporaneity of Neanderthal and modern man.

F. Clark Howell went one step further in 1958 when he went through all the documents relating to the Levant and showed that the Tabun and Skhul fossils were contemporary with European Neanderthal man. He showed, furthermore, that there were indeed two distinct populations in the Near East Mousterian period: Neanderthal man at Tabun and modern man at Skhul and, most likely, at Qafzeh. The similarities between the latter and the Cromagnons of Europe prompted him to name them Proto-Cromagnons.

Since this time, excavations have confirmed this interpretation. Between 1953 and 1960, R. Solecki at Shanidar, in Iraq, discovered seven new skeletons at Mousterian levels. They were described in 1983 by E. Trinkaus, who placed them with Neanderthal man. Other skeletons of this population, together with several isolated bone pieces, were excavated from the Amud cave by H. Suzuki in 1961 and from the Kebara cave in 1982 by a team led by O. Bar Yosef and myself. These latter two deposits are located in Israel.

At the same time, excavations in the Qafzeh cave were undertaken again, between 1965 and 1981, which resulted in the discovery of additional skeletal remains, several of which were in the Mousterian levels. This brought the anthropological series of this deposit up to over 20 individuals. Research on these skeletal remains (Vandermeersch 1981) confirmed that they belonged to a population with modern morphology.

Over 30 individuals with modern morphology have thus been found in the Near East Mousterian levels, issuing from the two deposits of Skhul and Qafzeh in Israel and from an equivalent number of Neanderthal men brought to light in Israel and Iraq.

Questions have been raised concerning the relative and absolute chronological position of these two series of fossils.

Two conflicting interpretations have been put forward: according to some people (Jelinek 1982), modern men belonged to the final part of the Mousterian period. They therefore came after the Near East Neanderthal men and were descended from them. According to others (Bar Yosef and Vandermeersch 1981), the Qafzeh modern men were extremely ancient, far more ancient than the Tabun Neanderthal. No phylogenetic relationship between these two populations could thus exist. I will not outline here the arguments developed in support of each of these interpretations as a series of absolute datings have just shed new light on this matter.

The Qafzeh modern men have been dated through use of thermoluminescence (Valladas *et al.* 1988) and ESR (Schwarcz *et al.* 1988) at between 90 000 and 100 000 years and the Skhul men through ESR at an equivalent age (Stringer *et al.* 1989).

Only two Neanderthal deposits have been the subject of dating: Kebara, which yielded an age of 55 000 years for the level from which skeleton issued and, secondly, Tabun, whose complete skeleton would appear to be aged at over 100 000 years (Stringer *et al.*, in press). These results show that the presence of Neanderthal men in the Near East was spread over a long period of time. These results tend towards proof of their contemporaneity with modern populations.

Another important result is that recent discoveries and datings show that modern man is as ancient in the Near East as he is in subSaharian Africa.

Europe

A completely different image is shown in Europe where modern man makes his appearance later, with the Cro-Magnons of the Upper Palaeolithic period. It is possible that they issue from the Proto-Cro-Magnons of the Near East and that they formed an immigrant population. Their earliest traces are to be found around 40 000 years in Eastern Europe. On our continent, they constituted a rival population with the Neanderthal men of the region who gave way to them little by little. Within a few million years, the latter had disappeared (Vandermeersch 1988). Europe is thus the only area in the world where we may note, in the Upper Palaeolithic period, both a biological and technological replacement.

The origin of modern man

In summary, in the Near East we have at our disposal fossils with modern morphology dated at 100 000 years and several African specimens that might also be of the same age. In Asia, the first modern men are, for the moment, dated as more recent: most likely between 50 000 and 100 000 years. In the Near East and in Africa, these populations were preceded by fossils whose cranial structure was already modern. However, they still demonstrated numerous archaic characteristics, especially at the level of the super-structures. They are aged between 100 000 and 200 000 years. Both the skull of Zuttiyeh and that of Ngaloba in East Africa may be highly rated. Generally they are referred to as archaic *Homo sapiens*.

There are skulls in Africa and Asia with a morphology midway between that of *Homo erectus* and *Homo sapiens*, which date from even more ancient periods, for example the skull from Dali in China and the skull from Ndutu in Tanzania. While their age is imprecise, they would seem likely situated between 200 000 and 400 000 years. We are thus led to acknowledge the fact that the important stages in the transformation which started from *Homo erectus* and ended with modern man are represented in extensive areas of the old world, with the exception of Europe. These data provide a strong argument in support of the idea of continuity in population settlement.

It may be possible to go further along these lines: anthropologists have pointed out that certain morphological characteristics are found significantly more frequently in South-East Asia, as compared to continental Asia from the period of *Homo erectus* to present day populations (Agner 1976; Larnach and MacIntosh 1976; Wolpoff *et al*. 1984). The fact that these characteristics persisted on a regional level would support the idea of a certain genetic continuity. These affinities had already been recognized by F. Weidenreich, who put forward the hypothesis of phyletic continuity in Asia in 1940.

This does not mean that modern populations of the diverse regions of Africa and Asia result uniquely from regional evolution movements and travelling by groups of men probably occurred. It is also likely that one region would have played a more important role than another in the origin of modern humanity, through the particular dynamics of its population; however, I do not believe that the theory of population replacement may be applied everywhere.

To conclude, I should like to make a few more remarks on the subject of the Near East, concerning the contemporariety of Neanderthal men and modern man in this region. Basically, there is a dilemma on how to interpret two populations that are very different from a morphological point of view. For a long time, Neanderthal men were considered as a species, *Homo neanderthalensis*; it was only from the 1940s that they were taxonomically placed at the level of a subspecies. For example, the beginning of this century, the famous paleontologist, M. Boule, wrote in his work on man at Chapelle-aux-Saints that they were only 'just above the level of the animal' and it is still thought by some that they did not possess articulated language. While it is true that the Neanderthal brain cavity had a volume comparable to that of present day man, its long, wide and flattened form was different and the incline of the occipital area was far less marked. These differences were supposed to reflect functional differences. In the Near East we can find, at the same epoch, both populations in the same region. There is a distance of 40 km at the most, between the Neanderthal deposits at Tabun and Kebara and those of Qafzeh and Skhul where remains of modern man were found. Present day chronological data would seem to indicate that this contemporaneity lasted several dozen millennia. Furthermore, the stone industries fall within the province of the same Mousterian culture. Also certain socio-cultural practices, such as voluntary burial, have been noted in the two types of population. When a Near Eastern Mousterian layer is excavated, it is impossible at the present time to determine who was at the origin of its formation. The European model is not to be found in the Near East. This refers to the rapid replacement of Neanderthal man by modern man, which is considered as the expression of the latter's intellectual superiority. If modern man and Neanderthal man shared the same culture for such a long time, it means that their differences were not as marked as their morphology would lead us to believe.

References

Bar Yosef, O. and Vandermeersch, B. (1981). Note concerning the possible age of the Mousterian layers in Qafzeh cave. In *Préhistoire du Levant*, pp. 281–6. CNRS, Paris.

Beaumont, P.B. (1980). On the age of Border Cave Hominids 1–5. *Palaeontol. Afr.* **23**, 21–33.

Boulder, S. (1992). *Homo sapiens* in Southeast Asia and the antipodes: archaeological versus biological interpretations. In *The evolution and dispersal of modern human in Asia* (ed. T. Akazawa, K. Aoki, and T. Kimura), pp. 559–90. Hokusen-sha Publishing Co., Tokyo.

Bräuer, G. (1984). A craniological approach to the origin of anatomically modern *Homo sapiens* in Africa and implications for the appearance of modern Europeans. In *The origin of modern humans: A world survey of the fossil evidence* (ed. F.H. Smith and F. Spencer), pp. 327–410. Alan R. Liss, New York.

Bräuer, G. (1989). The evolution of modern humans: recent evidence from Southwest Asia. In *The human revolution: behavioural and biological perspectives in the origins of modern human* (ed. P. Mellars and C. Stringer), pp. 123–54. Edinburgh University Press, Edinburgh.

Cook, H.B.S., Malan, B.D., and Wells, L.H. (1945). Fossil man in the Lebombo-Mountains, South Africa: 'The Border Cave', Ingwawnma District, Zululand. *Man* **45**, 6–13.

Coon, C.S. (1939). *The races of Europe*. Macmillan, New York.

Day, M.H. and Stringer, C.B. (1982). A reconsideration of the Kibish remains and the *erectus–sapiens* transition. In *L'Homo erectus et la place de l'homme de Tautavel parmi les Hominidés fossiles* (ed. H. de Lumley), pp. 814–46. CNRS/Louis Jean, Nice.

Grun, R., Smackleton, N.J., and Deacon, H.J. (1990). ESR dating of tooth enamel from Klasies River Mouth cave. *Curr. Anthropol.* **31**, 427–32.

Howell, F. and Clark, (1958). Upper Pleistocene men of the Southwest Asian Mouserian. In *Hundert Jahre Neanderthalensis*, pp. 185–98. Böhlam Verlag, Koln–Graz.

Hublin, J.J. (1978). *Le torus occipital transverse et les structures associées*. Thèse, Paris.

Jelinek, A. (1982). The Tabun Cave and Paleolithic in the Levant. *Science* **216**, 1369–75.

Larnach, S.L. and MacIntosh, N.W.G. (1970). *Craniology of the aborigines of North Queensland*. Oceania Monograph no.15, University of Sydney.

McCown, T.D. and Keith, A. (1939). *The stone age of Mount Carmel. Vol. II. The fossil human remains from Levallois Mousterian*. Clarendon Press, Oxford.

Schwarcz, H.P., Grün, R., Van, B., Bar Yosef, O., Valladas, H., and Tchernov, E. (1988). ESR dates for the hominid burial site of Qafzeh in Israel. *J. Human Evol.* **17**, 733–7.

Trinkaus, E. (1983). *The Shanidar Neandertals*. Academic Press, New York.

Valladas, H., Reyss, J.L., Joron, J.L., Valladas, G., Bar Yosef, O. and Vandermeersch, B. (1988). Thermoluminescence dating of Mousterian "Proto-Cro-Magnon" remains from Israël and the origin of modern man. *Nature* **331**, 614–16.

Vandermeersch, B. (1981). *Les Hommes Fossiles de Qafzeh (Israël)*. CNRS, Paris.

Wu Xinzhi, (1992). The origin and dispersal of anatomically modern humans in East and Southeast Asia. In *The evolution and dispersal of modern human in Asia*, pp. 373–8. Hokusen-sha Publishing Co., Tokyo.

2

Image of the human fossil brain: endocranial casts and meningeal vessels in young and adult subjects
ROGER SABAN

Introduction

In modern man, each of the envelopes of the brain (pericranium, cranium, meninges) possesses a specific vascular system for the cranial vault (Fig. 2.1):

(1) superficial arteriovenous system (fronto-temporo-occipital) for the pericranium;

(2) diploic venous system (fronto-temporo-occipital) for the cranium; and

(3) meningeal arteriovenous system (anterior and middle) for the dura mater, while the brain possesses its own cerebral arteriovenous system (anterior and middle).

All of these systems are superimposed and communicate with each other via the venous sinuses of the dura mater, constituting a regulating device for drainage of the cerebral blood. The dura mater sinuses communicate directly with:

(1) the superficial venous system via the emissary veins (parietal and mastoid emissary veins); and

(2) the cerebral system, via the superior (Trolard) and inferior (Labbé) anastomotic veins; while a diploico-meningeal system (Saban 1984) is derived from these same sinuses, characterized by the development of numerous intracranial anastomoses over the entire path of these vessels.

The system of sinuses (principally the superior sagittal sinus and the transverse sinuses) and the meningeal veins ensure the drainage of cerebral blood into the two jugular systems: the sinuses drain into the internal jugular vein posteriorly, and the meningeal veins drain into the external jugular vein anteriorly (Fig. 2.2).

Because of the intimate relations between the dura mater and the endocranium, the meningeal vessels (sinuses and veins) deeply groove their passage in the inner table of the cranial bones. The osteovenous contact is present at birth, as the veins follow the homonymous arteries over their entire path (Saban and Grodecki 1979). It is therefore possible, by means of endocranial

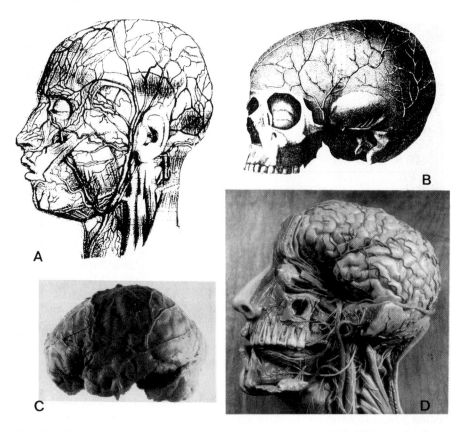

Fig. 2.1. The four vascular systems of the head. A, superficial arteriovenous system (adapted from Bonamy and Broca 1841); B, diploic venous system (adapted from Breschet 1830); C, meningeal arteriovenous system; D, cerebral arteriovenous system, an anatomical wax cast by Carlo Calenzuoli 1820. (Comparative anatomy laboratory of Museum, Paris.)

casts, to demonstrate the entire meningeal venous system in adults and to follow its changes during growth.

Endocranial casts of fossils reveal the functional importance of this venous system, which reflects cerebral development during human evolution in adults as well as in juvenile subjects.

The meningeal system of adult man

This system consists of the dura mater sinuses and the meningeal veins composed of two networks, an anterior network and a middle network which form the very dense vascular 'grid' in contact with the cranial vault.

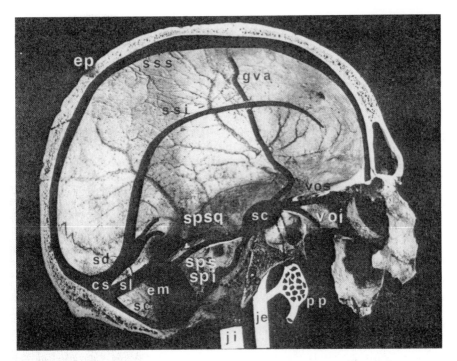

Fig. 2.2. The dura mater venous sinuses; left hemicranium of an adult man. cs, confluence of the sinuses; em, mastoid emissary vein; ep, parietal emissary vein; gva, great anterior vein; je, external jugular vein; ji, internal jugular vein; pp, pterygoid venous plexus; sc, cavernous sinus; sd, straight sinus; sl, transverse sinus; so, occipital sinus; spi, inferior petrosal sinus; sps, superior petrosal sinus; spsq, petrosquamous sinus; ssi, inferior sagittal sinus; sss, superior sagittal sinus; voi, inferior ophthalmic vein; vos, superior ophthalmic vein.

The dura mater sinuses (Fig. 2.2)

Divided into two territories, depending on whether they belong to the vault or the base of the skull, these sinuses are large venous channels with rigid walls formed by folds of the dura mater lined by endothelium. Of the 23 identified sinuses, some of which are inconstant, we shall only deal with those which have direct relations with the meningeal veins, i.e.

For the vault:

(a) the single superior sagittal sinus, which extends along the sagittal suture inside the falx cerebri between the *crista galli* in the frontal bone and the confluence of the sinuses in the occipital region, where it receives the straight sinus enlarged by the inferior sagittal sinus:

(b) the paired transverse sinuses, continuous with the superior sagittal sinus and terminating in the internal jugular vein behind the petrous temporal bone; and

(c) the paired occipital sinuses, a kind of diversion of the transverse sinuses, are situated on the occipital bone.

For the base:

(a) the paired cavernous sinus, situated on the sides of the *sella turcica*, are generally composed of a group of plexiform channels (remains of the embryonic pro-otic plexus), containing the internal carotid artery and branches of the trigeminal nerve. These sinuses, which receive the ophthalmic veins anteriorly, drain into the transverse sinus via the superior petrosal sinus, which runs over the superior surface of the petrous bone, within the tentorium cerebelli, and the inferior petrosal sinus, which travels on the inferior surface of the petrous bone;

(b) the paired petrosquamous sinuses, which normally disappear in the 99 mm fetus, but which may sometimes persist in adult life. They follow the internal petrosquamous suture and connect the transverse sinus to the common trunk of the middle meningeal veins which drains into the pterygoid venous plexus; and

(c) the inconstant paired sphenoparietal sinuses receive the anterior branch of the middle meningeal vein (great anterior vein) and connects the superior sagittal sinus to the cavernous sinus via the superficial middle cerebral vein.

Meningeal veins (Fig. 2.3)

The system of meningeal veins comprises two networks:

 (i) an anterior network which drains the frontal region; and
(ii) another network which drains the temporofrontal region.

As seen on endocranial casts (Fig. 2.4), the anterior meningeal veins form, in the frontal region, on either side of the longitudinal fissure, a convergent

---

Fig. 2.3. Impression of the meningeal veins on an endocranial cast in adult man (Comparative anatomy laboratory of Museum, Paris). A, right lateral view; B, topographical diagram on right lateral view (the great anterior vein is indicated with a dotted line): C, superior view; a, meningeal anastomosis; ad, diploic anastomosis; b, bregma; ba, anterior branch; bm, middle branch; bp, posterior branch; bp2, secondary posterior branch; c, cerebellum; eft, fronto-temporal incisure; etc, temporocerebellar incisure; f, frontal; fp, arachnoid granulation; lf, frontal lobe, lo, occipital lobe; lp, parietal lobe; lt, temporal lobe; λ, lambdoid suture; ma, anterior meningeal venous network; mm, middle meningeal venous network; o, occipital; p, parietal; rf, frontal branch of the middle meningeal veins; sc, coronal suture; sce, central sulcus; se, squamous suture; sl, transverse sinus, sla, lateral sulcus; sp, superior temporal sulcus; spe, parieto-occipital sulcus; spo, petro-occipital sinus; sps, sphenoparietal suture; spsq, petrosquamous sinus; ss, sagittal suture; ssp, sphenoparietal sinus; sss, superior sagittal sinus; st, inferior temporal sulcus; t, temporal.

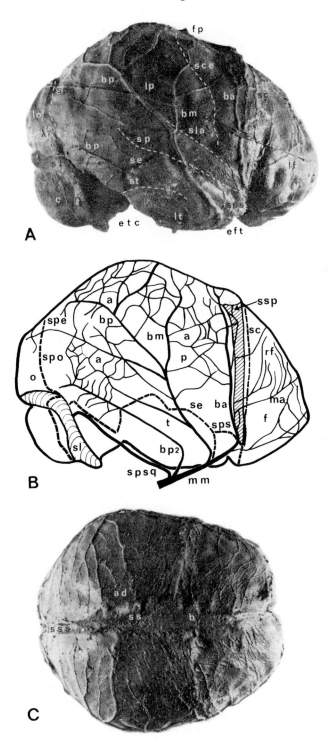

A

B

C

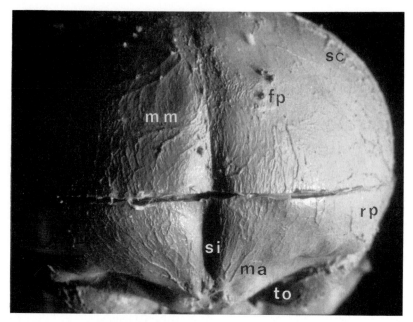

Fig. 2.4. Endocranial cast of the frontal region and anterior meningeal network. fp, arachnoid granulation; ma, anterior meningeal venous network; mm, middle meningeal venous network; rp, pterion region; sc, coronal suture; si, longitudinal fissure; to, orbital roof.

vascular network which drains into the superior ophthalmic vein via the cribriform plate of the ethmoid. Adjacent to the coronal suture and pterion, this network anastomoses with branches of the middle meningeal network forming a dense vascular grid over the frontal lobes.

The middle meningeal veins (Fig. 2.3) consist of a bilateral system of three principal branches, anterior, middle, and posterior, forming a common trunk which, on leaving the cranium, drains into the pterygoid venous plexus. The first and last of these branches unite in the temporal region, while the second drains indifferently into either of them. Each of these branches, derived from venous lacunae which accompany the superior sagittal sinus, receives a very large number of tributaries which anastomose with each other to form a very dense vascular plexus, principally located over the parietal region of the cranial vault. The ramifications of the anterior branch extend on to the frontal region and a large frontal branch is present in the pterion region. This system has two inconstant features, corresponding to the persistence of archaic features. The first, situated in the anterior branch, where the most anterior vein can reach enormous proportions, constitutes what has been incorrectly named as 'Breschet's sinus' (Fig. 2.5), but which corresponds to the great anterior vein. The second feature is the persistence of the petrosquamous sinus which connects the common trunk to the transverse sinus. This sinus is

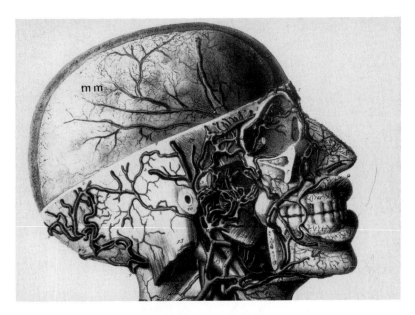

Fig. 2.5. The middle meningeal veins according to Breschet (1830). mm, middle meningeal venous network; gv, great anterior vein.

constant in all primates, in which it constitutes a communication between the transverse sinus and the external jugular vein, via the jugular foramen.

The system of meningeal veins in young subjects

As we have previously shown (Saban 1987), the system of anterior meningeal veins travels inside the longitudinal fissure and therefore does not form any grooves on the endocranial surface of the frontal bone. From the fifth year of life onwards, a short groove indicating the passage of the anterior meningeal vein close to the cribriform plate, appears on either side of the endofrontal crest. The complete network of convergent tributaries only appears on the endocranial surface during adulthood (see Fig. 2.4). In contrast, from birth onwards, the middle meningeal venous system (Saban 1986) shows a constant progression in the formation of the vascular grooves involving all of the parietal surface.

In the neonate, on the right side (Fig. 2.6A), the three branches, anterior, middle, and posterior, with a limited number of tributaries, form grooves on the parietal bone, while on the left side (Fig. 2.6B) the topography of the system is greatly simplified with no tributaries associated with the middle and posterior branches.

In the 40-day-old infant (Fig. 2.7), the topography of the three branches, although still only slightly ramified, shows the first anastomoses between the

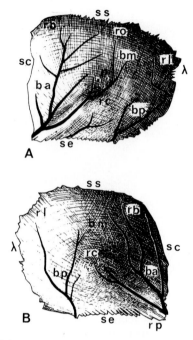

Fig. 2.6. Impressions of the middle meningeal veins on the endocranial surface of the parietal bone in the neonate. A, right parietal bone; B, left parietal bone. p, parietal tuber; rb, bregma region; rc, central region; rl, lambda region; ro, obelion region; rp, pterion region. (See Fig 2.3 for the other abbreviations.)

middle branch and the anterior branch as well as with the posterior branch. As before, there is a greater number of branches on the right side. On the superior view, none of the tributaries of the anterior meningeal vein cross the coronal suture, very widely open on either side of the anterior fontanelle, while the metopic suture persists.

The complexity of the network is considerably increased in the 1-year-old infant (Fig. 2.8). All of the branches receive numerous tributaries and the number of anastomoses between the three branches is increased, principally on the right side (Fig. 2.8B). The frontal region shows the presence of a large frontal tributary belonging to the anterior meningeal branch, located adjacent to the fronto-temporal incisure.

By the age of 2½ years, at the same time as the appearance of the first impressions of the parietal gyri, the middle meningeal network is completed by numerous anastomoses, as demonstrated by the grooves on the parietal bone (Fig. 2.9). The permanent venous network continues to progress until the age of 5 years (Fig. 2.10). All of the tributaries of the three middle meningeal branches are punctuated with small diploic foramina, as it is at this age that the diploic venous network starts to develop between the two

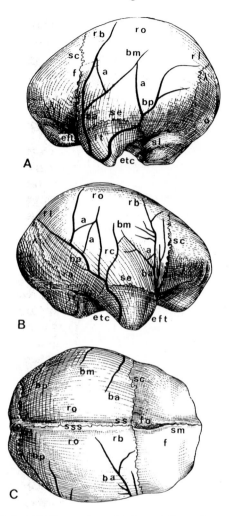

Fig. 2.7. Drawing of the middle meningeal veins on the endocranial cast of a 40-day-old infant. A, left lateral view; B, right lateral view; C, superior view. ra, asterion region; rb, bregma region; rc, central parietal region; rl, lambda region; ro, obelion region. (See Fig. 2.3 for the other abbreviations.)

bone tables of the cranial vault and channels entering the diploe can be seen in the coronal, sagittal, and lambdoid sutures.

Finally, from the age of 6½ years onwards (Fig. 2.11), the permanent vascular network appears to be developed, at least in terms of the middle meningeal veins, although it will only become fully operational during adulthood. The organization of the anastomoses becomes more elaborate. The tributaries of the anterior branch now cross the coronal suture and anastomose with the frontal branch, as can be seen on the left side (Fig. 2.11A and C).

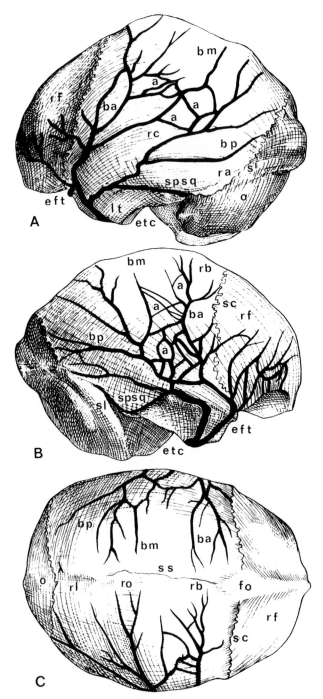

Fig. 2.8. Drawing of the middle meningeal veins in the 1-year-old infant. A, left lateral view; B, right lateral view; C, superior view. fo, fontanelle; rb, bregma region; rc, central parietal region; rl, lambda region; ro, obelion region. (See Fig. 2.3 for the other abbreviations.)

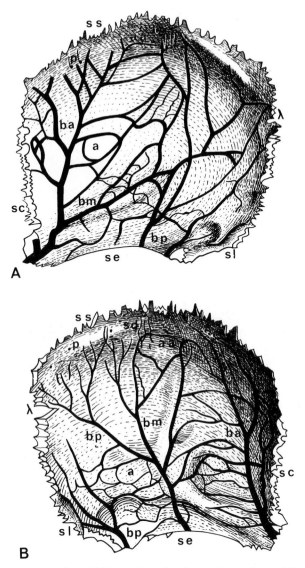

Fig. 2.9. Impressions of the middle meningeal veins on the endocranial surface of the parietal bone in the 2½-year-old child. A, right parietal bone; B, left parietal bone; gs, sagittal groove; p, diploic foramen. (See Fig. 2.3 for the other abbreviations.)

The zone of active growth of the coronal suture appears to constitute, up until this age, an obstacle to the penetration of vessels on to the frontal bone. The anterior meningeal system, composed of a single vessel up until the age of 5 years, also starts to show the first ramifications at the site of the first frontal gyrus, on either side of the longitudinal fissure (endofrontal crest). This

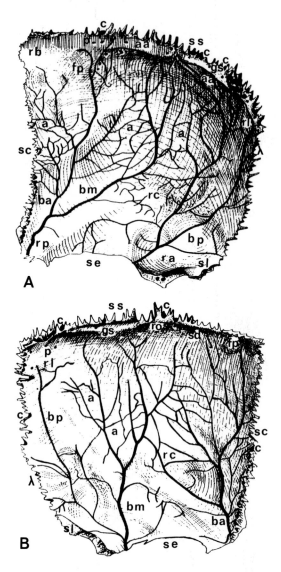

Fig. 2.10. Impressions of the middle meningeal veins on the endocranial surface of the parietal bone in the 5-year-old child. A, right parietal bone; B, left parietal bone. c, diploic channel; gs, sagittal groove; p, diploic foramen; ra, asterion region; rb, bregma region; rl, lambda region; ro, obelion region; rp, pterion region; s, diploic sulcus. (See Fig. 2.3 for the other abbreviations.)

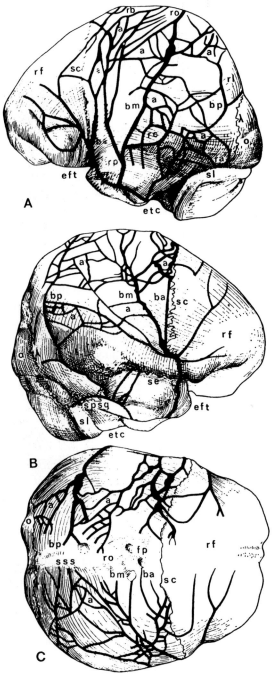

Fig. 2.11. Drawing of the middle meningeal veins on the endocranial cast of a 6½-year-old child. A, right lateral view; B, left lateral view; C, superior view. rb, bregma region; rc, central parietal region; rl, lambda region; ro, obelion region. (See Fig. 2.3 for the other abbreviations.)

system only becomes clearly demarcated in adults, at the same time as the appearance of the anastomoses with the frontal branches of the middle meningeal system.

The middle meningeal venous system in human fossils

Although the dura mater sinus system does not reveal any major variations in fossil forms from *Australopithecus* onwards, the meningeal veins reveal a progressively increasing complexity of their network in the course of human evolution (Fig. 2.12). As we shall demonstrate, there is a succession of the various stages observed above in the course of modern human development.

The meningeal veins in adult fossils

In the pre-human phase, about 2 million years ago, the middle meningeal venous system (the anterior meningeal venous system is not seen) was extremely simple in *Australopithecus*. *Gracilis* forms were characterized by the absence of the middle branch of the middle meningeal veins (Fig. 2.13A), while this branch is present in *robustus* forms (Fig. 2.13B), probably because of the marked increase in the volume of the brain (an average of more than 100 ml). However, in both forms, there is persistence of the archaic features conserved from simians, such as the presence of the petrosquamous sinus and a large frontal branch.

The simplified topography of the system in the parietal region in *Australopithecus robustus* recalls the arrangement observed in modern neonates (see Fig. 2.6), while that of *A. gracilis* is even more primitive.

The concomitant arrival of the *Homo* genus, in which the first forms had a cerebral capacity exceeding 700 ml, was associated with a new step in the complexity of the middle meningeal system, while the anterior meningeal system was still absent. With this new increase in cerebral capacity, the new evolutionary leap is characterized by the presence of the first anastomoses between the three branches of the middle meningeal veins (Fig. 2.14A and Fig. 2.15A). At the same time, the archaic features tend to fade with the loss of the frontal branch, while the petrosquamous sinus is still present. The topography of the system is therefore very similar to that observed in the modern 40-day-old infant (see Fig. 2.7), in which, however, the petrosquamous sinus, one of the last remains of archaicity in primates, has now disappeared.

The following step, in which the anterior meningeal system was still not present, was marked by the development, 1 million years ago, of *Pithecanthropus*, of which two forms have been identified (Saban 1980): an archaic form (*H. erectus*) with a cerebral capacity less than 1000 ml and a more advanced form (*H. palaeojavanicus*) with a cerebral capacity greater than 1000 ml. The cast of the middle meningeal network of *H. erectus* contains

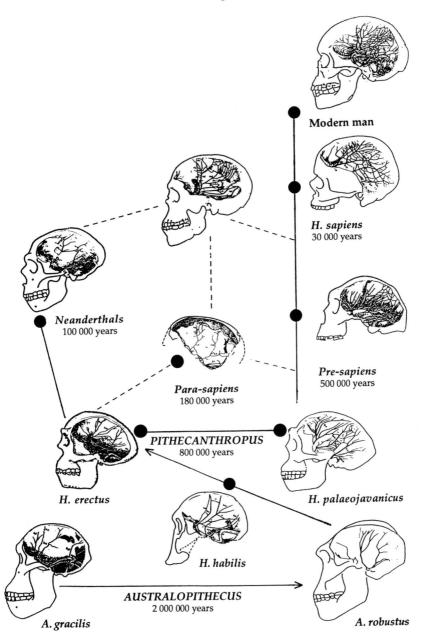

Fig. 2.12. The morphological stages of the meningeal vessels in human evolution.

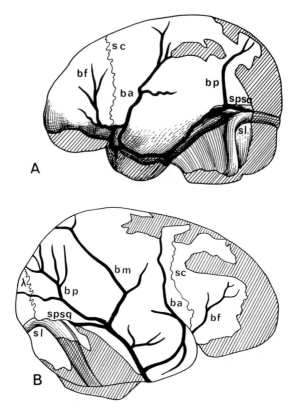

Fig. 2.13. Drawing of the middle meningeal veins on the endocranial cast of *Australopithecus*. A, *gracilis* form (Sts 60), left lateral view; B, *robustus* form (SK 1585), right lateral view, bf, frontal branch. (See Fig. 2.3 for the other abbreviations.)

three poorly ramified branches (anterior, middle, and posterior) as well as a frontal branch (Fig. 2.14B and Fig. 2.15B), while *H. palaeojavanicus* shows numerous anastomoses between the tributaries of the middle and posterior branches (Fig. 2.17A). The topography of the middle meningeal system of this latter form presents numerous similarities with that of the modern 1-year-old child (see Fig. 2.8). The two archaic and advanced forms of *Pithecanthropus* share common features in the organization of the system: reappearance of the frontal branch, predominance of the middle and posterior branches, persistence of the petrosquamous sinus, presence, in certain individuals, of the great anterior vein.

Examination of the meningeal vessels has led us to identify two morphological types in the evolution of man: an archaic type and an advanced type. The first type includes, from *H. erectus* onwards, Neanderthal man and the para-*sapiens*, which became extinct before the arrival of *H. sapiens*, and the second, continuous with *H. palaeojavanicus*, was continued by the pre-*sapiens* up to *sapiens* and modern man.

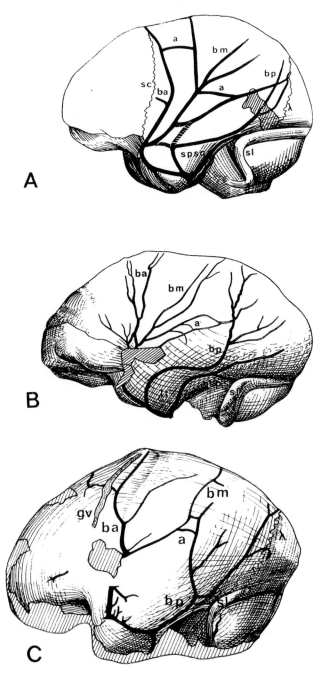

Fig. 2.14. The meningeal vessels of archaic adult forms of the human line. Endo-cranial cast, right lateral view. A, *Homo habilis* (ER 1470); B, *H. erectus* (Sangiran VI); C, Neanderthal (La Ferrassie). bf, frontal branch of the middle meningeal veins; gv, great anterior vein. (See Fig. 2.3 for the other abbreviations.)

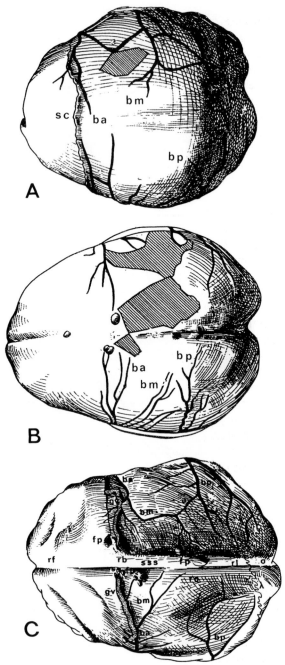

Fig. 2.15. The meningeal vessels of archaic adult forms of the human line. Endocranial cast, superior view. A, *Homo habilis* (ER 1470); B, *H. erectus* (Sangiran VI); C, (Neanderthal man); gv, great anterior vein; rb, bregma region; rf, frontal region; rl, lambda region; ro, obelion region.

In Neanderthal man, who lived 100 000 years ago, the middle meningeal venous system (Fig. 2.14C and Fig. 2.15C) retains many of the archaic features observed in *H. erectus*, despite the marked increase in brain volume (average of 1500 ml): poorly ramified branches, with virtually no anastomoses, persistence of the petrosquamous sinus, and, most importantly, constant presence of the great anterior vein on both sides. The anterior meningeal system first appears over the first two frontal gyri, on either side of the longitudinal fissure, as demonstrated by the endocranial cast of Moustier man (Fig. 2.16A). This network, still very poorly developed, only has a very small number of anastomoses with the frontal tributary of the anterior branch of the middle meningeal veins. Because of its unusual features, the middle and anterior meningeal network does not correspond to any of the morphological stages of this network observed in modern infants. The same applies to the para-*sapiens*, still very archaic, but much older than Neanderthals, as they first appeared almost 200 000 years ago. They are characterized by the multiplication of the ramifications of the middle branches, the absence of the anterior network, and the loss of the great anterior vein and the petrosquamous sinus, which distinguishes them from archaic *Pithecanthropus* (*H. erectus*).

In contrast, in the pre-*sapiens*, which were a continuation of the evolutionary model of *H. palaeojavanicus* and which first appeared at the beginning of the Acheulian age, about 500 000 years ago, the middle meningeal system (Fig. 2.17 and Fig. 2.18) was characterized by the marked increase in the number of anastomoses between the three branches. This is observed even in the earliest forms, such as Tautavel man and Swanscombe man (Fig. 2.18B), indicating, already at this time, a marked advance in sapienization. However, these anastomoses are preferentially observed in the superior region of the vault, as in Swanscombe man (Fig. 2.18B) and Omo II man (Fig. 2.17B and Fig. 18A), which is much more recent (about 100 000 years), comparable with the appearance observed in the modern 5-year-old child (see Fig. 2.10).

In the other even more recent forms (about 40 000 BC), such as La Quina V (Fig. 2.18C), the anastomotic network is situated in the central parietal region, as in the 6½-year-old child (see Fig. 2.11). Moreover, numerous tributaries of the anterior branch cross the coronal suture, over the frontal region. The presence of the petrosquamous sinus is encountered sporadically (Omo II, La Quina V, Brno, Combe Capelle). The anterior meningeal network, always very simplified, can be seen over the first frontal gyrus (see Fig. 2.16B), recalling the features observed in modern man at the age of 5 years, with several additional lateral anastomoses for the frontal branch of the middle meningeal veins, but also, in the later forms (La Quina V), branches in the coronal region.

At the end of the last Ice Age, at the time of the appearance of *H. sapiens*, about 30 000 years ago, the middle meningeal network reached its maximum complexity. The middle meningeal venous network constitutes a vascular grid

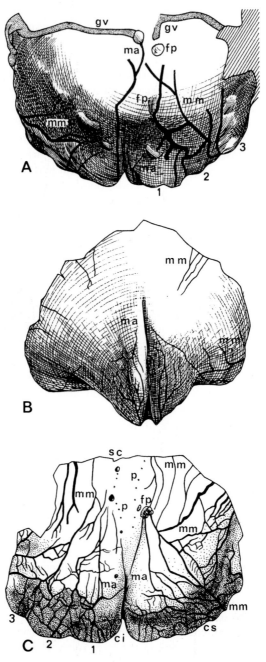

Fig. 2.16. The anterior meningeal venous system in adult fossil forms. Endocranial cast, anterior view. A, Neanderthal (Le Moustier); B, pre-*sapiens* (Tautavel man); C, *sapiens* (Hahnöfersand). 1, 2, 3; first, second, and third frontal gyri; gv, great anterior vein; p, diploic foramen. (See Fig. 2.3 for the other abbreviations.)

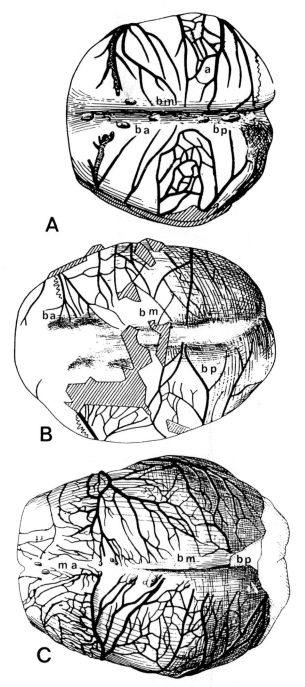

Fig. 2.17. The meningeal vessels of advanced adult forms of the human line. Endo-cranial cast, anterior view. A, *Homo palaeojavanicus* (Sangiran VII); B, pre-*sapiens* (Omo II); C, *sapiens* (Cro-Magnon III). (See Fig. 2.3 for the other abbreviations.)

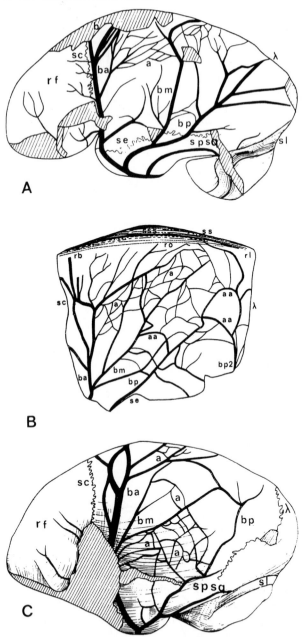

Fig. 2.18. The middle meningeal network in the pre-*sapiens*. A, Omo II (100 000 years), corresponding to the modern 3-year-old child; B, Swanscombe (400 000 years), corresponding to the modern 5-year-old child; C, La Quina V (40 000 years), corresponding to the modern 6½-year-old child. rb, bregma region; rf, frontal tributaries of the anterior branch of the middle meningeal veins; rl, lambda region; ro, obelion region. (See Fig. 2.3 for the other abbreviations.)

which involves all of the parietal region in Cro-Magnon III man, for example (Fig. 2.17C). The anterior network also forms a vascular grid, as illustrated by the endocranial cast of Hahnöfersand man (see Fig. 2.16C), demonstrating the multiplication of the anastomoses with tributaries of the anterior branch of the middle meningeal veins in the coronal region and, laterally, with the frontal branch. The vascular grid is therefore complete over the entire fronto-parietal cranial vault (see Fig. 2.2 and Fig 2.3).

The middle meningeal veins in juvenile fossil forms

The few juvenile specimens we have been able to study include Neanderthal, pre-*sapiens*, and *sapiens* forms from the time of birth until the age of 5 years.

Neanderthal infants

We have studied three specimens: the neonate from the Hortus cave, the La Ferrassie child with an estimated age of 2 years, and the Gibraltar II child, aged 5 years.

The two fragments of right and left parietal bone of the Mousterian neonate of the Hortus cave revealed the topography of the middle meningeal venous network. The anterior branch draining the bregma region can be seen on the anterior portion of the right parietal bone (Fig. 2.19A). It is very simple and receives a minimally ramified posterior tributary. The left parietal bone (Fig. 2.19B) shows the tract of the posterior branch which receives two bifurcated tributaries from the lambda region. This system is simpler than that of modern neonates (see Fig. 2.6), especially in terms of the anterior branch, while the posterior branch is situated more anteriorly. There is no middle branch.

The fragment of the right parietal bone of La Ferrassie child 8, which has been estimated to be about 2 years old, corresponds, *grosso modo*, to the anterior half of the bone. The grooves corresponding to the three branches of the middle meningeal system can be seen (Fig. 2.19C). Behind the coronal suture, the anterior branch, the most ramified receives, anteriorly, several tributaries derived from the frontal region and, posteriorly, a group of three tributaries derived from the bregma region. The central parietal part is crossed obliquely by the middle and posterior branches; the middle branch has a single ramification. This system, devoid of any anastomoses, appears to be very simplified in comparison with the modern 2½-year-old child (see Fig. 2.9).

The almost complete left parietal bone of the Gibraltar II Mousterian child, evaluated to be 5 years old, is limited by the coronal, sagittal, lambdoid, and squamous sutures. It reveals the traces of a relatively simple middle meningeal system (Fig. 2.19D). There are only a few anastomoses between the tributaries of the anterior and middle branches. The posterior branch descends from the

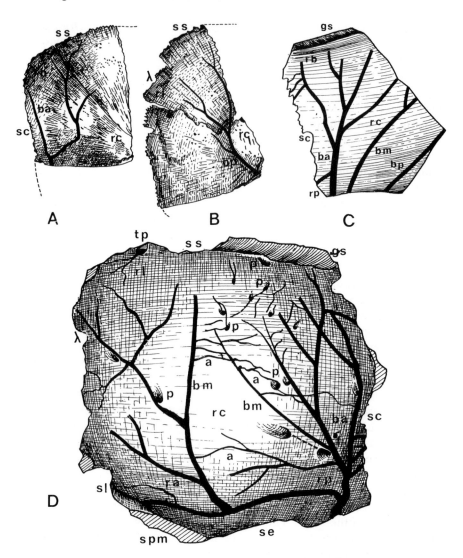

Fig. 2.19. The middle meningeal veins from fragments of parietal bones of Neanderthal child. A, anterior portion of the right parietal bone of the neonate from the Hortus cave; B, posterior portion of the left parietal bone of the neonate from the Hortus cave; C, anterior portion of the right parietal bone of the 2-year-old La Ferrassie child; D, left parietal bone of the 5-year-old Gibraltar II child. gs, sagittal groove; p, diploic foramen; ra, asterion region; rb, bregma region; rc, central parietal region; rl, lambda region; spm, petro-mastoid suture; tp, parietal foramen. (See Fig 2.3 for the other abbreviations.)

lambda region to as far as the region of the squamous suture. In view of its simplicity and the small number of anastomoses, this system differs from that of the modern 5-year-old child (see Fig. 2.10). The small number of anastomoses and the large quantity of diploic foramina concentrated in the bregma region appear to indicate the development of the diploic venous system, as in the modern 2½-year-old child (see Fig. 2.9).

As in the two previous fossils, it was impossible to identify the presence of the great anterior vein, characteristic of adult Neanderthal man. Moreover, the appearance of this vessel in children, when it is subsequently present in adults, is unknown. In Neanderthal man, this enormous vein, which acts like a sinus, may only have developed during adulthood, in the same way as it is sporadically observed in modern man as well as in *Pithecanthropus* and pre-*sapiens*.

Pre-sapiens *and* sapiens *children*

Amongst the pre-*sapiens*, we have selected, as an example, the fragment of left parietal bone from Rissian 15 child from La Chaise de Vouthon (Suard Cave) estimated to be 1 year old, corresponding to almost the entire anterior half of the bone. Strongly modelled by the impressions of the parietal cerebral gyri, while these impressions are not observed in modern children before the age of 2½ years, this parietal portion reveals a relatively simple middle meningeal system (Fig. 2.20A) with an anterior branch well endowed with tributaries, but with only three anastomoses, two anteriorly between the coronal tributaries and the other posteriorly with the middle branch. The posterior branch, divided into two, follows the squamous suture. The topography of the network recalls that observed in modern children of the same age. However, in this case the difference concerns the impressions of the cerebral gyri which correspond to the findings observed in a modern 5-year-old child.

Our example of *Homo sapiens* consisted of the examination of the parietal bones of the neolithic child in the Kitsos cave, near Lavrion (Greece) and the child from the Mungo Lake I grave (Australia), dated to be about 25 000 BP, with respective ages of approximately 2 and 5 years.

The two fragments of the right parietal bone of the Kitsos child 68.193/1950 (Fig. 2.20B), approximately 2 years old, correspond to the anterior portion of the bone (bregma and central regions) and a small superior portion of the obelion region. The anterior branch of the middle meningeal system, present on the anterior fragment, already constitutes a network well supplied with anastomoses, representing the early phase of the vascular grid observed in adults, as seen in the modern child of the same age (see Fig. 2.9). Deep grooves accompanied by diploic foramina can be seen adjacent to the sagittal groove, crossing the grooves of the middle meningeal veins. These grooves correspond to the path of the diploic anastomoses which have not yet been embedded inside the bone.

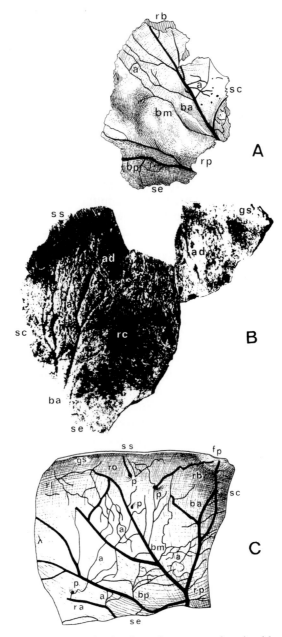

Fig. 2.20. The middle meningeal veins from fragments of parietal bones of advanced fossil forms of children of the human line. A, pre-*sapiens*, anterior portion of the left parietal bone of the 1-year-old child, La Chaise de Vouthon no. 15 (Suard cave); B, *sapiens*, anterior portion of the right parietal bone of the 2-year-old Kitsos child (Greece) no. 68.193/1950; C, left parietal bone of the 5-year-old child from Mungo Lake I (Australia). gs, sagittal groove; p, diploic foramen; ra, asterion region; rb, bregma region; rc, central parietal region; rl, lambda region; ro, obelion region; rp, pterion region. (See Fig. 2.3 for the other abbreviations.)

In modern man, the diploic venous system develops around the age of 2 years. Consequently, there no longer seems to be any difference in the establishment of the diploico-meningeal system between the neolithic child and the modern child, as both present the same stage of development of the meningeal blood supply.

The left parietal bone of the Mungo Lake I child, aged 5 years, presents a topography of the middle meningeal venous system (Fig. 2.20C) similar to that observed in the modern child of the same age (see Fig. 2.10).

There is a large network of anastomoses between the tributaries of the three branches. The frontal branch receives a large number of tributaries across the coronal suture, from the frontal region, as in the modern child.

Overall, the meningeal vasculature therefore reflects the complexity of human evolution. At each stage, marked, up until the appearance of *Homo sapiens*, by the development of a further complexity of the vascular system, there is a coexistence of archaic forms and advanced forms suggesting the existence of two types of evolution: one progressive or mental and the other adaptive or mechanical.

We have observed, in the development of meningeal networks, a correspondence in fossils with the postnatal development of this vascular system which, in modern man, regulates the drainage of cerebral blood. This again confirms the validity of the law of recapitulation of phylogenesis in ontogenesis expressed by Von Baer (1828) and defined by Haeckel (1866).

Due to the extreme simplicity of the middle meningeal system, which lacks the middle branch, *Australopithecus gracilis* would therefore correspond to the present-day antenatal state, while the acquisition of the middle branch in *A. robustus* recalls the arrangement seen in the modern neonate. The first anastomoses encountered in the first representative of the genus *Homo* (*H. habilis*) correspond to the stage observed in the 40-day-old infant. The multiplication of the number of ramifications of the three branches, characteristic of archaic forms of *Pithecanthropus* (*H. erectus*), is very similar to the findings in the 1-year-old infant. The constitution of a real network by the multiplication of the anastomoses becomes effective in the more advanced forms of *Pithecanthropus* (*H. palaeojavanicus*). This network continues to progress in the pre-*sapiens*, in which we find an increasingly dense meshwork, which is encountered in the modern 3-year-old (Omo III), 5-year-old (Swanscombe), and 6½-year-old (La Quina V) children, to eventually form, in *Homo sapiens*, the vascular network observed in adults (Cro-Magnon III).

The archaic juvenile forms appear to present, at the same age, a retardation in the ramification of the network, as for example in the Hortus cave neonate, in which the system corresponds to a modern antenatal stage because of the absence of the middle branch. This difference is also found in the Gibraltar Neanderthal child which is estimated to be 5 years old, on the basis of numerous features, while its meningeal venous network corresponds to that of a modern 2½-year-old child. This difference tends to fade amongst the pre-*sapiens* (La Chaise, Suard cave) and totally disappears in *sapiens*, in which the topography of the system corresponds to the same type in fossils and in modern man (Mungo I, Kitsos).

The anterior meningeal system, visible in modern man from the fifth year of life onwards, does not constitute a real anastomotic network until adulthood. Its delayed appearance is also observed in pre-*sapiens* fossils (Tautavel), in which it corresponds to the modern infantile stage, while in *sapiens*, it reaches its full development (Hahnöfersand) comparable to that observed in modern adults.

References

Bonamy, C., Broca, P., and Beau, E. (1841). *Atlas d'anatomie descriptive du corps humain*, **2**. Victor Masson, Paris.

Breschet, G. (1830). *Recherches anatomiques physiologiques et pathologiques sur le système veineux et spécialement sur les canaux veineux des os*. Paris.

Haeckel, E. (1866). *Generelle Morphologie der Organischen Formenwissenchaft, mecanisch begrundet durch die von Charles Darwin Reformite Descendanztheorie*. G. Reimer, Berlin.

Saban, R. (1980). Le système des veines méningées moyennes chez *Homo erectus* d'après le moulage endocranien. *C.R. 105e Congr. Soc. Sav. Caen*, Sciences (animal biology) **3**, 61–73.

Saban, R. (1984). *Anatomie et évolution des veines méningées chez les hommes fossiles*. Comité des Travaux Historiques et Scientifiques (Mémoires de la Section des Sciences **11**), 289 p. Paris.

Saban, R. (1986). Les rapports des veines méningées moyennes avec la paroi endocrânienne chez l'Homme, au cours de la croissance. *Nova acta Leopoldina* **58, 262**, 425–39.

Saban, R. (1987). Empreintes vasculaires du frontal et hominisation *Soc. Etudes Rech. Préhist. Les Eyzies, Bull*. **36**, 53–61.

Saban, R. and Grodecki, J. (1979). Rapports des vaisseaux méningés avec la paroi endocrânienne chez l'Homme. *C.R. 104e Congr. Soc. Sav. Bordeaux*, Sciences, (animal biology) **2**, 67–80.

Von Baer, K.E. (1828). *Ueber Entwicklungsgeschichte der Thiere. Beobachtung und Reflexion*. Bozntzäger, Koenigsberg.

DISCUSSION

Participants: L. de Bonis, Y. Coppens, R.L. Holloway, R. Saban, P.V. Tobias

Tobias: Your research has focused on the drainage system of the brain. But has your work thrown any light on the arterial side; that is, on the relative contribution to the blood supply of the brain by the carotid system and by the vertebral-basilar system, the two major systems that compete in providing blood to the brain?

Saban: I have limited my work on the drainage system because the marks on the endocranial walls are made by the meningeal veins. I have shown in a series of studies of the human fetus that the arterial branches are continuously covered by a great vein or, more often, by two veins. Nevertheless, the arterial blood pulse does certainly accentuate the process of imprinting.

Holloway: Did you find any correlation within the variations of the meningeal pattern and did you try to find any correlation between the total length of the anastomoses and cranial capacity?

de Bonis: You have mentioned two very distinct forms of the meningeal pattern among the fossils from Sangiran. Would it be possible to attribute these differences to within-population variations and not to two different forms or species?

Saban: In many fossils I have noticed some differences or asymmetries between both sides of the brain, but nothing systematic. To date, I was interested by the formation and the density of the network, but I did not try to quantify it. I think it will be possible and worth doing this with the new technical means of investigation. In all modern populations the meningeal pattern looks the same. It is characterized by a very dense and tight network, especially over the parietal region. The frontal part of the network is also well developed with a large amount of anastomosis. I do not see any very significant differences between the modern population. My purpose is to give different morphological types, without attempting to deal with hominid phylogeny. The fossils are simply grouped according to the characteristics of their vascular system. The density of the drainage network has a functional significance and is associated with brain size. From gracile to robust australopithecines we observe a brain size increase of 100 ml and the acquisition of middle branch. Between robust australopithecines and *Homo habilis* the cranial capacity increases from 520 ml to 770 ml and the meningeal pattern is more complicated. In *Homo erectus* the meningeal pattern remains fairly simple with the persistence of the petrosquamous sinus. When the cranial capacity reached over 1000 ml, it seems that it has a strong effect on the development of the middle meningeal network.

Tobias: Roger Saban's communication has added a very important dimension to our thinking, and that is the question of ontogenetic changes. In the modern child, between the age of 40 days and 2½ years the brain triples in size. Obviously, the bigger the brain, the bigger the number of branches and anastomoses per unit area of the surface of the brain. So, do the ontogenetic and phylogenetic changes indicate a progressive complexity of the vascular system? The reason why age changes are so important is that 50 per cent of all of early hominid specimens are juvenile and the great majority of them are suitable for studying patterns of convolution and meningeal vessels.

Coppens: To complete this topic, it appears that the meningeal network in great apes is more developed over the rear part of the brain, while in humans it is the anterior part of the brain. Can you comment on that?

Saban: The process of imprinting the different parts of the brain on the endocranial walls is very dependent on the position of the fetus *in utero* and acquisition of bipedalism in the young child. At about the age of 5 years, the brain undergoes dramatic growth together with a change in cranial morphology. In great apes, the meningeal pattern is important, but less dense than in hominids, and characterized by the dominance of the posterior region. We have the same basic pattern in orang-utan, on one hand, and in chimpanzee and gorilla, on the other. So, we observe the same pattern in two different lineages separated by a long time period. It is worth noting that in gorilla the anterior region of the network is more developed.

de Bonis: I do not understand what you mean when you talk about pre-*sapiens* and para-*sapiens* and what are their phylogenetic positions. I should point out that the fossils from Tautavel and Swanscombe display clear Neanderthal characteristics. They are pre-*neanderthals* and not pre-*sapiens*.

Coppens: The terminology you have used is very confusing, even if you are concerned with typology and not with phylogeny. I think that it might be advisable to call these types A, B, C, or D for the purpose of grouping, and to avoid any reference to the theory of pre-*sapiens* which is clearly refuted today. Besides. I am sure that the conclusions of your very fine study have a real interest for systematics and phylogeny. It seems that Neanderthals have a very particular meningeal pattern. Can you give more details about this?

Saban: For what I have called the pre-*sapiens*, including *Homo palaeo-javanicus*, the meningeal pattern is characterized by large amount of anastomoses. In the para-*sapiens* the meningeal network is rich, but with less anastomoses. Concerning Neanderthals, we observe the same pattern on the six or eight skulls known. The network remains simple and is characterized by the presence of a great anterior vein, which is a modification of one of the two veins running along the anterior branch of middle meningeal artery. The most anterior branch is large to form a sinus and sinks into the cavernous sinus. I should add that the anterior branch characteristic of Neanderthals is also observed in some recent *Homo erectus* (Sangiran 7) and sometimes in modern humans. This characteristic is not specialized and represents the survival of a primitive pattern.

3

Toward a synthetic theory of human brain evolution

RALPH L. HOLLOWAY

Introduction

This chapter presents a general synthesis of the evolution of the hominid brain. Elsewhere (Holloway 1968, 1979, 1983*a,c*) I have indicated why brain mass alone is not a sufficient variable to explain the evolution of human behaviour and the brain, and offered a number of different models relevant to synthesizing mass with reorganization and hierarchy (ontogenesis). Here, I argue that a major *reorganizational* change from a pongid to hominid pattern occurred some 3–4 million years ago in the posterior parietal, anterior occipital, and superior temporal portions of the cerebral cortex. This occurred *before* frontal lobe (Broca's area) reorganization in *Homo habilis*, about 1.8 million years ago. The doubling of brain size from roughly 750 cc in *H. habilis* to 1400 cc in modern *H. sapiens* occurred *after* (at least) three major *reorganizational* changes evolved in earlier hominid brains. These were:

1. Reduction of primary visual striate cortex (area 17 of Brodmann) and relative enlargement of extrastriate parietal cortex (areas 18 and 19), angular gyrus (area 39), and supramarginal gyrus (area 40) of the inferior parietal lobe. This represents the first major change from pongid to human brain organization, and occurred between 3 and 4 million years ago.

2. A reorganization of the frontal lobe occurred mainly involving the third inferior frontal convolution, known as Broca's area, toward a definitely human external morphology most fully in evidence in the KNM-ER 1470 specimen of *H. habilis*. This probably occurred between 2.5 and 1.8 million years ago.

3. In addition to brain enlargement (mostly as an allometric relation to body size), there was a change from a pongid pattern of minimal cerebral asymmetries to a human pattern involving a higher degree of left-occipital and right-frontal petalial enlargements, associated with greater hemispheric specialization. This appears most clearly in *H. habilis*, but may have occurred earlier in *Australopithecus africanus*, 2 to 3 million years ago (Holloway and de Lacoste-Lareymondie (1982)).

Regarding brain volume changes integrated with *reorganization*, I suggest the following sequence:

1. There was a small allometric increase in brain size from early *A. afarensis* to later *A. africanus*, i.e. from roughly 400 cc to perhaps 460 cc.

2. From *A. africanus* to *Homo habilis* there was an increase in brain size from about 450 cc to 700–750 cc in *H. habilis*, an increase of 250–300 cc. As there is little solid evidence for any major increase in body size between the two species, we cannot be certain that this significant increase was allometrically related to body size. (That is, specimens of habilines such as KNM-ER 1813 and KNM-ER 1805, or OH 7, or OH 8, were surely smaller than KNM-ER 1470 in both brain and body size.)

3. From *H. habilis* to *H. erectus*, there was a modest increase of brain size from roughly 750 cc to perhaps 900 cc which was most likely an allometric one, as the KNM-WT 15 000 'strapping youth' postcranial elements suggest a significant increase in body size, but only a small increase in brain size.

4. Following *H. erectus*, I would argue that most of neural evolution was a gradual increase of brain volume and increasing refinement or asymmetrical organization that was not allometrically related. There is very little evidence for any major change of body size from *H. erectus* to archaic *H. sapiens* that would provide an allometric basis (at least as we conceive it) for some 400–500 cc increase in brain size. Elsewhere (Holloway 1985*b*) I have suggested that the higher cranial capacities of Neanderthals (still *H. sapiens* to me) were related to lean body mass and possibly, adaptation to cold.

5. There has been a small and gradual reduction of brain size based on allometry from the Upper Palaeolithic to modern times. (See Table 3.1 for a summary of reorganizational and size changes.)

Evidence from comparative neuroanatomy

Excellent reviews of primate parietal cortex, both anatomically and behaviourally, have been provided by Hyvarinen (1982), Kaas and Huerha (1988), Kaas and Pons (1988), Pandya *et al.* (1988) Yin and Medjbeur (1988), and Pandya and Yeterian (1990). For detailed reviews of human brain function, see: Goldman-Rakic (1987), Nass and Gazzaniga (1987), and Ojemann and Creutzfeldt (1987). For the evolution of neocortex, see Allman (1990).

As I indicated in earlier publications (Holloway 1976, 1979, 1983*a,c*, 1988*b*) the quantitative data regarding volumes of true primary visual striate cortex shows that there has been a dramatic relative decrease of the volume of this tissue in *Homo sapiens*. The actual volume in *Homo* is 121 per cent *less than expected* for a primate of its brain weight (Holloway 1983*a*, 1988*b*). The obvious interpretation is that there was a sufficient amount of primary visual striate cortex for human vision *but* that there also was a relative *increase* in posterior parietal association cortex. The data for demonstrating this fact was provided by Stephan *et al.* (1981). This conclusion is strongly reinforced by the fact that the volume for the human lateral geniculate

Table 3.1. *Summary of reorganizational and size changes in the evolution of the hominid brain*

Brain changes	Taxa	Time (myr)	Evidence
1. Reduction area 17 (primary visual striate cortex) and relative increase of posterior parietal cortex	*A. afarensis* mean brain = *c.* 400 cc	*c.* 3	AL 162-28 has posterior position for IP and lunate sulci
2. Small increase in brain size, probably allometric	*A. africanus* mean brain = 440–450 cc	2.0–2.5	South African sample of brain endocasts
3. Reorganization of frontal lobe and an increase in cerebral asymmetries	*H. habilis* brain = 752 cc	1.8–2.0	KNM-ER 1470
4. Modest allometric in brain size, and increase in cerebral asymmetries	*H. erectus* brain = 800–1000 cc	1.5–0.5	*H. erectus* brain casts and postcranial bones e.g. KNM-WT 15 000
5. Gradual increase and refinements in cortical organization to modern *Homo* pattern	*H. sapiens neanderthalensis* 1200–1700 cc	0.1	Brain size, archaic *Homo* and Neanderthal endocasts
6. Small reduction in brain size among modern *H. sapiens*	*H. sapiens*	0.01 to present	Modern cranial capacities

nucleus falls some 146 per cent less than expected for a primate of its brain weight. The thalamus as a whole, however, is very close to predicted values for *Homo*, as are most other brain structures (e.g. cerebellum, cortex, hypothalamus, hippocampus). (We still need more quantitative studies on the distributions of cortical tissues, as well as other structures, within the order primates. Quantitative differences regarding dendritic branching and synapses would be most welcome. As cytoarchitectonic regions vary in different primate species, it is obvious that more than brain size has evolved within the order. Thus, notions such as 'extra cortical neurones', as used by Jerison (1973) and more recently Tobias (1987) are actually generating fictional numbers. (See Holloway 1966, 1968, 1974, 1979, for critiques.)

The volumetric differences (completely ignored by Jerison 1990, pp. 288–9) indicating a significantly lower proportion of area 17 in *Homo* is matched by the surface convolutional patterns which show that *all* pongids have an anterior location of the lunate ('*Affenspalte*') sulcus (e.g. Connolly 1950). The lunate sulcus is the anterior boundary of area 17. When the lunate is present in *Homo*, the sulcus is located in a quite posterior position (see Holloway 1985*a* for some of the history regarding this sulcus). Elsewhere (Holloway 1988*a*) I have provided measurements showing the relative anterior position of the lunate sulcus in 32 cerebral hemispheres of the chimpanzee and compared this position to the possible location of a lunate sulcus on the Taung *A. africanus* brain endocast.

A further point should be stressed. The intraparietal (IP) sulcus, which usually divides the parietal cortex into superior and inferior portions, *always* abuts (and terminates) against the lunate sulcus in all pongid species. This fact is of particular importance as the fossil endocasts (see next section) seldom show an unambiguous lunate sulcus, *whereas one can sometimes see the posterior end of the intraparietal sulcus. If the posterior end of the intraparietal is in a relatively posterior position, any lunate sulcus present on the specimen* must *appear posterior to it* (see Fig. 3.1).

Why is so much emphasis being placed on the posterior parietal cerebral cortex in this chapter? Posterior parietal cortex has been demonstrated to be of great importance in mediating the perception of spatial relationships among objects (see Kolb and Winshaw 1985, for a general review). The analysis and integration of sensory information (i.e. cross-modal matching) is upset when lesions are produced in this region (e.g. Mishkin *et al.* 1982). Geschwind's famous 1965 article on "disconnexion syndromes" made these points when analysing posterior lesions and receptive aphasias, and suggested this was an important foundation or prerequisite for language. It would not be a necessary correlate of language since chimpanzees have shown cross-modal matching (Davenport *et al.* 1973), as have some monkey species (Jarvis and Ettlinger 1977). In any event, the posterior parietal cortex is best known for its multi-modal processing of visual, auditory, and somatic sensory information, its relation to social communication and visuospatial integration (see Mishkin, this volume).

I suggest that a species newly adapting to a growing savannah-like environment (e.g. Chapter 6 and references therein) and utilizing a relatively recent adaptation (i.e. bipedal striding gait), benefited from some neural reorganization that enhanced an appreciation of visuospatial integration, at least with regard to its patchy distribution of food and water resources. (After all, bipedal locomotion probably correlates with increased uses of the hand in all sorts of activities from tool-using, collecting, carrying, scavenging, to flea-picking.) Furthermore, increased competence in social communication (not necessarily language) would surely have been an advantage for smaller bands of early hominids. (In addition, we should also expect some brain reorganization sufficient to service the musculoskeletal changes inherent in bipedal locomotion, but it is unclear where, beyond the postcentral sulcus, it would be in the parietal cortex.) Additionally, the inferior parietal lobe, particularly in its more left–right asymmetric functioning in *Homo*, plays a major role in the recognition of social behavioural complexity (including facial recognition), with the left side being more analytic than the Gestalt or so-called 'emotional' right side. Such an animal adjusting to a new locomotory adaptation and perhaps social existence ('social intelligence') was *Australopithecus afarensis*.

Still, it is wise to remember that the extant comparative evidence (i.e.

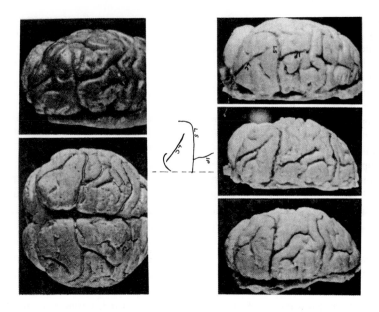

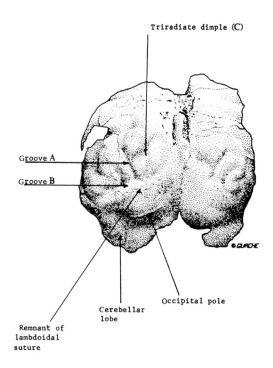

Triradiate dimple (C)

Groove A

Groove B

Occipital pole

Cerebellar
lobe

Remnant of
lambdoidal
suture

Fig. 3.1. (a) A composite photograph of five *Pan* troglodytes chimpanzee brain casts showing the invariant orientations of the intraparietal sulcus (IP), the lunate sulcus (LS), and the lateral calcarine sulcus (LC) as schematically diagrammed in the centre space. The lunate sulcus is the anterior boundary of primary visual striate cortex, area 17. The posterior part of IP always ends at the LS in pongids. (b) A stipple drawing of the brain endocast of the Hadar 162-28 *Australopithecus afarensis* specimen. The endocast is oriented in a dorsal and posterior view similar to the bottom two *Pan* brain casts in (a). Groove A is the intraparietal sulcus agreed upon by the author and Falk. Groove B could be either the lunate sulcus or a groove caused by the posterior lip of the parietal bone. In any event, the distance from the posterior part of IP to the occipital pole (OP) is approximately half the same distance in chimpanzee brain casts which are also smaller than the Hadar fossil endocast (see Table 3.2 for measurements). The posterior placement of IP is not a pongid feature but one more closely related to *Homo*. (Modified after Holloway 1983*b*.)

---

between *Homo* and *Pan*) is only indirect. Indeed, most of recent neuro-anatomy has studied the macaque monkey rather than chimpanzee, and it would appear that there are asymmetries in many other animals, and that the macaque cortex is almost a perfect homologue of human cerebral cortex. The direct evidence for such proposed changes in neural cortical organization must be verified from the fossil record. This leads to the next subject, palaeoneurology.

Evidence from palaeoneurology

In his 1987 article, Tobias neglected to mention that I have always indicated that the lunate sulcus could not be unambiguously seen on the Taung specimen of *Australopithecus africanus*. In 1981, I said the following: 'We cannot prove where the lunate sulcus is located, but only demonstrate where it is not' (Holloway 1981*a*, p. 49).

My arguments with Falk regarding the possible anterior position of the lunate sulcus in a chimpanzee position should indicate just how controversial and difficult it is to ascertain this landmark on fossil endocasts. Nevertheless, my sample of 32 chimpanzee hemispheres shows that their lunate sulcus is located more than 2 standard deviations *posterior* to where Falk (1980) suggests it is located on the Taung specimen. I clearly indicated in my 1985*a* contribution to the Dart volume edited by Tobias (1985) that neither Clark (1947) nor I were convinced that the Taung specimen even had a lunate sulcus. It would be a remarkable reversal of evolutionary process if the Taung specimen's lunate sulcus were in a position even more primitive (i.e. more anterior) than in *Pan* (e.g. *Macaca*), which is where Falk places it.

Earlier in time, after the split between pongids and hominids (? 6–10 million years ago), hominids lost the pongid anterior placement of the lunate sulcus. That is, a posterior placement (or none at all) became a derived

Table 3.2. *Some measurements comparing the posterior cortex of* Pan troglodytes *(chimpanzee) and* Australopithecus afarensis

Specimen	Volume (ml)	Distance (OP–IP, mm)
A. afarensis		
AL 162-28	375–400	15–16
Pan troglodytes		
Welker	400	35.5
YN70-119	290	30
YN73-74	407	37
YN80-7	320	30
B60-11	270	29
B60-7	300	36
Paris	295	30
B57-2	250	29
BMNH	275	30
JH1	207	24.5
JH2	237	32.5
JH3	297	31
JH4	209	32
JH5	200	30
JH6	247	32.5
JH7	202	29
JH8	281	35
JH9	214	30
Infant	140	25.5
Average	271	31.2
Standard deviation (s.d.)	20.7	3.08
Correlation = 0.65, P = 0.0032		

OP, occipital pole; IP, intraparietal sulcus.

Number of s.d.s of AL 168-28s IP is from the *Pan* average: s.d. 5.1.

Arc measurements are shown for occipital pole (OP) to the most caudal end of the intraparietal sulcus (IP). The measurements were made on fixed brains and brain casts. The infant values were *not* included in the calculations. Notice that if one accepts Falk's (1980) placement for the lunate sulcus on the Taung *A. africanus* specimen, the OP–IP measurement will be at least 42 mm (Holloway 1988*a*). Note the similarity of these measurements and statistics to those in Holloway (1985*a*, p. 57). Notice, also, that the chimpanzee infant has an OP–IP arc of 25.5 mm, considerably larger than the Hadar hominid AL 168-28 with 15.5 mm. The figures strongly suggest reorganization of posterior parietal and occipital cerebral cortex in early *Australopithecus*, with a reduction of primary visual striate cortex, e.g. area 17 of Brodmann, and a relative increase in inferior parietal cortex.

character state in *Australopithecus* or *Homo* (autapomorphy?), rather than a shared primitive retention with *Pan*. Fortunately, the AL 162–28 posterior cranial fragment of *A. afarensis* preserves enough morphology to distinguish the posterior end of the intraparietal sulcus discussed earlier. On this crucial matter Falk (1985) and I agree.

Our disagreement stems from how the fragment is oriented (Holloway 1983*b*; Falk 1985, 1986; Holloway and Kimbel 1986). Falk (1985) misplaced the endocast almost 40 degrees in an anticlockwise direction, which elevated the asterionic point above the small remnant of squamosal suture on the parietal bone. Measurements by my student and colleague, Jill Shapiro, indicate that in *Pan*, the superior margin of the squamosal suture of the parietal is often 10 mm or more superior to asterion when the skull is in the Frankfurt Horizontal plane (Holloway and Shapiro, 1992). In hominids, from *A. afarensis* on, I cannot find a *single* instance in which the squamous suture is not clearly superior to asterion. Falk's (1985) orientation of the cast places the asterionic point well superior to the squamous suture, thus achieving what only 2 out of 36 chimpanzee endocasts in my collection show: cerebellar lobes projecting more posteriorly than the occipital poles of the cerebral cortex! Falk's (1986) response to our criticism was to suddenly define two previously unidentified, and practically invisible sulci, the superior parietal, and the superior temporal, suggesting that their orientation would only fit in her orientation. Alas, these sulci cannot be identified on the AL 162–28 specimen with any confidence. It is more probable that these tiny dimples could be either s. sublunatus or s. postlunatus (see Shantha and Manocha 1969). In any event, one should not attempt to determine cranial orientation from insecure brain sulci. It is better to properly orient the endocast first. Thus the matter of relative locations of the asterion and squamosal suture, in addition to the indices I published earlier (Holloway 1985a and expanded herein) provide strong evidence that any lunate sulcus on the AL 162-28 would have been in a very posterior position compared to all pongids. Falk's measurement's of the occipital pole to posterior intraparietal sulcus are awaited. This is the best direct evidence available to show that posterior parietal and anterior occipital cortex underwent an important reorganizational process at least a million or so years *before* the appearance of *H. habilis*, and perhaps *A. africanus* as well. (See Table 3.2 for measurements.)

Behaviour and natural selection

Earlier (Holloway 1975, 1976, 1979, 1981b, 1983a,c) I argued that the human brain evolved under strong natural selection for enhanced social behaviour emphasizing communication. Four behavioural aspects were involved in the early reorganization of posterior parietal cortex:

1. Communication involving visual, auditory, and sensorimotor *integration* synthesis (Pandya and Yeterian 1990), i.e. multi-modal processing. Expansion of posterior parietal cortex, particularly the inferior parietal lobule, laid the '*fondation neurologique*' for language reception.
2. Social competence and social complexity involved greater skills in verbal and non-verbal communication.
3. Visuospatial integration related to: (a) tool-use and making, throwing objects with force, and accuracy; (b) more sophisticated longer-term

memory of spatial locations of self and others, as well as objects and resources (trees, water holes, etc.).

4. As natural selection favoured (2) and (3), asymmetrical parcellation of these functions probably increased, becoming more lateralized, with possible development of sex differences between the two hemispheres as distributed through the corpus callosum (Holloway 1983a, 1990). All of these are part and parcel of what is often called 'social intelligence'.

In sum, reorganization may first have involved the parietal lobe, which integrated with the first and second allometric brain increases between *A. afarensis* and *H. habilis*. In addition, reorganization of the frontal lobe occurred prior to *H. habilis*. Primitive language now had its major neurological foundations, and became increasingly elaborate throughout the rest of hominid evolution beginning with *H. erectus*, and reflected neurologically as a substantial increase in brain size, some of it related to body size, but most of it to behavioural complexity (see Holloway 1967, 1981b). I mention these speculative possibilities because we are really at an impasse regarding either the *degree* or *kind* of subtle differences that separate the human brain from those of other higher primates (e.g. Holloway 1968, 1979, 1983a,c).

Macroscopically, what we see are small reorganizational changes (e.g. sensorimotor and association cortex) and a tripling of brain size. I completely disagree with Gould (1988) that human language was simply a 'spandrel of the brain', a mere accidental epiphenomenon of *Homo*'s large brain size (see also Pinker, this volume). The strong communicative proclivities of higher primates, the clear-cut formation of a true *Homo*-like Broca's area in a small-brained hominid (KNM-ER 1470), the strong cortical asymmetry, and the presence of stone tools made to a standardized pattern all suggest some social communicative competence beyond that of any chimpanzee. Adding on reorganizational changes in the posterior parietal cortex as well as hemispheric asymmetry (thus specialization) makes the argument for any early onset of primitive language more probable, and surely communicative competence was under the direction of natural selection.

Conclusion

I wish to make some predictions about future genetic DNA findings in *Homo* and *Pan*. The difference between these two species' brains at the genetic level, while minor, will rest with the regulatory genes that control the rates of hyperplasia and hypertrophy of neural nuclei and regional cortical areas, as well as the whole brain in relation to timing and rate of growth. *Pan* and *Homo* are more than 99 per cent identical in their nuclear DNA, and their behaviour, apart from language, is very similar. Given the extraordinary degree of homology between macaques and *Homo* as discussed by Galaburda (1984a,b) and Deacon (1988a,b), very few differences in structural genes should be discernible. At least two gene groups will be found which relate to regulatory functions (either as true regulatory genes or as genes suppressing or ending portions of regulatory interactions with structural genes). One group will play a role in the relative reduction of Brodmann's area 17 (perhaps through fetal cell death), shutting off hyperplasia earlier in *Homo* and a compensatory increase in posterior parietal cortex (areas 7, 39, 40), i.e. an increased time or rate of hyperplasia and hypertrophy. This may well include differences between ontogenetic growth rates of right and left cortices, with the left being slowest, the right more precocious (e.g. see Scheibel 1990).

A second group will play a role in the overall growth of the central nervous system, extending the periods for both hyperplasia and hypertrophy of the whole brain. A third group will probably be found that controls the development and integration of the third inferior frontal convolution which is so heavily involved in human language production, despite the region's homology with that in the macaque (see Deacon 1988, and, in particular, Galaburda's 1984*a* review). Again, rates will differ between left and right areas, in accordance with Scheibel's (1990) finding regarding increased dendritic branching in left Broca's areas.

These predictions were suggested in an earlier paper (Holloway 1979, 1983*a,c*) based on studies of the residuals between expected and actual allometric volumes of various regions of the primate brain in *Homo*. Yet a fourth group might play a role in the differential hypertrophy of left and right hemispheric regions as a basis for task specialization.

Acknowledgements

Some of the data discussed were collected under NSF Grant BNS-84-18921. I am also grateful to Phillip Tobias, Alun Hughes, Bob C. Brain, R.E.F. Leakey, and Alan Walker, Don Johanson, William Kimbel, and Yves Coppens for allowing me to study the early fossil hominids. I especially thank Alan Mann for his encouragement, and Daisy Dwyer for her patience.

I also thank Jill Shapiro for her critical and helpful comments on this paper, and Joyce Monges and Gabrielle Malka for their translation of earlier versions of this paper into French.

References

Allman, J. (1990). Evolution of neocortex. In *Cerebral cortex* (ed. E.G. Jones and A. Peters), Vol. 8A, pp. 269–83. Plenum, New York.

Allman, J. and McGuiness, E. (1988). Visual cortex in primates. In *Comparative primate biology* (ed. H.D. Steklis and J. Erwin), Vol. 4, pp. 279–326. Liss, New York.

Clark, W.E.L. (1947). Observations on the anatomy of the fossil *Australopithecinae*. *J. Anat.* **81**, 300–33.

Connolly, C.J. (1950). *The external morphology of the primate brain*. C.C. Thomas, Springfield, Illinois.

Davenport, R.K., Rogers, C.M., and I.S. Russell (1973). Cross-modal perception in apes. *Neuropsychologia* **11**, 21–8.

Deacon, T.W. (1988*a*). Human brain evolution: 1. Evolution of language circuits. In *Intelligence and evolutionary biology* (ed. H.J. Jerison and I. Jerison), NATO Series ASI, Series G, Vol. 17, pp. 368–82. Springer, Berlin.

Deacon, T.W. (1988*b*). Human brain evolution: 2. Embryology and brain allometry. In *Intelligence and evolutionary biology* (ed. H.J. Jerison and I. Jerison), NATO Series ASI, Series G, Vol. 17, pp. 383–416. Springer, Berlin.

Falk, D. (1980). A reanalysis of the South African australopithecine natural endocasts. *Am. J. Phys. Anthropol.* **53**, 525–39.

Falk, D. (1985). Hadar AL 162–28 endocast as evidence that brain enlargement preceded cortical reorganization in hominid evolution. *Nature*, **313**, 45–7.

Falk, D. (1986). Reply to Holloway and Kimbel. *Nature* **321**, 536–7.

Galaburda, A.M. (1984*a*). The anatomy of language: lessons from comparative anatomy. In *Biological perspectives on language* (ed. D. Caplan, A.R. Lecours, and A. Smith), pp. 290–302. MIT Press, Cambridge, Mass.

Galaburda, A.M. (1984*b*). Anatomical asymmetries. In *Cerebral dominance: the biological foundations* (ed. N. Geschwind and A.M. Galaburda), pp. 11–25. Harvard University Press, Cambridge, Mass.

Geschwind, N. (1965). Disconnexion syndromes in animals and man. *Brain* **88**, 239–94; 585–644.

Goldman-Rakic, P.S. (1987). Circuitry of primate prefrontal cortex and regulation of behavior by representational memory. In *The nervous system*, Section I of *Handbook of physiology* (ed. V.B. Mountcastle, F. Plum, and S.R. Geiger), Vol. 5, Part 1, pp. 373–418. American Physiology Society, Bethesda, Md.

Gould, S.J. (1988). *The evolution of the human brain*. James Arthur Lecture. Delivered at the American Museum of Natural History, New York.

Holloway, R.L. (1966). Cranial capacity and neuron number: critique and proposal. *Am. J. Phys. Anthropol.* **25**, 305–14.

Holloway, R.L. (1967). The evolution of the human brain: some notes toward a synthesis between neural structure and the evolution of complex behavior. *General Systems*, **XII**, 3–19.

Holloway, R.L. (1968). The evolution of the Primate brain: some aspects of quantitative relationships. *Brain Res.* **7**, 121–72.

Holloway, R.L. (1974). On the meaning of brain size. Book review of H.J. Jerison's *Evolution of brain and intelligence*. *Science* **184**, 677–9.

Holloway, R.L. (1975). *The role of human social behavior in the evolution of the brain*. 43rd James Arthur Lecture (1973), 45 pp. American Museum of Natural History.

Holloway, R.L. (1976). Paleoneurological evidence for language origins. In *Origins and evolution of language and speech. Ann. N.Y. Acad. Sci.* **280**, 330–48.

Holloway, R.L. (1979). Brain size, allometry, and reorganization: toward a synthesis. In *Development and evolution of brain size: behavioral implications* (ed. M.E. Hahn, C. Jensen, and B.C. Dudek), pp. 59–88. Academic Press, New York.

Holloway, R.L. (1981*a*). Revisiting the S. African Australopithecine endocasts: results of stereoplotting the lunate sulcus. *Am. J. Phys. Anthropol.* **56**, 43–58.

Holloway, R.L. (1981*b*). Cultural symbols and brain evolution: A synthesis. *Dialect. Anthropol.* **5**, 287–303.

Holloway, R.L. (1983*a*). Human paleontological evidence relevant to language behavior. *Hum. Neurobiol.* **2**, 105–14.

Holloway, R.L. (1983b). Cerebral brain endocast pattern of *A. afarensis*. *Nature* **303,** 420–2.

Holloway, R.L. (1983c). Human brain evolution: a search for units, models, and synthesis. *Can. J. Anthropol.* **3,** 215–32.

Holloway, R.L. (1984). The Taung endocast and the lunate sulcus: a rejection of the hypothesis of its anterior position. *Am. J. Phys. Anthropol.* **64,** 285–8.

Holloway, R.L. (1985a). The past, present, and future significance of the lunate sulcus in early hominid evolution. In *Hominid evolution: past, present, and future* (ed. P.V. Tobias), pp. 47–62. Liss, New York.

Holloway, R.L. (1985b). The poor brain of *Homo sapiens neanderthalensis*: see what you please . . . In *Ancestors: the hard evidence* (ed. E. Nelson), pp. 319–24. Liss, New York.

Holloway, R.L. (1988a). Some additional morphological and metrical observations on *Pan* brain casts and their relevance to the Taung endocast. *Am. J. Phys. Anthropol.* **77,** 27–33.

Holloway, R.L. (1988b). The brain. In *Encyclopedia of human evolution and prehistory* (ed. I. Tattersall, E. Delson, and J. van Couvering). pp. 98–105. Garland, New York.

Holloway, R.L. (1990). Sexual dimorphism in the human corpus callosum: its evolutionary and clinical implications. In *Apes to angels: essays in anthropology in honor of Phillip V. Tobias* (ed. G. Sperber), pp. 221–8. Wiley–Liss, New York.

Holloway, R.L. and Kimbel, W.H. (1986). Endocast morphology of Hadar hominid AL 162–28. *Nature* **321,** 536.

Holloway, R.L. and LeLacoste-Lareymondie, M.C. (1982). Brain endocast asymmetry in pongids and hominids: some preliminary findings on the paleontology of cerebral dominance. *Am. J. Phys. Anthropol.* **58,** 101–10.

Holloway, R.L. and Shapiro, J. S. (1992). Relationship of squamosal suture to asterion in pongids (*Pan*): relevance to early hominid brain evolution. *Am. J. Phys. Anthropol.* **89,** 275–82.

Hyvarinen, J. (1982). The posterior parietal lobe of the primate brain. *Physiol. Rev.* **62,** 1060–129.

Jarvis, M.J. and Ettlinger, G. (1977). Cross-modal recognition in chimpanzees and monkeys. *Neuropsychologia* **15,** 499–506.

Jerison, H.J. (1973). *Evolution of brain and intelligence*. Academic Press, New York.

Jerison, H.J. (1990). Fossil evidence on the evolution of the neocortex. In *Cerebral cortex* (ed. E.G. Jones and A. Peters), pp. 285–309. New York, Plenum.

Jerison, H.J. and Jerison, I. (1988). *Intelligence and evolutionary biology*. NATO Series ASI, Series G, Vol. 17. Springer, Berlin.

Kaas, J. and Huerta, M.F. (1988). The subcortical visual system of primates. In *Comparative primate biology* (ed. H.D. Steklis and J. Erwin), Vol. 4, pp. 327–92. Liss, New York.

Kaas, J. and Pons, T.P. (1988). The somatosensory system of primates. In *Comparative primate biology* (ed. H.D. Steklis and J. Erwin), Vol. 4, pp. 421–68. Liss, New York.

Kolb, B. and Whishaw, I.Q. (1985). *Fundamentals of human neuropsychology* (2nd edn). Freeman, New York.

Milner, A.D., Ockleford, E.M., and W. Dewar (1977). Visuo-spatial performance following posterior parietal and lateral frontal lesions in stumptail macaques. *Cortex* **13,** 350–60.

Mishkin, M., Lewis, M.E., and Ungerleider, L.G. (1982). Equivalence of parietopreoccipital subareas for visuospatial ability in monkeys. *Behav. Brain Res.* **6,** 41–55.

Nass, R.D. and Gazzaniga, M.S. (1987). Cerebral lateralization and specialization in human central nervous system. In *The nervous system*, Section 1 on *Handbook of Physiology* (ed. V.B. Mountcastle, F. Plum, and S.R. Geiger, Vol. 5, Part 1, pp. 701–62. American Physiological Society, Bethesda, Md.

Ojemann, G.A. and Creutzfeldt, O.D. (1987). Language in humans and animals: contributions of brain stimulation and recording. In *The nervous system*, Section 1 of *Handbook of physiology* (ed. V.B. Mountcastle, F. Plum, and S.R. Geiger), Vol. 5, Part 1, pp. 675–99. American Physiological Society, Bethesda, Md.

Pandya, D.P., Seltzer, B., and Barbas, Helen (1988). Input-output organization of the primate cerebral cortex. In *Comparative primate biology* (ed. H.D. Steklis and J. Erwin), Vol. 4, pp. 39–80. Liss, New York.

Pandya, D.P. and Yeterian, E.H. (1990). Architecture and connections of cerebral cortex: implications for brain evolution and function. In *Neurobiology of higher cognitive function* (ed. A.B. Scheibel and A.F. Wechsler), No. 29, UCLA Forum in Medical Science, pp. 53–83. Guilford, New York.

Scheibel, A.B. (1990). Dendritic correlates of higher cognitive function. In *Neurobiology of higher cognitive function* (ed. A.B. Scheibel and A.P. Wechsler), No. 29, UCLA Forum in Medical Science, pp. 239–68. Guilford, New York.

Shantha, T.R. and Manocha, S.L. (1969). The brain of chimpanzee. I. External morphology. In *The chimpanzee* (ed. G.H. Bourne), Vol. I, pp. 187–237. University Park Press, Baltimore.

Stephan, H., Frahm, H., and Baron, G. (1981). New and revised data on volumes of brain structures in Insectivores and Primates. *Folia Primatol.* **35,** 1–29.

Tobias, P.V. (ed.) (1985). *Hominid evolution: past, present and future*. Liss, New York.

Tobias, P.V. (1987). The brain of *Homo habilis*: A new level of organization in cerebral evolution. *J. Hum. Evol.* **16,** 741–62.

Yin, Tom C.T. and Medjbeur, S. (1988). Cortical association areas and visual attention. In *Comparative primate biology* (ed. H.D. Steklis and J. Erwin), Vol. 4, pp. 393–422. Liss, New York.

DISCUSSION

Participants: L. de Bonis, J.-P. Changeux, Y. Coppens, R.L. Holloway, M. Mishkin, P. Picq, S. Pinker, F.H. Ruddle, P. V. Tobias, L. Weiskrantz

Weiskrantz: Can you see on the casts the difference in the slope of the fissure of Sylvius between the left and right side of the human brain, which is also correlated with a larger supratemporal plane on the left, and when do you first see that in a cast or in any species in human evolution?

Holloway: There is some slight evidence of asymmetry in chimpanzee and macaques, but it is very hard to see it on human endocasts. The hominid endocasts are too poor to be certain of this asymmetry, but more research has to be done.

Coppens: I just want to say again how your analysis of the reorganization of the *Australopithecus* brain is consistent with the important changes in anatomy, locomotion, diet, and behaviour. In comparison with African apes, *Australopithecus* shows small canines molarized premolars, thick enamel, a different diet, a short face, a bipedal mode of locomotion, a human pattern of parturition, and the first evidence of tool-making. It is difficult to think that such a creature kept an ancestor's brain. As far as language is concerned, I have something of environmental obsession. It think that the movement of the larynx is part of the adaptation of the upper respiratory apparatus to the drought 3 million years ago. In other words, we speak because it was dry.

Pinker: When you say that there is evidence for Broca's area in *Homo habilis*, do you see trace of the sulci that define the convolution?

Holloway: The frontal lobe is broadened in KNM-ER 1470 and is more human looking. There is a relief or a bump that makes a good homologue of area 45. In *Homo* this is a weak evidence, but it is better developed than in *Australopithecus*.

Pinker: Some research has been done on the basicranium in order to make speculations about the positioning of the pharynx. Does the basicranium of *Homo habilis* display the modern pattern?

Holloway: This work has been done by Jeffrey Laitman and his colleagues. Obviously, one cannot ignore the peripheral tissues and one has to allow for the possibility of what occurs at the base of the cranium. However, I disagree with their conclusions, especially about Neanderthals. Even if there are strong differences in basicranial anatomy between Neanderthals and modern humans, which may influence the complexity and the resonance of vocal cord production, I do not think it really restricts in any neurological way the production of sounds.

Picq: The work done by Jeffrey Laitman on basicranial anatomy remains very controversial. I recall the contribution of Bernard Vandermeesrch. He and his team have found a Neanderthal hyoid bone which presents all the modern human characteristics. Thus, the differences in basicranial anatomy are not reflected on the hyoid bone, which is involved in the complex physical process of sound production.

de Bonis: Do you find in the history of the human brain something which would be like the ontogenetic stage of the human brain like, for example, meningeal vessels discussed by Saban?

Holloway: We do not have many fossil specimens, and most of them are of adults. I really do not know the ontogenetic stages of brain development within any one particular fossil hominid group. Anyway, I do not think that recapitulation was involved in the process of human brain evolution.

Picq: I would like to have your comments about brain size and body size evolution or brain evolution and allometry, because the study of heterochrony is of growing importance in palaeoanthropology. What we know about brain size and body size evolution might be explained, as a process, in the light of the recent advances in genetics, especially homeotic and regulatory genes. In comparison with *Australopithecus*, *Homo habilis* had a significantly larger brain, but there is no clear evidence of body size increase and their growth period was apparently similar. Then, we could have mutations affecting the rate of growth of some stage of brain ontogeny. From *H. habilis* through *H. erectus*, and perhaps Neanderthals, we have evidence of both brain and body size increase, along with probably a longer growth period. These changes might be the result of selective pressures acting on body size with mutations of some regulatory genes. Brain size may have changed partly in following the allometric relationships between the brain and the body. Finally, modern humans, in comparison with archaic *H. sapiens*, appear more gracile and also have a slightly smaller body and brain. Several authors spoke about neoteny or hypomorphosis. What do you think about this very general and simplistic scheme, even if it puts aside important aspects of brain reorganization?

Holloway: About brain evolution and genetics, I have made some predictions and I have presented them. I will comment first on the relationship between brain and body size. In fact, there is a high correlation between brain size, body size, and stature, particularly in males, for a sample of over 1000 specimens of modern humans. The correlation was in the order of 0.3, which seems not that much. But it is very high in a population of roughly 5 billion or so in which accessibility to proteins affect stature and growth. There is a way in which the brain could be reduced a little bit in size. I think that the hypothesis about modern human gracilization is quite valuable because Neanderthal brain size is about 100 to 150 cc larger than our own on average. The muscle insertions on the skeleton suggests more limb body mass, which would seem to be highly correlated to brain size. For example, we find the heaviest brain and higher cranial capacities in Eskimos, who are among the largest people of the world, but not relative to stature.

With *H. habilis*, it is much more difficult. The best fossil we have for the brain is KNM-ER 1470. But we do not have the postcranial skeleton. On the other hand, the best postcranial skeleton comes from Olduvai, the very diminutive OH 62. There are also the skulls KNM-ER 1813 and 1805, both very small with cranial capacities of about 510 and 585 cc, respectively. The body size seems to me, at least in early *H. habilis*, possibly very small. Another possibility is that there was a tremendous sexual dimorphism in that population, KNM-ER 1470 being a male and the smaller ones females. I am afraid no further discoveries can clear this up.

Coppens: The fossil skeleton OH 62 found by Don Johanson at Olduvai has been described as a *H. habilis*. I am more inclined to think that it is an *Australopithecus*. In Eastern Africa there is the same sort of evolution as in South Africa with *A. africanus* and *A. robustus*, I mean the sequence OH 62 and *A. boisei*. I think that *Australopithecus* emerged in Eastern Africa and then moved or developed its geographical distribution with, subsequently, a type of parallel evolution in both regions. In fact, after meeting Don Johanson and Tim White about OH 62, they told me that because it is a gracile form, and gracile forms at Olduvai are *H. habilis*, they deduced that this skeleton was a *H. habilis*. Anyway, it remains that there is too large a gap in size between KNM-ER 1470 and OH 62, even if *H. habilis* was very dimorphic, and I agree with you that KNM-ER 1813 is a *H. habilis*.

Holloway: I think that the question of tempo and mode of evolution of the brain, and the relationship between brain size, body size, and growth are going to be very complex. My point is that I strongly object to the hypothesis that the brain evolved last, i.e. first we have bipedal locomotion, then stone tools, and suddenly the brain develops. The business of reorganization and increase in size has a very complex inter-digitation through time, varying in its rate and duration, together with the social aspects of hominids themselves. For several years I have been striving for proof that there was reorganization of the brain. My efforts should come up with the scheme that there are four possible ways in which the brain could be reorganized, with three having no effect on size. In the first case, called A, you can have brain size increase without any change between the size of the components or the connections between them. This change could occur isometrically or allometrically. The second case, B, does not involve any change in absolute brain size. This is the situation in which changes somewhere in the genome, or in a regulatory gene, will limit the development of one region and induce some hyperplasia in the other. This would produce, for instance, the expansion of the parietal region and the posterior position of the lunate sulcus, which happened about 3 to 4 million years ago. Hypothesis C is more difficult and it is called a hierarchical or mutational change. The fibre systems are maturing at different rates and/or increasing in number between the different cortical regions through the corpus callosum. But because of Pasko Rakic's views, I might now withdraw it. Finally, in case D we now have asymmetries and no change in overall brain size. There, one had the left-occipital petalia and a right-frontal petalia. That is the torque that appeared about 1.8 to 2 million years ago in *Homo*. We have empirical palaeoneurological evidence for cases A, B, and D. In two cases the organization is totally different and the size remains exactly the same.

Changeux: May I ask Frank Ruddle whether you consider some regulatory genes that could account for changes A and B? For in the last case, D, it is known that there are mutations in snail and mouse which invert the left and right distribution of organs. Is it known, in *Drosophila*, for instance, if some of these homeotic mutations you talked about could change the symmetry of organization?

Ruddle: I do not know of any homeotic genes which behave that way, and the genes you have mentioned are not part of that particular family. I would not rule it out, but I do not think there is any evidence as yet.

Mishkin: You gave us one of your reasons, among many, for the moving backward of the lunate sulcus as being the result of evolutionary pressure because of the necessity for greater spatial abilities. That was probably based, I suppose, from what we have known for some time abut the role of the inferior parietal lobule. Actually, it was our hypothesis based upon our work on the monkey. Therefore, we were rather surprised to find that the activation, in our particular way of studying spatial ability in man, activated the superior parietal lobule. It does not mean that this evidence is in any way in contradiction with your supposition. I agree with you about the expansion of the parietal lobe in man. I am only wondering about the relationship between the inferior and superior lobules. The issue may depend on a better understanding of an intermediate brain between monkey and man, namely chimpanzee. We need to look at the spatial functions in chimpanzee, using non-invasive techniques, combined with increased efforts to try to find out what will activate the inferior parietal gyrus in man. So, could the reorganization have been so profound as to have pushed a functional area that far dorsally as a result of the enlargement of the part of the brain necessary to handle whatever is activated in the inferior parietal lobule?

Holloway: Unfortunately, we lack sufficient information on the chimpanzee. Obviously, there is a problem with the great increase in the posterior parietal cortex. I mean there must be some allometric relationship in the process of enlargement. Clearly, we need more research on the chimpanzee.

Characteristics of cranial endocasts

Holloway: I have brought some endocasts to give you an idea of the quality of the characters studied and their variability. For the purpose of comparison, there is a hominid endocast from Hadar (AL 162-28, an australopithecine) and two endocasts of chimpanzee, one gyrencephalic, one lissencephalic. There is always a constant pattern of morphology between the lunate sulcus, which is the anterior limit of area 17 or the primary visual striate cortex, and the intraparietal fissure, separating the superior and inferior parietal lobules, which tend to go from the lunate sulcus to the somatosensory following a particular angle. This pattern is very uniform in all anthropoids and particularly pongids. In all apes the lunate sulcus always occupies an anterior position, the intraparietal fissure abuts in the posterior part of the sulcus. In humans, when we can observe the lunate sulcus, it has a very posterior position. The human pattern is the result of brain allometry. If we plot the volume of area 17 against brain size in primates, we have very high correlations. The log-log plotting of data from 46 species of primates is very tight if we omit modern humans. If we include the human, it is 121 per cent less than that would be expected for a primate of this size. We obtain similar results when plotting the size of the lateral geniculate against brain size. This residual, which indicates a very negative deviation relative to the interspecific prediction, has a meaning. The comparative evidence indicates major changes in the parietal region. The cerebral morphology is very much in favour of changes affecting the parietal lobes, hence my great interest in the lunate sulcus. I am trying to use these constant features to see if there is any new pattern or autapomorphy in hominids, especially the position of the lunate sulcus relative to the other features used as landmarks. These landmarks are very important and must be oriented correctly; a major point of controversy between Dean Falk and myself. If we orient correctly the endocast of Hadar hominid AL 168–22, we can see that the lunate sulcus does *not* have an anterior position.

Tobias: About the two chimpanzee endocasts, I do not think we have a lissencephalic brain, but instead a lissencephalic endocast. We know that the marks on the inside of the cranium disappear with age. But we do not know what leads to the smoothing of the endocranium. We assume that there is some osteoclastic resorption. As an example, I spoke about the poor fit of the medula oblongata in the foramen magnum. It is interesting to note that in children of 5 to 6 years of age, the cranial capacity has attained virtually 100 per cent of its adult size. But when we get older the brain becomes smaller, although we have no real evidence of reduction of the cranial capacity.

Mishkin: What you have demonstrated about chimpanzee endocasts is very interesting. It will also be very interesting to do some experimental work on monkeys concerning the relationships between the endocranium and the pattern of brain fissuration.

Tobias: We have carried out a series of studies in my laboratory. We have brains that derive from skulls and we had the endocranial casts. In general, we have found a remarkably good correlation. We can say that when there are impressions on the endocasts, they do reflect true morphology. But not all the brain morphology reflects itself on the endocasts. So, the absence of impressions is not significant, but the presence of impressions is significant. I wonder why Ralph Holloway focuses so much of his attention on the posterior part of the brain and the lunate sulcus. Because hominids are upright walking creatures, the impressions are better on the frontal part of the brain than near the top where the lunate sulcus should be. For this reason, Dean Falk's investigations must be taken very seriously. She studied not merely the amount of fissuration on the lateral aspect of the frontal lobe, but also the pattern of fissuration. She found clear differences between living apes and modern humans. With regard to early hominids, *Australopithecus* was in the ape pattern and *H. habilis* (KNM-ER 1470) the human. The frontal area is surely one of the most favourable regions to study the pattern of cerebral reorganization in hominid evolution. An important reorganization also took place in the inferior part of the parietal which leaves more obvious impressions on endocasts.

Coppens: There is not only the problem of the lunate sulcus. According to your earlier work on *Australopithecus* endocasts, there are several important modern features, such as the height of the cerebellum, the round tips of the temporal lobes, the number of gyri and sulci in the frontal area, among others.

Changeux: What is the frequency of the characteristics appearing on the endocasts, for example, on this lissencephalic endocast?

Holloway: In chimpanzee, you find that the mean values for the cerebral height is much less than it seems to be in Taung, a gracile australopithecine. But you will find occasional chimpanzees with a higher cerebral height. Now, in robust australopithecines you have rather modern values. The endocast KMN-ER 1585, with a capacity of 530 cc, shows a beautiful left petalia, unfortunately the right frontal is missing. It is the same for OH 5 described by Phillip Tobias. These features are interesting with regard to Sussman's work describing tools and possible manipulative abilities of robust australopithecines.

When we want to study the pattern of fissuration of the frontal lobe, we contend with more problems. What has been stated derives mainly from the description of Sts 60 from Sterkfontein. That frontal lobe appears to have a great deal of fissuration, but it is very difficult to pinpoint each one of those gyri and sulci. I am not sure that the advance features described by Dean Falk are totally certain, because when I look at a large sample of chimpanzee braincasts, I am impressed by a fair amount of fissuration. I remain more convinced about major changes in the posterior part of the brain than in the anterior because comparatively speaking, the ape lunate sulcus does not appear in modern humans. When was it lost?

4

The brain of the first hominids

PHILLIP V. TOBIAS

Introduction

Palaeoneurology is a field beset with relatively few facts but many theories. This was the situation early in this century, when the palaeontologically unsupported expectation that encephalic enlargement had occurred early in hominid evolution led to the acceptance by some investigators of the forged remains of the large-brained Piltdown skull (Spencer 1990; Tobias 1990), and to the rejection of the small-brained Taung child, the first australopithecine to be discovered in Africa (Dart 1925). It remains a feature of the situation close to the end of the century, although it must be admitted that more facts have accumulated, even as the speculations have multiplied.

The story of early hominid brains has to be read from carefully dated, well-identified, fossilized calvariae, or from endocranial casts formed within them —by nature or in the laboratory. Such materials confine the Hercule Poirot, who would read 'the little grey cells' of early hominids, to statements about the size, shape and surface impressions of such similitudes of ancient brains. To this limiting factor must be added another major limitation in palaeo-neurological studies, namely the paucity of fossil remains suitable for such studies.

Which were the first hominids? (6–4 million years BP)

Molecular biological studies set the emergence of the family Hominidae as a late Miocene phenomenon, about 6.4 to 5 million years before the present (myr BP). Shortly before that time, it is postulated, an African hominoid lineage existed which comprised common ancestors to the chimpanzee and human beings. The gorillas had already stemmed from an earlier part of the hominid lineage (9–6 myr BP). In the final million years of the Miocene epoch, the chimpanzee-human line of descent underwent a split or divergence into the proto-chimpanzees and the proto-humans (or hominids *sensu stricto*).

Fossil calvariae or endocranial casts from those times (6.4–5 myr BP) would be extremely useful as presenting a possible starting point for our study of the evolution of the hominid brain. Unfortunately, the only hominoid fossils from the critical time, immediately before or after the postulated last divergence, are a molar tooth (Lukeino, 6 myr BP), a partial mandible (Lothagam,

5.5–5 myr BP), another partial mandible (Tabarin, 4.9–4.15 myr BP), a fragment of humerus (Chemeron, 5.07–4.15 myr BP), several fragments from Sahabi (*circa* 5 myr BP), another fragment of humerus (Kanapoi, *circa* 4 myr BP), a femur fragment (Maca, 4.0–3.5 myr BP), and a part of a frontal bone (Belohdelie, *circa* 4 myr BP). The only cranial specimens among these are a parietal fragment from Sahabi in Libya and the frontal part from Belohdelie in Ethiopia. Both are too incomplete to provide much information about the brain, save that the frontal is part of a small cranium, the endocranial capacity of which would evidently have been low, as in apes and as in early hominid specimens attributed to the genus *Australopithecus* (Hill and Ward 1988).

The earliest available hominid endocasts (4–3 million years BP)

Although it is often assumed that the hominids were upright and bipedal from their earliest emergence, it is only some 2 to 2.5 million years later that suitable cranial material is available. Thus, for 2 million years or more, after the hominid lineage is believed to have diverged from that of the chimpanzees, we have no direct information about the hominid brain.

The oldest specimens that have permitted some endocranial detail to be detected are several dated to about 3.5 to 3.2 million years ago from Hadar in Ethiopia and Laetoli in Tanzania. The entire population of hominid remains from Hadar and Laetoli has been assigned by Johanson *et al.* (1978) to a proposed new species, *Australopithecus afarensis*, although some investigators have been led to conclude that two different hominid species are represented (e.g. Senut and Tardieu 1985). On the basis of the diagnostic features originally proposed for the new species, Tobias (1980) drew attention to a number of marked resemblances between the Hadar and Laetoli series and the Transvaal fossils of *A. africanus*; while Falk (1990) has suggested that at least one of the Laetoli hominids is on a different lineage from those of Hadar. Irrespective of the systematics involved, the endocrania of the *A. afarensis* specimens show the following features:

1. Small endocranial capacity: one Hadar specimen has a capacity estimated at between 485 and 500 cm^3; the other two 'suggest small brain sizes, well within *Pan* limits (i.e. less than 400 ml)' (Holloway 1988). In other words there is no evidence that the population(s) represented had a mean absolute brain size greater than that of the African great apes of today (Fig. 4.1).

2. The presence of grooves suggesting enlargement of the occipital and one or both marginal venous sinuses. This variant is present in the South African 'robust' australopithecine, *A. robustus*, and the East African 'hyper-robust' australopithecine, *A. boisei*, and attention was first drawn to it in both species by the author (Tobias 1967, 1968). The feature, which is an infrequent variation in *A. africanus*, *H. habilis*, *H. erectus* and *H. sapiens*, occurs in six out of six Hadar specimens, including AL 288 (nicknamed 'Lucy'). The

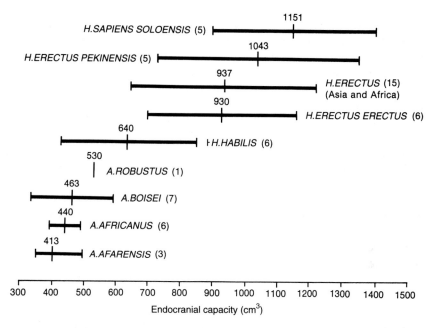

Fig. 4.1. The mean values and population ranges of absolute endocranial capacities of early hominid taxa belonging to the genera *Australopithecus* (A.) and *Homo* (H.). For each taxon or fossil series, the thick horizontal bar represents the 95 per cent population limits; the vertical line across the central part of the bar shows the value of the sample mean.

Homo erectus erectus: an Indonesian subspecies of *H. erectus*; *H. erectus pekinensis*: a Chinese subspecies of *H. erectus*; *H. erectus* (Asia and Africa): this series comprises the Indonesian and Chinese subsets, together with several African specimens of *H. erectus*.

presence of this feature in *A. afarensis* and in *A. boisei* and *A. robustus* is a major reason why some investigators have inferred that *A. afarensis* is ancestral to the last two species (Falk and Conroy 1983; Holloway 1988).

The only relevant specimen from Laetoli, LH 21, differs from the Hadar specimens in showing clear grooves for transverse and sigmoid sinuses, but no grooves for enlarged occipital and marginal sinuses. Thus, the Laetoli specimen shows the pattern which is most common in *A. africanus, H. habilis, H. erectus* and *H. sapiens*. Falk (1990) has used this evidence to place the Laetoli hominid on a different lineage from the Hadar hominids. Walker and Leakey (1988) believe that the evidence of the more recently discovered *A. boisei* specimens east and west of Lake Turkana supports the attribution of Omo L338y-6 to *A. boisei*, although the Omo specimen does not possess enlarged occipital-marginal sinus grooves.

3. For *A. afarensis*, little evidence is available on the morphology of the brain surface, though in AL 288 of Hadar, the frontal fragment suggests

narrowness (Holloway 1988). Holloway (1975) has long insisted that signs of 'cortical reorganization' were evident in the fossil hominid record, *before* brain enlargement had become apparent. The main evidence he has adduced has been the supposed posterior placement of the lunate sulcus, to which Dart (1925) had first drawn attention in the endocranial cast of the Taung child. Indeed, Holloway (1975, p. 404) has gone so far as to claim that the supposed position of the lunate sulcus in the Taung child's endocast was 'perhaps the most conclusive proof for reorganisation of the brain from a pongid to human pattern'. Although the recognition of the lunate sulcus on endocasts has been contested and denied by many investigators (Keith 1931; Weidenreich 1936; Bonin 1963; Clark 1964; Falk 1980, 1989; Tobias 1981, 1983, 1987), Holloway (1988) has again referred to the putative position of this sulcus in support of a claim that the Hadar AL 162-28 endocast shows cortical reorganization. A careful reading of the evidence adduced by Holloway (1988, p. 101) in support of his claim reveals no convincing argument for such reorganization. Nor is the author able to accept Holloway's claims that his measurements from the occipital pole to the alleged lunate sulcus impression 'show that some cortical *reorganisation* took place early in hominid evolution, involving a reduction in the lateral extent of primary visual striate cortex'; nor that '... there are strong suggestions that this occurred in "robust" australopithecines' (Holloway 1988, p. 102).

It is the author's conclusion that the earliest available hominid endocasts, dated to between 4 and 3 myr BP, reveal only that: (i) the size of the brain of the Hadar hominids was no bigger, in absolute terms, than that of the African great apes; and (ii) the pattern of cranial venous sinus drainage was occipital-marginal in 6/6 Hadar crania and transverse-sigmoid in 1/1 from Laetoli. Apart from the curious venous sinus drainage pattern, there is no convincing evidence that these early East African hominids showed any differences of brain form or size from those characterizing the apes. On this basis it may be possible for us to draw the provisional conclusion that, for 2 to 2.5 million years after hominids came into being as a separate zoological family, presumably possessed of upright stance and bipedal gait, we have as yet no convincing evidence for externally manifest changes in the brain, either of size, form, or sulcal or gyral pattern. Further discoveries from these distances in time may lead us to alter this interpretation.

Hominid crania and endocasts (3–1.8 million years BP)

Before 3 myr BP, the available hominid specimens, all from East Africa, are considered by many investigators to belong to a single lineage and a single species, *Australopithecus afarensis*, although some investigators (e.g. Olson 1981; Ferguson 1983; Senut and Tardieu 1985; Falk 1990) consider that there is evidence favouring the existence of more than one hominid species at that time.

After 3 myr BP and in the final 1.2 million years of the Pliocene epoch, *all* investigators are agreed that a more complex picture of the hominids is apparent. From about 2.6 to 2.5 myr BP onwards, at each time level for which data are available, there is indisputable evidence of the coexistence of at least two different species of hominid. From 3 to about 2.3 myr BP, we have evidence in South Africa of the existence of the type species of *Australopithecus*, namely *A. africanus*. In East Africa, from about 2.5 to about 1.8 myr BP, there is testimony to the existence of what the author called the 'hyper-robust' lineage. In this line most investigators recognize a single species, *A. boisei*, although several hold that the earliest members of the lineage should be identified as a distinct species, *A. aethiopicus*. The author agrees with Walker and Leakey (1988) that at present it seems unwise to separate this lineage into successive species.

These two hominid lineages — those of *A. africanus* and of *A. boisei* — may be presumed to have stemmed from a common hominid ancestral form. At some stage earlier than 2.5 myr BP this earlier hominid lineage is inferred to have undergone a cladogenetic split into the postulated two later hominid lines of descent. The brains of these 3–1.8 myr BP hominids were all of small absolute size. See Table 4.1 for some examples.

The values given in Table 4.1 show that, by and large, the mean absolute

Table 4.1. *Some classified endocranial capacities of hominids 3.2–1.8 myr BP*

Species	Dating (myr BP)	Specimen	Capacity (cm³)	Source
A. afarensis	c.3.2	AL 162-28	400	Holloway (1988)
(Hadar)	c.3.2	AL 333–105	400	Holloway (1988)
	c.3.2	AL 333-45	485–500	Holloway (1983)
		Mean	*circa* 413.5	Tobias (1987)
A. africanus	c.3.0	MLD 37/38	425	Conroy *et al.* (1990)
	3.0–2.5	Sts 60	428	Holloway (1975)
		Sts 71	428	Holloway (1975)
		Sts 19	436	Holloway (1975)
		Sts 5	485	Holloway (1975)
	?c.2.2	Taung 1	440	Holloway (1975)
		Mean	440.3	Conroy *et al.* (1990)
A. robustus	1.8–1.7	SK 1585	530	Holloway (1972)
A. boisei	2.5	KNM-WT 17000	410	Walker *et al.* (1986)
	2.1	Omo L338y-6	427	Holloway (1981*a*)
	c.2.0	KNM-ER 13750	450–480	Holloway (1988)
	1.8	KNM-WT 17400	?*circa* 390–400	Holloway (1988)
	1.8	OH 5	530	Tobias (1963)
	c.1.5	KNM-ER 406	510	Holloway (1973)
	c.1.5	KNM-ER 732	506	Holloway (1973)
		Mean	463.3	This chapter

endocranial capacities of *A. africanus* (440 cm^3) and *A. boisei* (463 cm^3) are very similar. However, Holloway (1988) has claimed that there was 'an important increment in size' from earlier to later members of the *A. boisei* lineage. To test this claim, the sample of *A. boisei* may be divided into two subsets: specimens from 2.5 to 2 myr BP have capacities ranging from 410–?480 cm^3 and a mean of 434 cm^3 ($n=3$), whereas the later specimens (1.8–1.4 myr BP) have a sample range of ?390–530 cm^3 and a mean of 485 cm^3 ($n=4$), i.e. 11.75 per cent greater. When allowance is made for the tiny samples and the probable presence of specimens of both sexes among the seven crania, and not necessarily in the same ratio in the two subsets, this early–late difference of the means must be deemed as being of doubtful or limited importance. Within the *A. africanus* subsets, although the lowest value happens to belong to what seems to be the oldest specimen (MLD 37/38), no trend appears to be displayed within the *circa* 0.8 myr BP lineage assumed to be represented.

In other words, the mean values for four australopithecine taxa listed in Table 4.1 — namely, ?413.5, 440.3, 463.3, (530) — all show no real advance on the mean values for extant great apes.

The mean values for large pooled samples of male and of female great apes (Tobias 1971*a,b*, 1975) are shown in Table 4.2. Since the fossil samples are most likely to include male and female specimens, we may compare their means with the weighted means of pooled male plus female data for the apes (see Table 4.3). In relation to these pooled mean values for apes, the estimated mean values for fossil hominid taxa (excluding *A. robustus* with an available sample of one) show the percentage ratios indicated in Table 4.4.

When we consider that the modern human inter-racial mean is of the order

Table 4.2. *Mean endocranial capacities of male and female great apes*

Species	Male (cm^3) (n)	Female (cm^3) (n)	Sexual dimorphism (%)
Pan troglodytes	398.5 (163)	371.1 (200)	6.9
Gorilla gorilla gorilla	534.6 (414)	455.6 (254)	14.8
Pongo pygmaeus	434.4 (203)	374.5 (199)	13.8

Table 4.3. *Mean endocranial capacities of great apes (male plus female pooled data)*

Species	n	cm^3	Source
Pan troglodytes	363	383.4	Tobias (1971*a*)
Pan paniscus	11	343.7	This chapter
Gorilla gorilla gorilla	668	504.6	Tobias (1971*a*)
Pongo pygmaeus	402	404.8	Tobias (1971*a*)

Table 4.4. *Mean endocranial capacities of three australopithecines species expressed as percentage ratios of mean endocranial capacities of three great ape species*

Fossil taxon	Mean (cm³)	Pan troglodytes	G. gorilla	Pongo pygmaeus
A. afarensis	c.413.5	?108	?82	?102
A. africanus	440.3	115	87	109
A. boisei	463.3	121	92	114

of 1350 cm³, which is 3.52 times that of the chimpanzee, 2.68 times the gorilla value, and 3.33 times the orang-utan value, the comparative inter-hominoid index values for *Australopithecus* species are seen to have hardly increased at all. In comparison with the chimpanzee mean values, *Australopithecus* species show a small but definite increase in mean absolute capacity, ranging from an increment of ?8 per cent in *A. afarensis*, through 15 per cent in *A. africanus*, to 21 per cent in *A. boisei*. In comparison with the orang-utan mean values, the corresponding australopithecine increments are ?2, 9, and 14 per cent, respectively. Only in relation to the gorilla mean values do the australopithecine data show shortfalls, the respective decrements being ?18, 13, and 8 per cent.

When several indices of encephalization are employed to relate estimates of mean endocranial capacity to estimated body size, these confirm that the various australopithecine species were somewhat more encephalized (i.e. possessed a somewhat higher mean relative endocranial capacity) than the chimpanzee (Jerison 1973; Holloway and Post 1982; McHenry 1982; Tobias 1987).

To conclude on the available data for three species in the genus *Australopithecus*, they show absolute and relative endocranial capacities (or approximations to brain sizes) which, on average, are slightly smaller than those of the gorilla, but slightly larger than those of the chimpanzee and the orang-utan.

If the molecular data are correct in assigning modern chimpanzee and modern man as the most closely related extant hominoids, then it probably follows that the comparison of hominid features with those of the chimpanzee should be considered most relevant. In such comparisons, *the australopithecines—even the smallest-brained taxon—show a small but definite advance over the chimpanzee in both absolute and relative brain size.*

It is pertinent to enquire whether there are any other morphological differences between the endocasts of australopithecines, of whichever lineage, and those of apes.

Australopithecine endocasts from East and South Africa have several structural features in common, which distinguish them from those of the extant apes:

1. In gross pattern, the impression of the australopithecine foramen magnum, which is regarded as reflecting the position of the brainstem, reveals a

somewhat more anteriorly implanted brainstem than in the brains of extant apes, towards, although not as far forwards as, the position that pertains in later humans. In modern humans, there is however an appreciable discrepancy between the small diameters of the brainstem and the large diameters of the foramen magnum (Tobias and Symons 1992); hence it is important to allow for this loose fit of the brainstem as it traverses the foramen magnum, in the assessment of the brainstem's position.

2. The parietal lobe of the cerebrum appears well developed (Holloway 1988).

3. The cerebellar hemispheres are underslung (Tobias 1967), so that the occipital poles of the cerebrum generally form the most posterior part of the endocast (to which generalization, the oldest *A. boisei* endocast (that of KNM-WT 17 000) and the second oldest (that of Omo L338y-6) may be exceptions — see Holloway (1981*a*, 1988).

4. Most of the australopithecine endocasts show the combination of right fronto-petalia and left occipito-petalia, a combination which Galaburda (1984) describes as the most common in modern man, and which Holloway (1988, p. 98) states is not found in the apes, 'even the highly asymmetrical endocasts of *Gorilla*'. As in modern man, there are exceptions to this combination among the early hominids (LeMay 1976, 1984; Holloway and De LaCoste-Lareymondie 1982; LeMay *et al.* 1982; Tobias 1987; Holloway 1988).

Some investigators, most notably Schepers (1946) and more recently Holloway (1970, 1974, 1975, 1985, 1988), have urged a fifth set of criteria upon us, namely the pattern of sulci, especially with regard to the position of what is taken to represent the lunate sulcus: Holloway sees this in early hominid endocasts as in a human-like posterior position, rather than in an ape-like anterior position. Others, chiefly Falk (1989) and Falk *et al.* (1989), see only ape-like sulcal patterns in the australopithecine endocasts. As far as the lunate sulcus is concerned, the author agrees with Clark (1964) that it is really not possible to identify the lunate sulcus with certainty from the impressions on the early hominid endocasts.

The above four gross rearrangements of the brain of various australopithecines might have been related to underlying cytoarchitectonic alterations, but in the absence of agreement on changes in the sulcal and gyral patterns, we do not have even pointers in that direction. Suffice to say that the evidence of these gross rearrangements, as listed above, and of a relatively small increase in absolute and relative endocranial capacity, constitutes the *total available information on differences between the brains of australopithecines and of present-day chimpanzees*. These represent the sum of the externally detectable encephalic changes that we may infer had occurred in the 2.0–3.0 myr that elapsed between the chimpanzee–human divergence and the dates of the *A. afarensis* and early *A. africanus* endocasts — provided we assume that minimal change had occurred between the brain of the last common ancestor before the chimpanzee–human divergence and the extant chimpanzee brain.

Two brain-related anatomical features of early hominids

Two brain-related but not directly neurological sets of structural alterations made their appearance in some of the early hominids: (1) venous drainage, and (2) shape of the foramen magnum.

1. *Variant of venous drainage*

One structural alteration is the occurrence of venous drainage by the occipital and marginal sinuses (instead of, or in addition to, the lateral or transverse-sigmoid sinus drainage pattern). This occipital-marginal sinus pattern attained a very high incidence (up to 100 per cent) in the Hadar hominids assigned to *A. afarensis* (Holloway 1981*a*; Kimbel 1984), and in *A. boisei* and *A. robustus* (Tobias 1967, 1968), while showing a low frequency in *A. africanus* and in species of *Homo* (Tobias 1987). This subject has been developed in detail recently by Falk and Conroy (1983), Kimbel (1984), Falk (1986, 1988, 1990), and Tobias and Falk (1988).

In the occipital-marginal sinus variant, virtually the entire or a substantial part of the drainage of the superior sagittal and straight sinuses is by way of a system of enlarged occipital and marginal sinuses, the drainage thus bypassing or supplementing the transverse-sigmoid sinus system. The occurrence of a 'mixed' system of drainage on the Taung endocast of *A. africanus* has been reported by Tobias and Falk (1988).

The author has tended to see in the two patterns of drainage a polymorphism, with varying frequencies of the variants in different hominid taxa (Tobias 1968, 1991). This view does not gainsay the possibility that the trait possessed adaptive significance. Kimbel (1984, p. 261) concluded that the alternate states of the drainage pattern are adaptively equivalent and neutral and of 'no valence in broader taxonomic and phylogenetic contexts'. In sharp contrast, Falk and Conroy (1983) and Falk (1986, 1988) have built on earlier work by Conroy (1980*a,b*, 1981) and proposed that there are decided functional and adaptive aspects of the occipital-marginal sinus system. They draw attention to the important role of the vertebral venous plexus in regulating cerebral venous pressure and cerebrospinal fluid pressure in the upright position of the human body. They point out, moreover, that the occipital-marginal sinuses drain not only into the internal jugular vein but also, and especially, into the vertebral venous plexuses. They propose the hypothesis that 'selection for bipedalism placed an increasing cerebral venous drainage load on the vertebral plexus. By 3 million years ago, early robust-like hominids had already responded to this selection pressure by developing to a very high frequency the enlarged accessory sinus system which has numerous connections to the vertebral plexus' (Falk and Conroy 1983, p. 780).

In her later work, Falk (1986, 1990) has proposed alternative routes for the delivery of blood to the vertebral venous plexuses. These routes are a host of emissary foramina, especially the mastoid and parietal foramina, the frequencies of which, she claims, show an increase in *A. africanus* and later hominids, alongside declining frequencies of enlarged occipital and marginal sinuses and of multiple hypoglossal canals. Falk (1990) has gone on to develop a more controversial theory. Pointing out that the brain is extremely sensitive to heat and that emissary veins have been shown to cool the brain under conditions of hyperthermia, she proposes that the network of emissary veins in the lineage leading to *Homo* 'acted as a radiator that released a thermal constraint on brain size'. She goes on to describe the radiator as a 'prime releaser' of brain size evolution. This new radiator theory has already elicited critical comment (see 'Open peer commentary' accompanying Falk 1990), which has led Falk to modify some aspects of her theory, but she has not abandoned it. Moreover, Zihlman (1990) has described Falk's theory as plausible and this author agrees with Zihlman that it is a useful hypothesis, in that it endeavours to synthesize data, relates venous drainage and pressure in the brain to the upright posture, formulates an adaptive framework, and highlights issues for further research and discussion.

2. *The shape of the foramen magnum*

A second brain-related feature is the acquisition of an extraordinary cardioid shape of the foramen magnum in *A. boisei*, to which the author first drew attention in his account of the type specimen, Olduvai hominid 5 (Tobias 1967). The foramina in *A. boisei* are absolutely short in anteroposterior diameter and absolutely broad in maximum transverse diameter, as compared with those of other early hominids. The small available samples of *A. robustus* suggest that there is a tendency towards broadening of the foramen magnum in this species, although the frequency of broadened foramina magna may be smaller than in *A. boisei*.

Among the Hadar hominids, the anatomical region about the foramen magnum is poorly represented. One specimen, AL 333-45, has the margin of the foramen sufficiently well preserved to enable Kimbel *et al.* (1982) to describe the foramen as 'an elongate oval', while that of AL 333-114 is said to be 'greatly arched' (Kimbel *et al.* 1982). In the published account, no mention is made of the shape of the foramen in AL 333-105. Casts of these three specimens were kindly sent to the author by Dr W. H. Kimbel and a new study has been made of them (Tobias 1994). We have observed that two of the characteristics of the cardioid foramen magnum were clearly present. First, the anterior margin took off transversely behind the front of the occipital condyle. Secondly, there was a marked hollowing of the lateral margin of the foramen behind the condyle. Thus, even on these incomplete specimens, the preserved morphology indicates that the Hadar hominids almost certainly possessed the cardioid foramen magnum.

The anteroposterior truncation of the foramen magnum in *A. boisei* led the author at first to consider the possibility that a posterior extension of the sphenoid air sinus in the heavily pneumatized cranium of *A. boisei* might have been responsible for the foreshortening. This explanation was excluded after the author examined the posterior extent of the sphenoid sinus in numerous hominoid crania. It was also considered unlikely that the shortening of the basicranium had produced the effect, since such shortening usually involves the basis cranii anterior to basion (Tobias 1967, p. 60).

A third possible causal factor is that the brainstem of *A. boisei* might have shown a peculiarly broadened shape, to which, during ontogeny, the four developmental anlagen of the foramen magnum margin would have moulded themselves (Tobias 1991). However, this does not seem probable, because a recent study by the author, assisted by J. Symons, clearly reveals that the brainstem does not fit closely within the foramen magnum (Tobias and Symons 1992). An examination has been made of six median sagittal sections and of nine coronal sections through the foramen magnum of modern human heads. In each of the six adult sagittal sections, as well as in one infant specimen, the anteroposterior diameter of the medulla oblongata has been determined in the plane of the foramen magnum. This has been expressed as a percentage of the anteroposterior diameter of the foramen, measured from basion to opisthion, in the same specimen. The resulting index may be designated the *Sagittal medullo-foraminal index*. Similarly, in each of the nine coronal sections through the foramen magnum, the transverse diameters of the medulla oblongata, and of the foramen itself have been determined in the plane of the section. From these two metrical characters on each specimen, a *transverse medullo-foraminal index* has been determined.

Table 4.5 shows the results obtained from the two samples of sectioned human heads. The table shows that, in the plane of the foramen magnum, the medulla oblongata occupies only about 23–33 per cent (mean 28 per cent) of the anteroposterior diameter (basion-opisthion) of the foramen and some 35–50 per cent (mean 38 per cent) of the transverse diameter. In the light of these figures, the size of the most caudal part of the medulla oblongata does not seem to play a major developmental role in determining the size and shape of the foramen magnum.

If we assume that, during ontogeny, the developing foramen magnum moulds itself around the soft tissue structures that traverse the foramen, it is

Table 4.5. *Results from two samples of sectioned human heads*

| Index | Sample | | Mean | Standard deviation |
	Size	Range		
Sagittal medullo-foraminal	6	22.7–33.3	27.7	4.834
Transverse medullo-foraminal	9	35.1–50.0	38.1	4.575

perhaps to the other contents that we should direct our attention in search of an explanation of the cardioid foramen magnum. As a preliminary examination, we may note the following structures that regularly traverse or frequently occupy the foramen magnum.

Neural:
The caudal part of medulla oblongata.
The inferior part of the vermis of cerebellum.
The tonsil of the middle lobe of each cerebellar hemisphere.
The spinal root of the nervus accessorius.
Rami communicantes of sympathetic nerve fibres around the vertebral arteries.

Meningeal
Sheath of dura mater.
Leptomeninges.
The highest triangular process of the denticulate ligament.
The upper part of cisterna magna.

Arterial
Vertebral arteries.
Meningeal branches of vertebral arteries.
Posterior inferior cerebellar artery.
Posterior spinal arteries.
Anterior spinal artery.

Venous
Anterior and posterior longitudinal spinal veins.
Marginal venous sinus (occasionally).
Venous channels connecting with vertebral veins.
Venous channels anastomosing with occipital and marginal sinuses, basilar plexus, internal vertebral venous plexus, external vertebral venous plexus, venous plexus of the hypoglossal canal, and condylar emissary veins.

Syndesmological
Membrana tectoria.
Longitudinal fascicle of the cruciform ligament of the atlas.
Apical ligament of the dens.
Alar ligaments.
Anterior atlanto-occipital membrane.
Posterior atlanto-occipital membrane.

There is little doubt, from the preliminary study by Tobias and Symons (1992), that both the cisterna magna (cisterna cerebellomedullaris) and parts of the cerebellum may occupy substantial although variable proportions of the adult foramen magnum. Thus, in the median sagittal plane, the cisterna magna may occupy 45–55 per cent of the anteroposterior diameter of the foramen magnum — appreciably higher percentages than those filled by the medulla oblongata. In the sagittal plane, also, the inferior vermis of the cerebellum may occupy 17–42 per cent of the anteroposterior diameter of the foramen. In the coronal or transverse plane, the tonsils of the left and right cerebellar hemispheres, or the tonsil of one or the other hemisphere alone, may occupy as little as 10 per cent, but usually in our series 24–46 per cent, and in one case 82 per cent, of the transverse diameter of the foramen magnum in the plane of the section. This figure varies somewhat according to whether the coronal plane of section is more anterior, midway, or more posterior along the length of the foramen magnum. The mean for six specimens (after allowance had been made for varying plane of section) was 38.9 per cent with a standard deviation of 17.031.

The author previously noted a striking lateral embayment of the lateral margin of the foramen magnum of A. boisei, immediately posterior to the occipital condyle, on each side. It was at this position that the maximum transverse diameter of the foramen was measured. The paired lateral embayments were clearly responsible, in the main, for the greater transverse diameter of the foramen in A. boisei than in other australopithecines.

Which of the soft structures enumerated above could have accounted for the lateral embayment noted by the author in the foramen magnum of A. boisei? In modern man, the structures prominently lodged in this area of the foramen on each side are the tonsil of the cerebellar hemisphere, the vertebral artery, the spinal root of the accessory nerve, and the major venous anastomoses between intracranial venous sinuses and channels, on the one hand, and the internal vertebral venous plexuses on the other.

It is tempting to relate the occurrence of well-developed venous anastomoses to the presence of enlarged occipital and marginal venous sinuses in A. boisei and to Falk's and Conroy's concept of marked venous drainage to the vertebral plexuses in these hominids. Could the postulated excessive venous anastomoses in A. boisei have led to the development of the prominent lateral embayments of the foramen magnum of that species? This interpretation appears to be supported by the presence, in A. afarensis and A. robustus, of apparent foramen magnum broadening, together with the presence of enlarged occipital-marginal sinuses.

Another anatomical feature, which could have been responsible for the development of the lateral embayment, is the tonsil of each cerebellar hemisphere. In modern man, our results have shown that the tonsil(s) may occupy 39 per cent, on the average, of the width of the foramen magnum. It could be postulated that, in A. boisei, the tonsil in many cases filled a larger percentage of the width of the foramen and that this appendage of the cerebellar hemisphere was causally related to the formation of the lateral embayments of the foramen. To test this hypothesis, evidence might be sought on the relative size of the cerebellar hemispheres in A. boisei. If there were signs of cerebellar expansion in A. boisei, part of the cerebellum might be expected to have lodged, regularly and appreciably, in the foramen magnum. In the author's study (Tobias 1967) of the endocranium of A. boisei, it was indeed found that the cerebellum showed signs of greater expansion than in *A. africanus*. This was shown, for example, by the relatively higher position (or 'ascent') of the internal occipital protuberance, marking the upper limit of the cerebellum in *A. boisei* (Tobias 1967, p. 56). It was even reasoned that the inferred rapid enlargement of the cerebellum might have been the key factor precluding the development of the common, hominid, trans-cerebellar route for the drainage of blood, from the sagittal and straight sinuses, to the jugular venous bulb, in favour of the retention into adulthood of the common foetal arrangement of enlarged occipital and marginal sinuses, as in *A. boisei* (Tobias 1967, pp. 70–1). The large size of the *A. boisei* cerebellum was revealed also by absolute measurements and by a number of encephalometric indices of the cerebellar impressions in the endocranial cast and it was commented, 'The cerebellum seems to have been better developed in [*A. boisei*] than in any of the other australopithecine brains' (Tobias 1967, p. 92). If this enlargement applied to all parts of the cerebellar hemispheres, it would follow that *A. boisei* would have possessed large cerebellar tonsils. The enlargement of the cerebellum probably effected relatively crowded conditions in the posterior cranial fossa. In consequence, an appreciable degree of herniation of the tonsil into the foramen magnum may be expected to have occurred in a percentage of individuals.

The evidence adduced earlier by the author that the cerebellum of *A. boisei* had undergone definite enlargement at first seemed to provide support for the hypothesis that tonsillar protrusion or herniation was the morphogenetic factor most likely to have been responsible for the conspicuous lateral embayment of the margin of the foramen magnum of *A. boisei*. However, the author has abandoned this view because of the inconstancy of tonsillar protrusion in modern humans (Tobias and Symons 1992).

Newer studies by the author have led him to conclude that well-developed venous anastomoses are most likely to be causally related to the formation of a lateral embayment: that the anastomoses must have been substantial in certain early hominids is testified to by the persistence of enlarged marginal sinuses (Tobias and Symons 1992).

Hominid endocasts (from 1.8 million years BP) *Homo habilis*

The earliest species attributed to the genus *Homo*, namely, *Homo habilis*, has endocasts testifying to major advances in the evolution of the brain. They have been described in detail recently (Tobias 1987, 1991), so a summary is all we need give here.

1. The first indications of a dramatic increase in *absolute* brain size in *H. habilis* are revealed by a mean capacity of 640 cm^3 which is 45 per cent greater than the mean of 440 cm^3 for *A. africanus* (Fig. 4.1).

2. By various measures of brain–body ratio, or indices of encephalization,

H. habilis shows an increase in *relative* brain size; it marks the first appearance in the fossil Hominidae of that hallmark of mankind, namely a disproportionately enlarged brain size.

3. From encephalometric features, the increase in the *H. habilis* brain is seen to involve a definite broadening (mainly of the frontal and parietal lobes of the cerebrum), and a moderate heightening, but not a sensible lengthening of the cerebral hemispheres.

4. The sulcal pattern of the frontal lobe is very similar to that of *H. erectus* and of modern *H. sapiens* and quite different from that of extant apes (Holloway 1978; Falk 1983).

5. The gyral impressions on the frontal lobe include a well-marked prominence in the position of the posterior part of the inferior frontal convolution: this corresponds to the position of Broca's area and its bulbosity exceeds that in the corresponding cortical zone of *A. africanus*.

6. There is a right fronto-petalia (assessed by anterior or rostral protrusion) in the few *H. habilis* crania from which it has been possible to produce tolerable endocasts including left and right frontal poles of the cerebral hemispheres. The posterior or caudal projection of the occipital pole is more variable: in the endocast of a presumptive male of *H. habilis*, left occipito-petalia is present, whilst, in the endocasts of two putative females, right occipito-petalia is manifest. In a modern human series reversal of the modal pattern of right fronto- and left occipito-petalia occurs more commonly in women, while the blend of right fronto-petalia with right occipito-petalia (as in one of our female *H. habilis* specimens) occurs in association with *non-right-handedness* (Bear *et al.* 1986).

7. On the impression of the parietal lobe, the superior parietal lobule is well developed (Holloway 1981*b*) and, in several endocasts of *H. habilis*, is asymmetrical with left predominance (Tobias 1987). The anterior part of the superior parietal lobule corresponds to Brodmann area 7: according to Mountcastle *et al.* (1975, 1984), judgements of shape and visual relations involve neuronal activity in area 7 (Eccles 1989). Moreover, Roland (1985), using the injection of radio-xenon, studied which areas of the brain showed an increase in regional cerebral blood flow when subjects were exposed to tasks requiring visuospatial judgement. *Inter alia*, they noted large increases in the posterior superior parietal areas — 36 per cent on the left and 30 per cent on the right. It would be useful to pursue the author's observation of superior parietal lobule asymmetry in *H. habilis*, to see whether it is present in larger series of early hominids and in modern *H. sapiens*, as this is a cerebral asymmetry which has not previously been reported (Tobias 1987). If the anatomical asymmetry is confirmed, it may provide evidence of a functional asymmetry in the representation of visuospatial discrimination and judgement.

8. The parietal lobe in *H. habilis* is well expanded transversely and the inferior parietal lobule is strongly developed — in contrast to the arrangement

in australopithecines and apes. Detectable impressions of the supramarginal and angular gyri, comprising the inferior parietal lobule, are present for the first time in the hominid lineages. Since this area forms part of the larger Wernicke's area or posterior speech cortex, it has been claimed that — with the anterior speech cortex of Broca present as well — H. habilis is the first species, in the history of the hominids, to show the two most important neural bases for language abilities. (The third area, the superior speech cortex, is part of the supplementary motor area; it lies on the superomedial surface of the cerebral hemisphere and therefore its presence cannot be detected on an endocast.)

9. One H. habilis endocast (Olduvai hominid 7) shows evidence of asymmetry of the sulcus lateralis (Sylvian fissure): the left sulcus ascends only slightly from anterior to posterior with a low termination, while that on the right rises steeply. This difference tallies with the position in later hominids including modern man (Cunningham 1892; Le May and Culebras 1972; Le May 1976, 1977)

10. The anterosuperior part of the occipital lobe of H. habilis is expanded transversely, in keeping with the expansion of the adjacent posterosuperior part of the parietal lobe. This parieto-occipital transverse expansion is more marked than the frontal transverse expansion and gives the endocast an ovoid rather than ellipsoid contour when viewed from above.

11. The middle meningeal vascular pattern of H. habilis is more beset with branches and anastomoses than are the patterns of the australopithecines so far described (Tobias 1967, 1987; Saban 1983).

12. Unlike the Hadar hominids and the 'robust' australopithecines, H. habilis endocasts show the transverse-sigmoid pattern of venous sinus drainage which is modal in A. africanus and H. sapiens. In two out of three specimens in which the area is preserved, the superior sagittal sinus groove drains to the right, whereas in OH 24 it drains to the left. This is the same specimen that shows right fronto- and occipito-petalia, which in modern man is associated with non-right-handedness. The author does not know of any published proven association between cerebral dominance or handedness, and the side to which the superior sagittal sinus drains, but the evidence of 'Twiggy' is interesting and suggestive. J. Symons, a medical science student in the author's department, has recently found that there is a strong, inverse association in modern man between the side showing occipito-petalia and the side to which the superior sagittal sinus drains: of 54 crania featuring left occipito-petalia, 50 showed evidence of drainage to the right; of 14 crania with right occipito-petalia, 9 showed drainage to the left. The left-left association occurred in only a single specimen out of 90 crania, and the right-right concurrence in 3 out of 90 crania, i.e. in only 4 out of 90 crania were the occipito-petalia and the direction of superior sagittal sinus drainage on the same side, whereas in 59 out of 90 crania they were on opposite sides. In the remaining 27 crania varying patterns of symmetry of petalia and/or of drainage were encountered (Henneberg and Symons 1991).

Conclusion

In the stages by which the presumably ape-like brain of the last common ancestor of man and chimpanzee was made over into the brain of mankind, it seems that relatively small changes accompanied the emergence of the earliest available and analysable hominids, the australopithecines. These comprised a minimal increase in absolute and relative size and some limited reorganization of the overall anatomical structure of the brain. As for neurologically important changes in the brain, there is scarcely any evidence of surface alterations in the sulcal and gyral patterns and what differences have been claimed, especially in regard to the expansion of the prestriate areas, are to say the least problematical on presently available evidence.

However, major expansion of the brain and critical cortical reorganization were striking features of the postulated transition from *A. africanus* (or an *A. africanus*-like form) to *H. habilis*. These changes included notable augmentation of the cerebrum, strong lateral expansion of the parieto-occipital region, the appearance of a human-like sulcal pattern, and the emergence for the first time of protuberances overlying what are interpreted as the anterior and posterior speech cortices. That these homologues of Broca's and Wernicke's areas were used by a speaking ancestor, *H. habilis*, was proposed by Tobias (1980, 1981, 1983), supported by Falk (1983) and by Eccles (1989).

Thus, it is with the appearance of the *H. habilis* brain that a gigantic step was taken to a new level of organization in hominid brain evolution.

Acknowledgements

My thanks are due to the conveners, Drs Changeux and Chavaillon, the Fondation Fyssen, Joel Symons, and Mrs Heather White.

References

Bear, D., Schiff, D., Saver, J., Greenberg, M., and Freeman, R. (1986). Quantitative analysis of cerebral asymmetries: fronto-occipital correlation, sexual dimorphism and association with handedness. *Arch. Neurol.* **43**, 598–603.

Bonin, G. von (1963). *The evolution of the human brain*, pp. 1–92. Chicago University Press.

Clark, W.E. LeGros (1964). *The fossil evidence for human evolution: an introduction to the study of palaeoanthropology* (2nd edn), pp. 1–201. Chicago University Press.

Conroy, G.C. (1980a). Evolutionary significance of cerebral venous patterns in palaeoprimatology. *Z. Morph. Anthrop.* **71**, 125–34.

Conroy, G.C. (1980b). Cerebral venous hemodynamics and the basicranium of *Cebus*. *Am. J. Phys. Anthropol.* **53**, 37–41.

Conroy, G.C. (1981). Cranial asymmetry in ceboid primates: the emissary foramina. *Am. J. Phys. Anthropol.* **55**, 187–94.

Conroy, G.C., Vannier, M.W., and Tobias, P.V. (1990). Endocranial features of *Australopithecus africanus* revealed by 2- and 3-D computed tomography. *Science* **247**, 838–41.

Cunningham, D.J. (1892). Contribution to the surface anatomy of the cerebral hemispheres. *Roy. Irish Acad. Sci. Cunningham Memoirs* **VII**. Dublin.

Dart, R.A. (1925). *Australopithecus africanus*: the man ape of South Africa. *Nature* **115**, 195–9.

Eccles, J.C. (1989). *Evolution of the brain: creation of the self.* Routledge, London.

Falk, D. (1980). A re-analysis of the South African australopithecine natural endocasts. *Am. J. Phys. Anthropol.* **53**, 525–39.

Falk, D. (1983). Cerebral cortices of East African early hominids. *Science* **221**, 1072–4.

Falk, D. (1986). Evolution of cranial blood drainage in hominids: enlarged occipital/marginal sinuses and emissary foramina. *Am. J. Phys. Anthropol.* **70**, 311–24.

Falk, D. (1988). Enlarged occipital/marginal sinuses and emissary foramina: their significance in hominid evolution. In *Evolutionary history of the "robust" australopithecines* (ed. F. Grine), pp. 85–96. Aldine de Gruyter, New York.

Falk, D. (1989). Ape-like endocast of "ape-man" Taung. *Am. J. Phys. Anthropol.* **80**, 335–9.

Falk, D. (1990). Brain evolution in *Homo*: the "radiator" theory. *Behav. Brain Sci.* **13**, 333–81.

Falk, D. and Conroy, G.C. (1983). The cranial venous sinus system in *Australopithecus afarensis*. *Science* **306**, 779–81.

Falk, D., Hildebolt, C., and Vannier, M.W. (1989). Reassessment of the Taung early hominid from a neurological perspective. *J. Hum. Evol.* **18**, 485–92.

Ferguson, W.W. (1983). An alternative interpretation of *Australopithecus afarensis* fossil material. *Primates* **24**, 397–409.

Galaburda, A.M. (1984). Anatomical asymmetries. In *Cerebral dominance: The biological foundations* (ed. N. Geschwind and A.M. Galaburda), pp. 11–25. Harvard University Press, Cambridge, Mass.

Henneberg, M. and Symons, J. A. (1991). Relationship of brain asymmetry to the lateralization of venous drainage from the cranial cavity. *Newsletter Anatom. Soc. S. Africa* **24**, 18.

Hill, A. and Ward, S. (1988). Origin of the Hominidae: the record of African large hominoid evolution between 14 my and 4 my. *Yrbk Phys. Anthropol.* **31**, 49–83.

Holloway, R.L. (1970). Australopithecine endocast (Taung specimen, 1924): a new volume determination. *Science* **168**, 966–8.

Holloway, R.L. (1972). New australopithecine endocast, SK 1585, from Swartkrans, South Africa. *Am. J. Phys. Anthropol.* **37**, 173–85.

Holloway, R.L. (1973). New endocranial values for the East African early hominids. *Nature* **243**, 97–9.

Holloway, R.L. (1974). The casts of fossil hominid brains. *Scient. Am.* **231**, 106–15.

Holloway, R.L. (1975). Early hominid endocasts: volumes, morphology and significance for hominid evolution. In *Primate functional morphology and evolution* (ed. R.H. Tuttle), pp. 393–416. Mouton, The Hague.

Holloway, R.L. (1978). Problems of brain endocast interpretation and African hominid evolution. In *Early hominids of Africa* (ed. C. Jolly), pp. 379–401. Duckworth, London.

Holloway, R.L. (1981a). The endocast of the Omo L338y-6 juvenile hominid: gracile or robust *Australopithecus*? *Am. J. Phys. Anthropol.* **54**, 109–18.

Holloway, R.L. (1981b). Exploring the dorsal surface of hominoid brain endocasts by stereoplotter and discriminant analysis. *Phil. Trans. R. Soc. Lond.* **B292**, 3–5.

Holloway, R.L. (1983). Cerebral brain endocast pattern of *Australopithecus afarensis* hominid. *Nature* **303**, 420–2.

Holloway, R.L. (1985). The past, present and future significance of the lunate sulcus in early hominid evolution. In *Hominid evolution: past, present and future* (ed. P.V. Tobias), pp. 47–62. Liss, New York.

Holloway, R.L. (1988). "Robust" australopithecine brain endocasts: some preliminary observations. In *Evolutionary history of the "robust" australopithecines* (ed. F.E. Grine), pp. 97–105. Aldine de Gruyter, New York.

Holloway, R.L. (1990). Falk's radiator hypothesis. *Behav. Brain Sci.* **13**, 360.

Holloway, R.L. and De LaCoste-Lareymondie, M.C. (1982). Brain endocast asymmetry in pongids and hominids: some preliminary findings on the palaeontology of cerebral dominance. *Am. J. Phys. Anthropol.* **58**, 101–10.

Holloway, R.L. and Post, D.G. (1982). The relativity of relative brain measures and hominid mosaic evolution. In *Primate brain evolution: methods and concepts* (ed. E. Armstrong and D. Falk), pp. 57–76. Plenum, London.

Jerison, H.J. (1973). *Evolution of the brain and intelligence*, pp. 1–482. Academic Press, New York.

Johanson, D.C., White, T.D., and Coppens, Y. (1978). A new species of the genus *Australopithecus* (Primates: Hominidae) from the Pliocene of eastern Africa. *Kirtlandia* **28**, 1–14.

Keith, A. (1931). *New discoveries relating to the antiquity of man*, pp. 1–512. Williams and Norgate, London.

Kimbel, W.H. (1984). Variation in the pattern of cranial venous sinuses and hominid phylogeny. *Am. J. Phys. Anthropol.* **63**, 243–63.

Kimbel, W.H., Johanson, D.C., and Coppens, Y. (1982). Pliocene hominid cranial remains from the Hadar Formation, Ethiopia. *Am. J. Phys. Anthropol.* **57**, 453–500.

Le May, M. (1976). Morphological cerebral asymmetry of modern man, fossil man and nonhuman primates. *Ann. N.Y. Acad. Sci.* **280**, 348–66.

Le May, M. (1977). Asymmetries of the skull and handedness. *J. Neurol. Sci.* **32**, 243–53.

Le May, M. (1984). Radiological, developmental and fossil asymmetries. In *Cerebral dominance: the biological foundations* (ed. N. Geschwind and A. M. Galaburda), pp. 26–42. Harvard University Press, Cambridge, Mass.

Le May, M., Billig, M., and Geschwind, N. (1982). Asymmetries of the brains and skulls of nonhuman primates. In *Primate brain evolution* (ed. E. Armstrong and D. Falk), pp. 263–78. Plenum, New York.

Le May, M. and Culebras, A. (1972). Human brain: morphologic differences in the hemispheres demonstrable by carotid arteriography. *New Engl. J. Med.* **287**, 168–70.

McHenry, H.M. (1982). The pattern of human evolution: studies on bipedalism, mastication and encephalization. *Ann. Rev. Anthropol.* **11**, 151–73.

Mountcastle, V.B., Lynch, J.C., Georgopoulos, A., Sakata, H., and Acuna, A. (1975). Posterior parietal association cortex of the monkey. *J. Neurophysiol.* **38**, 871–908.

Mountcastle, V.B., Motter, B.C., Steinmetz, M.A., and Duffy, C.J. (1984). Looking and seeing: the visual functions of the parietal lobe. In *Dynamic aspects of neocortical function* (ed. G.M. Edelman, W.E. Gall, and W.M. Cowan), pp. 159–93. Wiley, New York.

Olson, T.R. (1981). Basicranial morphology of the extant hominids and Pliocene hominids: the new material from the Hadar Formation, Ethiopia, and its significance in early human evolution and taxonomy. In *Aspects of human evolution* (ed. C.B. Stringer), pp. 99–128. Taylor and Francis, London.

Roland, P.E. (1985). Cortical activity in man during discrimination of extrinsic patterns and retrieval of intrinsic patterns. In *Pattern recognition mechanisms* (ed. C. Chagas, R. Gattass, and C. Gross), pp. 215–46. *Pontificiae Academiae Scientiarum Scripta Varia*, **54,** Vatican City, Rome.

Saban, R. (1983). Les veines meningees moyennes des australopitheques. *Bull. et Mem. de la Soc. d'Anthrop. de Paris* **10,** 313–24.

Schepers, G.W.H. (1946). The endocranial casts of the South African ape-men. In *The South African fossil ape-men: the australopithecinae* (ed. R. Broom and G.W.H. Schepers), pp. 153–272. *Transv. Mus. Mem.* **2.**

Senut, B. and Tardieu, C. (1985). Functional aspects of Plio-Pleistocene hominid limb bones: implications for taxonomy and phylogeny. In *Ancestors: the hard evidence* (ed. E. Delson), pp. 193–201. Liss, New York.

Spencer, F. (1990). *Piltdown: a scientific forgery*. Natural History Museum Publications, London and Oxford University Press.

Tobias, P.V. (1963). Cranial capacity of *Zinjanthropus* and other australopithecines. *Nature* **197,** 743–6.

Tobias, P.V. (1967). *Olduvai Gorge*, Vol. 2. *The cranium and maxillary dentition of* Australopithecus (Zinjanthropus) boisei, pp. 1–264. Cambridge University Press.

Tobias, P.V. (1968). The pattern of venous sinus grooves in the robust australopithecines and other fossil and modern hominoids. In *Anthropologie und Humangenetik* (K. Saller *Festschrift*), pp. 1–10. Gustav Fischer, Stuttgart.

Tobias, P.V. (1971*a*). *The brain in hominid evolution*, pp. 1–170. Columbia University Press.

Tobias, P.V. (1971*b*). The distribution of cranial capacity values among living hominoids. In *Proc. III Internat. Congr. Primat., Zurich, 1970*, Vol. 1, pp. 18–35.

Tobias, P.V. (1975). Brain evolution in the Hominoidea. In *Primate functional morphology and evolution* (ed. R. H. Tuttle), pp. 353–92. Mouton, The Hague.

Tobias, P.V. (1980). "*Australopithecus afarensis*" and *A. africanus*: critique and an alternative hypothesis. *Palaeont. Afr.* **23,** 1–17.

Tobias, P.V. (1981). *The evolution of the human brain, intellect and spirit*, pp. 1–70. University of Adelaide Press, Australia.

Tobias, P.V. (1983). Recent advances in the evolution of the hominids with especial reference to brain and speech. *Pontificiae Acad. Scient. Scripta Varia* **50,** 85–140.

Tobias, P.V. (1987). The brain of *Homo habilis*: a new level of organization in cerebral evolution. *J. Hum. Evol.* **16,** 741–61.

Tobias, P.V. (1990). Introduction to a forgery. In *Piltdown: A scientific forgery* (Frank Spencer), pp. viii–xii. Natural History Museum Publications, London and Oxford University Press.

Tobias, P.V. (1991). *Olduvai Gorge*, Vols 4A and 4B. *The skulls, endocasts and teeth of* Homo habilis. Cambridge University Press.

Tobias, P.V. (1994). The craniocerebral interface in early hominids. In *Integrated paths to the past: palaeoanthropological advances*. In honor of Francis Clark Howell (ed. R.S. Corruccini and R.L. Ciochon), pp. 185–203. Prentice Hall, New York.

Tobias, P.V. and Falk, D. (1988). Evidence for a dual pattern of cranial venous sinuses on the endocranial cast of Taung (*Australopithecus africanus*). *Am. J. Phys. Anthropol.* **76,** 309–12.

Tobias, P.V. and Symons, J.A. (1992). Functional, morphogenetic and phylogenetic significance of conjunction between cardioid foramen magnum and enlarged occipital and marginal venous sinuses. *Perspect. Human Biol. 2. Archaeol. Oceania* **27** (3), 120–7.

Walker, A. and Leakey, R.E.F. (1988). The evolution of *Australopithecus boisei*. In *Evolutionary history of the "robust" australopithecines* (ed. F. Grine), pp. 247–58. Aldine de Gruyter, New York.

Walker, A., Leakey, R.E., Harris, J.M., and Brown, F.H. (1986). 2.5-Myr *Australopithecus boisei* from West Lake Turkana, Kenya. *Nature* **322,** 517–22.

Weidenreich, F. (1936). Observations on the form and proportions of the endocranial casts of *Sinanthropus pekinensis*, other hominids and the great apes: a comparative study of brain size. *Palaeont. Sinica* **7** (4), 1–50.

Zihlman, A. (1990). The problem of variation. *Behav. Brain Sci.* **13,** 367–8.

DISCUSSION

Participants: Y. Coppens, R.L. Holloway, A. Roch Lecours, P. Picq,
P.V. Tobias

Brain asymmetries

Roch Lecours: There is a great deal of variability of the different asymmetries of the brain, especially for the parietal lobules. What is the size of your sample to confirm the torque of the brain, that is, left-occipital, right-frontal prominence, and do modern apes show asymmetries of this type?

Tobias: In my department we have a collection of about 3000 modern human skulls, and we have the brains of many of them. There are statistical correlations between the asymmetries, but not a one–one rigid correlation. When I said that the superior sagittal sinus goes to the right, it is commonly in association with left occipito-petalia and vice versa. For the fossils, the samples are very small: three endocasts for *Australopithecus afarensis*, six for *A. africanus*, and seven for *A. boisei*. So, in fact, these are minuscule samples, but to find six out of six showing the variant in one taxon, and six out of six showing the modern human pattern in another fossil taxon, that is surely significant.

Holloway: Concerning the pattern in modern apes, I can provide this according to the collection of 92 great apes' endocasts I have studied, which contains approximately 30 to 40 specimens per genus. Chimpanzees tend to show very little cerebral asymmetries, orang-utans almost none. Gorillas show strong asymmetries favouring the left occipital, but not for the right frontal. The torque pattern seems to be confined to *Homo*. Regarding brain petalias, one has to be careful because, as for the posterior occipital, it can describe its caudal projection also its lateral extent. For the right temporal, the main part is not the anterior projection, but its lateral expansion.

Tobias: It is important to be precise, because the classical concept of petalia is used to describe the rostral projection for the fronto-petalia and the caudal projection for the occipito-petalia.

Pattern of variation of brain evolution in hominids

Holloway: I think that *Australopithecus* does show much more of the human pattern for the posterior part of the brain.

Coppens: I think also that the brain of *Australopithecus* was better, that is, more human-like, than you said. Besides, I would like to have your opinion on the possibility of the existence of several species of *Homo* at the time of *Homo habilis*.

Picq: According to your discussion, there is no doubt that *H. habilis* represents a new level of brain organization. However, it seems that, according to recent works published by Ralph Holloway, there is some strong evidence for a dramatic evolution of the brain in the robust australopithecine lineage, as documented by the transition between 'the black skull' (a very robust australopithecine skull with dark colouration of the bone), which could represent a new species, and *A. boisei*. In other words, brain evolution, both in terms of increase in size and/or reorganization, was not a unique event in hominid evolution.

Tobias: I am not at all convinced that it is meaningful or helpful in the study of fossil hominids to multiply species. It is happening today, I believe, largely because of a preconceived philosophy of evolution. The rigid application of cladistics almost demands branching, and every branching calls for new species names. I believe that where authors are manufacturing different species, they are not taking into account sufficiently the intraspecific variability. The 'black skull' is certainly an *A. boisei*-like specimen that I believe simply expands our concept of variability within the species *A. boisei*. Modern humans form a highly variable species. *Homo erectus* is similarly marked by appreciable regional variability, and I believe such polytypy is there at the level of *Homo habilis*. It is a feature that was made possible by man's culture. Our speciation and evolution are different from most other species' by virtue of the cultural dimension. This has enabled us to diversify all over the world without speciating, unlike other creatures.

5

Evolution of neocortical parcellation: the perspective from experimental neuroembryology

PASKO RAKIC

The mechanism underlying the subdivision of neocortex into structurally and functionally distinct areas is central to our understanding of human brain evolution and the emergence of cognitive capacity. We are exploring the possible cellular and molecular mechanisms that may be involved in this parcellation, using as a model the developing cerebrum in the macaque monkey. Phylogeny of the neocortex is considered here in the context of the radial unit hypothesis of cortical development (Rakic 1988). According to this hypothesis, the mitotically active zone, situated at the surface of the cerebral ventricle, is depicted as a two-dimensional mosaic of proliferative units which constitute the rough protomap of prospective species-specific cytoarchitectonic areas. Each radial unit consists of several clones that produce a cohort of neurones sharing a common site of origin in the ventricular zone, a common migratory pathway in the intermediate zone and a final columnar deployment (ontogenetic column) in the cortical plate. We postulate that size of the cortical mantle, including the species-specific pattern of the cytoarchitectonic area, depends on the number of contributing radial units, while the thickness of the cortex depends on the magnitude of cell production within each unit. Although the initial number of units in a given species are likely to be set up early in embryogenesis by regulatory genes, organization and final size of each cytoarchitectonic area is established through interactions with appropriate afferents. The enlargement of existing or introduction of new cytoarchitectonic areas in evolution could initially occur through a heterochronic process that includes the modification of the rate and cessation of cell proliferation during the phase of radial unit formation. We suggest that the new set of radial units interact with afferent systems, creating an opportunity for the formation of novel input/target/output relationships which, if inheritable, may be subject to natural selection. Recent experiments on the embryonic brains of extant species, including induction of a novel cortical area by experimental manipulation of the developing thalamocortical input in Primates, support this hypothesis (Rakic *et al*. 1991; Rakic 1988).

Introduction

Traditionally, the problem of cerebral evolution has been studied by comparative anatomists (for example, Ariens Kappers *et al.* 1936; Armstrong and Falk 1982). However, developmental mechanisms by which the brain evolves and adapts to an ever-changing environment have not been sufficiently explored. In this chapter I will outline the possible cellular mechanisms that may underlie the evolution of distinct cytoarchitectonic areas in primates.

My view is based on the series of studies of cortical embryogenesis in macaque monkey and on comparisons of developmental events in this species with corresponding events observed in human embryos. Comparing features of the adult cerebral hemispheres in these two species reveals a difference in the size and pattern of cytoarchitectonic areas. These 'maps' generally reflect species-specific functional specializations at the cortical level. Since the time that humans and macaque monkeys diverged from a common ancestor approximately 20 million years ago (Fleagle 1988), the cerebral cortex of both species has undergone remarkable expansion and modification. However, the total surface area of cortex in humans is not only approximately 10 times larger than in macaque but, significantly, the relative proportion of various cytoarchitectonic areas is also quite different between these species (Blinkov and Glaser 1968). For instance, while some areas seem to have been conserved during the evolutionary process (for example, primary visual cortex or area 17), others exhibit enlargement (for example, secondary visual cortex, area 18); humans in general have a larger proportion of association cortex. Finally, new cytoarchitectonic subdivisions have emerged during evolution.

It seems reasonable to assume that these interspecies cytoarchitectonic differences (for example in overall size, relative proportions of existing areas, introduction of new areas) occur initially as structural modifications caused by gene mutations. I have started with the belief that information obtained from the study of embryonic development in living species may provide clues about possible cellular and molecular mechanisms that have occurred during evolution. Many examples from other fields of biology support the validity of this approach (Gould 1977). I will try to interpret cellular events that could occur during evolution in the framework of the protomap and radial unit hypotheses (Rakic 1988). These two hypotheses are derived from a series of neuroembryological studies starting approximately two decades ago (Rakic 1971, 1972). To illustrate the basic concept and how it applies to the neocortex, I will focus on the development of cytological and biochemical differences between the two visual cortical areas 17 and 18 (Fig. 5.1). In particular, I will deal with the emergence of distinct lamination, modular organization, separate thalamic input, and characteristic distribution of neurotransmitters, their receptors, and a mitochondrial enzyme cytochrome oxidase (CO) in

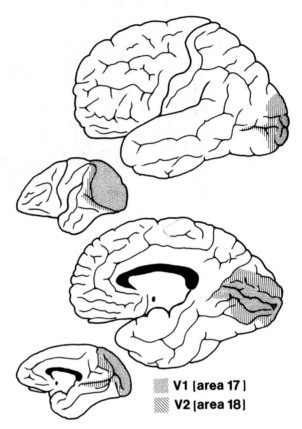

Fig. 5.1. Lateral and medial surface of cerebral hemisphere in the macaque monkey (left) and human (right) drawn at approximately the same scale. In the human cerebrum, which is approximately 10 times larger than the monkey's, the primary visual cortex (V1 or area 17) occupies only approximately 3 per cent (dotted region) of the total surface, whereas in the monkey the same cortex occupies approximately 15 per cent. In contrast, secondary visual cortex (V2 or area 18) is relatively smaller in monkey compared to man (illustration based on Brodmann (1909); data based on Blinkov and Glezer (1968) and R.W. Williams and P. Rakic (unpublished observations) for the macaque monkey).

these two areas (Hubel and Wiesel 1977; Livingstone and Hubel 1988; Rakic *et al.* 1988; Kuljis and Rakic 1990; Felleman and Van Essen 1991).

Early cellular events

The telencephalic wall in the mammalian embryo contains several cellular zones that do not have counterparts in the adult cerebrum (for example, Rakic 1982). The lining of the cerebral ventricle during the entire first half

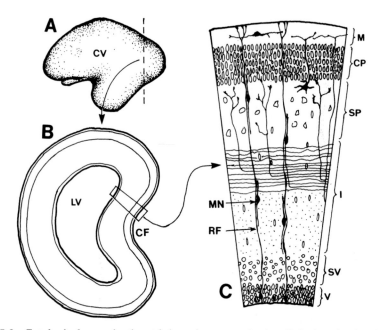

Fig. 5.2. Cytological organization of the primate cerebral wall during the first half of gestation. **A,** Cerebral vesicle of 60–65-day-old monkey fetuses is still smooth and lacks characteristic convolutions that will emerge in the second half of gestation. **B,** Coronal section across the occipital lobe at the level indicated by a vertical dashed line in A. The lateral ventricle at this age is still relatively large and only the identification of incipient calcarine fissure (CF) marks the position of the prospective visual cortex. **C,** A block of the tissue dissected from the upper bank of calcarine fissure. At this early stage one can recognize six embryonic layers from the ventricular surface (bottom) to the pial surface (top): ventricular zone (V), subventricular zone (SV), intermediate zone (I), subplate zone (SP), cortical plate (CP), and marginal zone (M). Note the presence of migrating neurones (MN) moving along radial glial fibres (RF) which span the full thickness of the cortex. The early afferents from the brainstem, thalamus, and other cortical areas invade the cerebral wall and accumulate in the subplate zone, where they make transient synapses before entering the cortical plate.

of gestation of both the macaque and human consists of proliferative cells that eventually produce all neurones of the neocortex. These precursors form the sheet of germinal cells or ventricular zone (Fig. 5.2). Progenitors of neuronal and radial glial cells in this zone are arranged in a pseudostratified manner and are attached to the ventricular surface by their endfeet (Rakic 1972). Many cells in this zone have radial processes that protrude towards the pial surface and into the outer cell-free marginal zone (Fig. 5.2C). A group of precursor cells that form a pseudostratified column separated by a palisade of glial cells has been termed the proliferative unit (Rakic 1978). Cohorts of cells produced in succession from the same proliferative unit migrate along radial glial fascicles and pass through the intermediate and subplate zones before entering the developing cortical plate (Fig. 5.2C).

The waves of neurones arrive successfully at the cortical plate and pass by each other (Rakic 1974), becoming arranged radially in the form of cell stacks which I call ontogenetic columns (Rakic 1982). The ontogenetic column can be defined as a cohort of neurones that originated from several progenitors situated in the same proliferative unit. Such columns can be easily recognized in the histological sections cut across the cortical plate during mid-gestation (for example, Fig. 5.1 in Rakic 1988). Although some postmitotic cells do not obey radial glial constraints (Rakic et al. 1974; Schmechel and Rakic 1979) and may move tangentially (Walsh and Cepko 1992; O'Rourke et al. 1992), the majority of neurones in a given column of the cortex are generated in the underlying sectors of the ventricular zone and strictly follow radial pathways (Rakic 1978; Nakatsuji et al. 1991; Tan and Breen 1993). The cell class that migrates radially is constrained in movement by an affinity for elongated glial fibres that span the entire foetal cerebral wall during the period of corticogenesis (Rakic 1972, 1978). Based on such observations, I suggest that the proliferative unit, migrating pathway, and ontogenetic column together form the radial unit that extends from the ventricular to the pial surface (Rakic 1988). Recent evidence indicates that the entire radial unit of migrating cells is coupled by junctions that provide intercellular communication (Lo Turco and Kriegstein 1992). According to the radial unit hypothesis, the neocortex can be considered as a mosaic derived from a large number of such radial units. These units may be considered as the ontogenetic basis of functional columns that are observed by anatomical and physiological methods in the adult cerebral cortex (Szenthagothai 1978; Mountcastle 1979; Eccles 1984).

Formation of radial units

Autoradiographic analysis of monkey embryos injected with ^{3}H-thymidine and sacrificed at short intervals shows that before the fortieth embryonic day (E40) all cells in the ventricular zone are still dividing. Most divisions during this phase are symmetrical; each progenitor produces two additional progenitor cells during each mitotic cycle (Rakic 1988). Thus, each extra round of mitotic divisions before E40 doubles the number of cells, resulting in an exponential increase in the number of progenitor cells in the ventricular zone (Fig. 5.3A). Conceivably, even a slight prolongation of this mode of proliferation could be indirectly responsible for the eventual surface enlargement of the cerebral cortex.

Around E40 some progenitor cells begin to produce neurones that leave the ventricular zone and will never divide again (Rakic 1974, 1988). The inhibition of DNA synthesis after this point in neuronal cell history is so powerful that they do not re-enter the mitotic cycle even under pathological

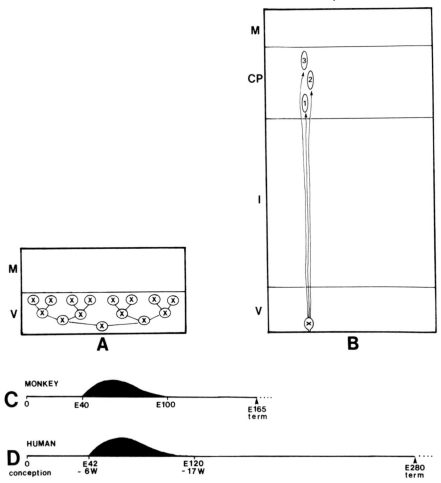

Fig. 5.3. (A) Schematic model of symmetrical cell divisions which predominate before the fortieth embryonic day (E40). At this early embryonic age the cerebral wall consists of only the ventricular zone (V), where all cells proliferate and the marginal zone (M), where some of them extend their radial processes. Symmetric division produces two progenitors during each cycle and causes rapid horizontal lateral spread. (B) Model of asymmetrical or stem division which becomes predominant in the monkey embryo after E40. During each asymmetrical division a progenitor produces one post-mitotic neurone which leaves the ventricular zone and another progenitor which remains within the proliferative zone and continues to divide. Postmitotic neurones migrate rapidly across the intermediate zone (I) and become arranged vertically in the cortical plate (CP) in reverse order of their arrival (1,2,3). (C) Diagrammatic representation of the time of neurone origin in macaque monkey. The data are obtained from the ^{3}H-thymidine autoradiographic analyses (Rakic 1974). (D) Estimate of the time of neurone origin in the human neocortex based on the number of mitotic figures within the ventricular zone, supravital DNA synthesis in slice preparations of fetal tissue and the presence of migrating neurones in the intermediate zone of the human fetal cerebrum. (Based on Rakic (1978) and Rakic and Sidman (1968).)

conditions that induce cell proliferation in other tissues (Rakic 1985). Analysis of the kinetics of cell proliferation in the ventricular zone indicates that, after E40 in the monkey embryo, many precursors begin to divide asymmetrically (Rakic 1988). This type of proliferation, also known as stem cell division, produces one daughter which is permanently postmitotic and the other which continues to divide or die. The postmitotic cell detaches its endfoot from the ventricular surface and migrates toward the pial surface, eventually becoming a cortical neurone (Fig. 5.3B). The other daughter cell remains attached to the ventricular zone by the endfoot and usually continues to divide for some time, giving each round an additional pair of unequal cells: one progenitor and one postmitotic neurone (Rakic 1988). This pattern of cell division in the monkey foetus proceeds during the next 30–60 days depending on the cortical area (Rakic 1974, 1976, 1982). Thus, mitotic activity in the ventricular zone of the macaque monkey can be divided into two broad phases:

(1) the phase of proliferative unit formation, which proceeds mainly by symmetrical division and occurs before E40;
(2) the phase of ontogenetic column formation, which proceeds mainly by asymmetric division and lasts between E40 and the completion of cortico-genesis in a given region.

These two phases may be triggered by the activation of different regulatory genes that control the type of mitotic division. Consequently, the duration of the first phase determines the size and duplication of cytoarchitectonic areas, whereas the second phase regulates the number and differentiation of neuro-nal phenotypes within each ontogenetic column (Rakic 1988).

Evidence for the radial organization of the developing cortex has come initially from light and electron microscopic analysis of the embryonic cerebral vesicle primates (Rakic 1971, 1972). The ventricular zone from the early stages of development shows a pseudostratified organization of the radially deployed germinal cells. These studies revealed the presence of columns of bipolar cells that are opposed to the elongated radial glial fibres. Dynamic cellular events in the primate embryonic cerebral wall are presented schemati-cally in Fig. 5.4 and further details can be found in primary references (for example, Rakic 1971, 1972, 1977, 1978, 1988; Rakic et al. 1974).

Formation of the cortical protomap

The most obvious relevance of the radial unit model for understanding corti-cal phylogeny is the corollary that the larger the number of proliferative units

Fig. 5.4. The relationship between a small patch of the proliferative, ventricular zone (VZ) and its corresponding area within the cortical plate (CP) in the developing primate cerebrum. Although the cerebral surface in monkey expands during prenatal development, resulting in a shift between the VZ and CP, ontogenetic columns (outlined by cylinders) remain attached to the corresponding proliferative units by the

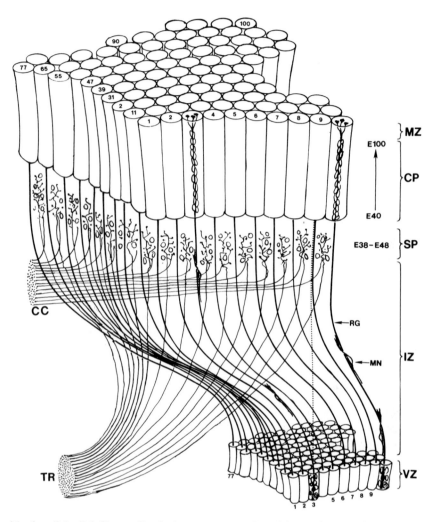

grid of radial glial fibres. Cortical neurones produced by a given proliferative unit between embryonic age 40 (E40) and E100 migrate in succession along common radial glial guides (RG) and form a single ontogenetic column. Each migrating neurone traverses the intermediate (IZ) and subplate (SP) zones, which contain 'waiting' afferents originated in the thalamic radiation (TR) and other cortical areas in the same or other hemisphere (CC). After entering the cortical plate, each wave of migrating neurones bypasses the population of previously generated neurones and assumes a position at the interphase between the CP and marginal zone (MZ). As a result of this process, proliferative units 1–100 give rise to ontogenetic columns 1–100 that are arranged in the same relative position to each other. The glial scaffolding prevents a mismatch between proliferative unit 3 and ontogenetic column 9 (dashed line). Thus, the specifications of topography and/or modality depend on the spatial distribution of proliferative units, while the radial (vertical) position of neurones within each unit depends on its time of origin. (From Rakic 1988.)

in a given species, the larger the surface of the cortex (for example, Fig. 5.4). Theoretically, the fate of both cells and units could be determined solely by cell lineage relationships, hence, this model is termed the 'fate map' hypothesis. Few scientists studying mammalian embryology subscribe to this view. Alternatively, all neurones and radial units that they form within the cortical plate could be initially identical or equipotential. According to this view, also popularly known as the '*tabula rasa*' hypothesis or the 'equivalency model', differentiation of cortex into cytoarchitectonic fields is imprinted on the uncommitted cells of the developing cortical plate by thalamic afferents connecting sensory receptors at the periphery (for example, Creutzfeldt 1977; O'Leary and Stainfield 1989; Killackey 1990). This hypothesis is attractive since there is little doubt that cortical afferents play an important role in shaping the size and character of cytoarchitectonic areas. However, recent studies have revealed molecular and cellular heterogeneity within the embryonic cerebrum, suggesting that information about the basic species-specific blueprint of the cortical areas may emerge before and independent of ingrowth afferents (see below). Since this basic blueprint can be modified by the interactions of various inputs, I proposed the 'protomap' hypothesis, which postulates that synergic action between factors both intrinsic and extrinsic to cortical cells is required for the formation of cortical cyto-architecture (Rakic 1988). The prefix 'proto' was introduced to underscore the malleable character of the primordial embryonic map.

The first line of evidence for the cortical protomap came from studies of normal brain development (Rakic 1972, 1978, 1988). A relatively well-preserved register between proliferative units within the ventricular zone and an array of ontogenetic columns in the cortical plate suggests a definite spatial relationship between two structures (Fig. 5.3). Moreover, ^{3}H-thymidine labelling of mitotic DNA replication shows clear regional differences in magnitude of cell proliferation. For example, a portion of the ventricular zone subjacent to the one area of the monkey cortex produces more neurones than an equivalent size portion of another area (Rakic 1976, 1982, 1988; Kennedy and Dehay 1993). The molecular heterogeneity of the cortical plate neurones is suggested by the finding that afferents from specific thalamic nuclei are attracted exclusively to some cortical areas, but not to others (Boltz *et al.* 1990). Indeed, histochemical and immunocytochemical analyses of the embryonic forebrain show that various macromolecules, including some proto-oncogenes, glycoconjugates, transcription factors, neurotransmitters, prohormones, intermediate filament proteins, and several types of adhesion molecules, are expressed as a gradient or in spatially restricted manner early or independently of thalamic input (Hutchins and Casagrande 1989; Johnson and Van der Kooy 1989; Schambra *et al.* 1989; Steindler *et al.* 1989; Barbe and Levitt 1991; Walther and Grauss 1991; Aramatsu *et al.*1992; LaMantia *et al.* 1992; Ferri and Levitt 1993; Cohen-Tannoudji *et al.* 1994). Early commitment and phenotype expression of individual cortical neurones before they reach their destination has been observed by a variety of other techniques (for example, McConnell 1988; Lo Turco *et al.* 1991; Schwartz *et al.* 1991). Finally, the presence of Primate-specific distribution pattern of cytochrome oxidase (CO), a molecular marker for cellular modules in the visual cortex involved predominantly with trichromatic vision, in monkeys subjected to prenatal binocular enucleation, provides additional evidence that at least certain species-specific molecular features of the visual cortex may develop independently of the information from the photoreceptors at the periphery (Kuljis and Rakic 1990).

Our recent studies suggest that each level of a given sensory system (for example, retina, lateral geniculate, visual cortex) may have its own protomap (for example, Rakic *et al.* 1991). For example, at the retina level, we have evidence that a species-specific photoreceptor mosaic develops before rods and cone subtypes develop any direct contacts with the retinofugal neuronal system (Wikler and Rakic 1991). Likewise, in early enucleated animals area 17 develops area-specific distribution of neurotransmitter receptors (Rakic and Lidow 1993) and cytochrome oxidase-enriched areas in the absence of the input from the photoreceptors (Kuljis and Rakic 1990). Therefore, proper maturation of the visual system requires articulating these protomaps into a functionally adequate system via synergic interactions that involve selective stabilization and elimination of unfit inputs, a general mechanism suggested for the formation of neural connections, including cerebral cortex (for example, Rakic 1981; Changeux and Dehane 1989). A test of the protomap hypothesis can be carried out by selectively diminishing the size of input to the cortex at critical embryonic stages.

Modifiability of the cortical protomap

As the term indicates, the protomap hypothesis does not require that all cellular and molecular details of the organization and final size of cytoarchitectonic areas be intrinsically and rigidly specified. Experiments in several species, including Primates, show that thalamic input plays an important role in regulating the size of a given cytoarchitectonic area. For example, in macaque monkey, prenatal bilateral enucleation diminishes the number of geniculocortical afferents, which in turn causes a proportional reduction in the size of area 17 which nevertheless retains normal thickness, layering pattern, synaptic density, and pattern cytochrome oxidase reactivity (Rakic 1988; Kuljis and Rakic 1990; Rakic *et al.* 1991; P. Rakic and M. S. Lidow, submitted). Thus, the number of ontogenetic columns devoted to a given area and, therefore, the final size is regulated by the number of thalamic afferents and, perhaps, also other classes of afferents that arrive at the critical developmental stages (Rakic 1988).

The developmental mechanism for rearrangement of the cortical area in enucleates is not understood, but several possibilities can be considered. Area

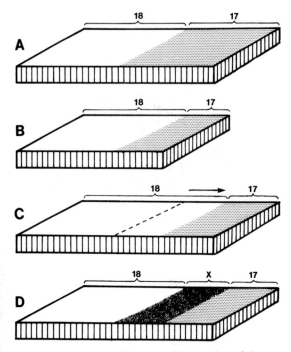

Fig. 5.5. Schematic representation of the possible modes of decrease in the size of area 17 caused by experimental reduction of thalamic input. (A) Relation between areas 17 and 18 in a normal animal. (B) Changes induced by differential cell death in area 17. (C) Encroachment of adjacent area 18 into the territory of area 17. (D) Formation of an abnormal hybrid cytoarchitectonic area (X) that consists of neurones intrinsically destined for area 17 but which receive input characteristics for area 18. (From Rakic 1988.)

17 can simply lose a number of ontogenetic columns (Fig. 5.5B). Alternatively, area 18, which normally receives input from the adjacent thalamic nucleus (pulvinar) and from the other parietal and temporal cortices, could take over some of the territory ordinarily destined for area 17 (Fig. 5.5C). Finally, a number of columns that were specified for area 17 (X in Fig. 5.5D) could, in the absence of normal afferents, receive input from the pulvinar and become a 'hybrid' cortex that is 'intrinsically' area 17 and 'connectionally' area 18 (Rakic 1988). Recently, we obtained evidence that enucleation during the midgestational period results in the formulation of a novel cytoarchitectonic area situated at the periphery or as an island in the middle of an otherwise normal-looking area 17 (Rakic *et al.* 1991). The experimentally induced 'hybrid' cortex, which we termed area X, has cytoarchitectonic features that are neither that of area 17 nor 18 (Fig. 5.6).

The experimental paradigm described above provides a glimpse of how novel cytoarchitectonic fields could be introduced during cortical evolution. In our experiment, specific thalamic input was diminished while its target

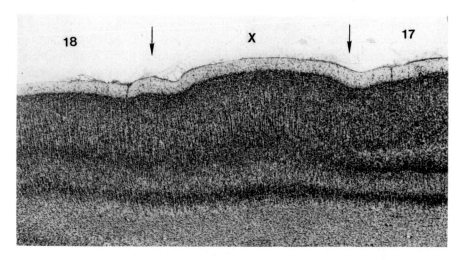

Fig. 5.6. A view of the cortex at the lateral surface of the occipital lobe in a 3 month old monkey that was bilaterally enucleated on the eighty-first embryonic day. The Nissl-stained section displays clearly abrupt changes in cytoarchitectonic patterns (arrows) that delineate area 17, area 18, and area X interposed between them. (From Rakic *et al.* 1991.)

cortical area, normally destined to become area 17, remained the same size. Thus, we experimentally created an imbalance of input and target ratio. During evolution, however, a similar imbalance of the input–target relationship could occur when a cortical area expands by an increase in the production of ontogenetic columns. The expanded cortical area could then serve as a substrate for competition among incoming afferent systems, including various thalamocortical fibres. Thus, adjustment between slightly altered protomaps from the periphery to the cortex could create new cytoarchitectonic combinations.

Implications for cortical evolution

There are many instructive examples of how the data obtained from studies of comparative neuroanatomy and neuronal plasticity can be used to elucidate principles of cortical evolution (for example, Ariens-Kappers *et al.* 1936; Armstrong and Falk 1982; Ebbeson 1984; Finlay *et al.* 1987; Kaas 1988; Allman 1990). A major drawback of this approach is the lack of adequate experimental evidence and controls; also it suffers from the problem of interpretation of cross-phyletic homology (for example, Northcutt 1990). Perhaps we cannot design a completely satisfactory experimental paradigm, nor should we expect to find an answer by searching for a 'magic' molecule that induces a new area. Since all known ion channels, transmitters, and receptors are present in the cortex of all Primates and even in the brains of all mammals (Hille 1984; Marc 1989), it is unlikely that the emergence of a single macromolecule could explain the emergence of a new cortical area in human. This may stand in contrast to the significant role that a small modification of some key molecules may have played in the evolution of other bodily systems (Medawar 1953). My strategy has been to search for possible cellular mechanisms of cortical evolution, using the principles of corticogenesis derived from the studies of embryogenesis in living Primates as guidelines. I do so with the sober realization that many assumptions may be highly speculative and quite difficult to prove.

I start with the conviction that several aspects of cerebral evolution are explicable in terms of the radial unit hypothesis. For example, the cortical surface in human is 10 times larger than in monkey, but its thickness remains relatively little changed. This differential expansion may be explained by the large increase in the number of proliferative units during the phase of unit formation. As evident from Fig. 5.3, during this early phase each round of symmetric cell divisions doubles the number of progenitor cells. However, during the second phase, when asymmetrical divisions predominate, each additional cycle adds only a single neurone to a given ontogenetic column. Therefore, an average of 3.2 extra rounds of cell division during the first phase could increase the surface of the cortex approximately 16 times, whereas three extra cycles at the end of corticogenesis would produce only a 3 per cent thicker cortex. Such a small change in the proliferative behaviour of cells in the ventricular zone may account for the difference in cortical size between the macaque monkey and *Homo sapiens* (Fig. 5.3A). Thus, large expansion in the total surface of the cortex could be attributed to a single heterochronic process. Obviously, this process alone would not be sufficient to explain the development of connectivity that ensues following formation of the larger cortical plate.

The model illustrated in Fig. 5.3 may be useful for understanding the mechanisms underlying the total expansion of cortical surface. However, how can one explain the uneven expansion of cytoarchitectonic areas? One possibility is that the heterochronic process that modifies the timing of proliferation kinetics in the mosaic of the ventricular zone produces additional ontogenetic columns only in certain regions. Such extra rounds of cell divisions could occur due to genetic mutation and natural selection. This is an attractive concept because it is in full harmony with ideas that language (and the cortical areas that are involved) probably has been introduced relatively fast rather than in a slow and graded manner (for example, Chapter 17). However, additional cortical columns composed of cells resulting from a single event could subsequently compete for the various afferent inputs and form a variety of novel input–output affiliations. One can further speculate that a given thalamic input which spreads to a larger number of columns in the cortex may retrogradely increase the size of the innervating population in this nucleus by spearing overproduced neurones. This could have a further cascading effect on the neurones and receptor cells at the periphery. Thus, protomaps at several levels of the brain could be modified. However, since selection pressures in evolution act upon the functional output, not on the structure itself, these connections must be validated through behaviour. The new synaptic relationships resulting from new neuronal interaction may be adverse, neutral, or may enhance behavioural adaptation. As pointed out by Jacob (1977), the new feature may not be optimal, but must be good enough. Only in such a case would it provide a survival advantage for the species over many generations that is fostered through natural selection. For example, by reducing adult mortality it could improve lifetime reproductive success.

The model presented illustrates how information obtained from studies of embryonic development of the cerebral cortex can be used to suggest possible cellular mechanisms underlying phylogenetic development of this structure. Such an approach provides a possibility of understanding the evolution of cerebral cortex and the ascent of human mental capacity. It has been suggested that evolutionary construction of the mammalian brain may require as many genes as were needed for all morphogenetic and metabolic functions in phyletic history (John and Micklos 1988). However, the difference between the genome of macaque monkey and human is thought to be relatively small and the explanation for cortical expansion rests predominantly upon the process of heterochrony followed by rearrangements of connections. Although the introduction of novelty in evolution by the process of hetero-chrony has been described for morphological features of non-neural organs of the body (Alberch *et al.* 1979), understanding the role of heterochrony in phylogenetic development of the brain presents special problems because of the complex interplay among multiple epigenetic growth factors which regulate gene expression during development (Changeux 1983; Easter *et al.* 1985; Edelman 1988; Purves 1988). During genesis of the cerebral cortex, such cellular interactions probably play a more significant role than in any other organ and this is perhaps, why progress has been slow in this field of research.

References

Allman, J. (1990). Evolution of neocortex. In *Cerebral cortex* (ed. A. Peters and E.G. Jones), Vol. 8. Academic Press, New York.

Alberch, P., Gould, S.J., Oster, G.F., and Wake, D.B. (1979). Size and shape in ontogeny and phylogeny. *Phyleobiology* **5**, 296–317.

Ariens Kappers, C.B., Huber, G.C., and Crosby, E.C. (1936). *The comparative anatomy of the nervous system of vertebrates including man.* Hofner, New York.

Arimatsu, Y., Miyamoto, M., Nihonmatsu, I., Hirata, K., Uratani, Y., Hatanaka, Y., and Takiguchi-Hoyashi, K. (1992). Early regional specification for a molecular neuronal phenotype in the rat neocortex. *Proc. Nat. Acad. Sci. USA* **89**, 8879–83.

Armstrong, E. and Falk, D. (ed.) (1982). *Primate brain evolution.* Plenum, New York.

Barbe, M.F. and Levitt, P. (1991). Early commitment of fetal neurons to the limbic cortex. *J. Neurosci.* **11**, 519–37.

Boltz, J., Novak, N., Götz, M., and Bonhoffer, T. (1990). Formation of target-specific organotypic slice cultures from rat visual cortex. *Nature* **346**, 359–62.

Blinkov, S.M. and Glezer, I.J. (1968). *The human brain in figures and tables, a quantitative handbook*. Plenum, New York.

Brodmann, K. (1909). *Vergleichende Localizationsationslehre der Grosshichinde*. Barth, Leipzig.

Changeux, J.-P. (1983). *L'homme neuronal*. Fayard, Paris.

Changeux, J.-P. and Dehane, S. (1989). Neuronal models of cognitive functions. *Cognition* **33**, 63–109.

Cohen-Tannoudji, M., Babinet, C., and Wassef, M. (1994). Early intrinsic specification of the mouse somatosensory cortex. *Nature*. (In press.)

Creutzfeldt, O.D. (1977). Generality of the functional structure of the neocortex. *Naturwissenschaften* **64**, 507–17.

Easter, S.S., Jr, Purves, D., Rakic, P., and Spitzer, N.C. (1985). The changing views of neuronal specificity. *Science* **230**, 507–11.

Ebbeson, S.O.E. (1984). Evolution and ontogeny of neural circuits. *Behav. Brain Sci.* **7**, 332–50.

Eccles, J.C. (1984). The cerebral neocortex. A theory of its operation. In *Cerebral cortex* (ed. E.G. Jones and A. Peters), Vol. 2, pp. 1–36. Plenum, New York.

Edelman, G.E. (1988). *Topobiology. An introduction to molecular embryology*. Basic Books, New York.

Fellman, D.J. and Van Essen, D.C. (1991). Distributed heirarchical processing in the primate cerebral cortex. *Cerebral Cortex* **1**, 3–54.

Ferri, T.R. and Levitt, P. (1993). Cerebral cortical progenitors are fated to produce region-specific neuronal populations. *Cerebral Cortex*, **3**, 187–98.

Finlay, B.L., Wikler, K.C., and Sengelaub, D.R. (1987). Regressive events in brain development and scenarios for vertebrate brain evolution. *Brain: Behav. Evol.* **30**, 102–17.

Fleagle, J.G. (1988). *Primate adaptation and evolution*. Academic Press, New York.

Gould, S.J. (1977). *Ontogeny and Phylogeny*. The Belknap Press, Cambridge, MA.

Hille, B. (1984). Ionic channels: evolutionary origins and modern roles. *Q. J. Exp. Physiol.* **74**, 785–804.

Hubel, D.H. and Wiesel, T.N. (1977). Ferrier lecture. Functional architecture of macaque monkey visual cortex. *Proc. R. Soc. Lond. B.* **198**, 1–59.

Hutchins, J.B. and Casagrande, V.A. (1989). Vimentin: changes in distribution during brain development. *Glia* **2**, 55–6.

Jacob, F. (1977). Evolution and tinkering. *Science* **196**, 1161–6.

John, B. and Miklos, G.L. (1988). *The eukaryote genome in development and evolution*. Allen and Unwin, London.

Johnson, J. and Van der Kooy, D. (1989). Protooncogene expression identifies a columnar organization of the ventricular zone. *Proc. Natl. Acad. Sci. USA*, **86**, 1066–70.

Kaas, J.H. (1988). Development of cortical sensory maps. In *Neurobiology of neocortex* (ed. P. Rakic and W. Singer), pp. 101–30. Wiley and Sons, New York.

Kennedy, H. and Dehay, C. (1993). Cortical specification of mice and men. *Cerebral Cortex* **3**, 171–86.

Killackey, H. (1990). Neocortical expansion, An attempt towards relating phylogeny and ontogeny. *J. Cog. Neurosci.* **2**, 1–17.

Kuljis, R.O. and Rakic, P. (1990). Hypercolumns in the primate visual cortex develop in the absence of cues from photoreceptors. *Proc. Natl. Acad. Sci. USA*, **87**, 5303–6.

LaMantia, A.S., Whitesides, J.G., Colbert, M., and Linney, E. (1992). Activated retinaic acid receptors define a distinct domain in the developing mouse forebrain. *Abst. Soc. Neurosci.* **18**, 1097.

Livingstone, M.S. and Hubel, D.H. (1988). Segregation of form, color, movement, and depth, anatomy, physiology and perception. *Science* **240**, 740–9.

Lo Turco, J.J., Blanton, M.G., and Kriegstein, A.R. (1991). Initial expression and endogenous activation of NMDA channels in early neocortex development. *J. Neurosci.* **11**, 792–9.

Lo Turco, J.J. and Kriegstein, A.R. (1992). Clusters of coupled neuroblasts in embryonic neocortex. *Science* **252**, 563–6.

McConnell, S.K. (1988). Fates of visual cortical neurons in the ferret after isochronic and heterochronic transplantation. *J. Neurosci.* **8**, 945–74.

Marc, R.E. (1989) Evolution of retinal circuits. In *Proceedings of the Second International Congress of Neuroethology*. Theme, Stuttgart.

Medawar, P.B. (1953). Some immunological and endocrinological problems raised by the evolution of vivparity in vertebrates. *Symp. Soc. Exp. Biol.* **7**, 320–38.

Mountcastle, V. (1979). An organizing principle for cerebral function: the unit module and the distributed system. In *The neurosciences, fourth study program* (ed. F.O. Schmitt and F.G. Worden), pp. 21–42. MIT Press, Cambridge, MA.

Nakatsuji, M., Kadokawa, Y., and Suemori, H. (1991). Radial columnar patches in the chimeric cerebral cortex visualized by use of mouse embryonic stem cells expressing β-galacto-oxidase. *Devel. Growth Differ.* **33**, 571–8.

Northcutt, R.G. (1990). Ontogeny and phylogeny. A re-evaluation of conceptual relationships and some applications. *Brain Behav. Evol.* **36**, 116–40.

O'Leary, D.M.D. and Stainfield, B.B. (1989). Selective elimination of axons extended by developing cortical neurons is dependent on regional locale. Experiments utilizing fetal cortical transplants. *J. Neurosci.* **9**, 2230–46.

O'Rourke, N.A., Dailey, M.E., Smith, S.J., and McConnell, S.M. (1992). Diverse migratory pathways in the developing cerebral cortex. *Science.* **258**, 299–302.

Purves, D. (1988). *Body and brain. A trophic theory of neural connections*. Harvard University Press, Cambridge.

Rakic, P. (1971). Guidance of neurons migrating to the fetal monkey neocortex. *Brain Res.* **33**, 471–6.

Rakic, P. (1972). Mode of cell migration to the superficial layers of fetal monkey neocortex. *J. Comp. Neurol.* **145**, 61–84.

Rakic, P. (1974). Neurons in rhesus monkey visual cortex. Systematic relation between time of origin and eventual disposition. *Science* **183**, 425–7.

Rakic, P. (1976). Differences in the time of origin and in eventual distribution of neurons in areas 17 and 18 of visual cortex in rhesus monkey. *Exp. Brain Res.* (Suppl. 1), 244–8.

Rakic, P. (1977). Prenatal development of the visual system in the rhesus monkey. *Phil. Trans. R. Soc. Lond. B* **278**, 245–60.

Rakic, P. (1978). Neuronal migration and contact guidance in primate telencephalon. *Postgrad. Med. J.* **54**, 25–40.

Rakic, P. (1981). Development of visual centers in primate brain depends on binocular competition before birth. *Science* **214**, 928–31.

Rakic, P. (1982). Early developmental events: cell lineages, acquisition of neuronal positions, and areal and laminar development. *Neurosci. Res. Prog. Bull.* **20**, 439–51.

Rakic, P. (1985). Limits of neurogenesis in primates. *Science* **227**, 154–6.

Rakic, P. (1988). Specification of cerebral cortical areas. *Science* **241**, 170–6.

Rakic, P. (1985). Limits of neurogenesis in primates. *Science* **227**, 154–6.

Rakic, P. (1988). Specification of cerebral cortical areas. *Science* **241**, 170–6.

Rakic, P. and Lidow, M.S. (1993). Distribution and density of neurotransmitter receptors in the absence of retinal input from early embryonic stages. *J. Neurosci.* (In press.)

Rakic, P. and Sidman, R.L. (1968). Supravital DNA synthesis in a developing human and mouse brain. *J. Neuropath. Exp. Neurol.* **27**, 246–76.

Rakic, P., Stensaas, L.J., Sayre, E.P., and Sidman, R.L. (1974). Computer-aided three-dimensional reconstruction and quantitative analysis of cells from serial electronmicroscopic montages of fetal monkey brain. *Nature (London)* **250**, 31–4.

Rakic, P., Gallager, D., and Goldman-Rakic, P.S. (1988). Areal and laminar distribution of major neurotransmitter receptors in the monkey visual cortex. *J. Neurosci.* **8**, 3670–90.

Rakic, P., Suner, I., and Williams, R.W. (1991). A novel cytoarchitectonic area induced experimentally within the visual striate cortex. *Proc. Natl. Acad. Sci. USA* **88**, 2083–7.

Schambra, V.B., Sulik, K.K., Petrasz, P., and Lande, J.M. (1989). Ontogeny of cholinergic neurons in the mouse forebrain. *J. Comp. Neurol.* **288**, 101–22.

Schmechel, D.E. and Rakic, P. (1979). A Golgi study of radial glial cells in developing monkey telencephalon. *Anat. Embryol.* **156**, 115–52.

Schwartz, M.L., Rakic, P., and Goldman-Rakic, P. (1991). Early phenotype expression of cortical neurons. Evidence that a subclass of neurons has callosal axons. *Proc. Natl. Acad. Sci. USA* **88**, 1354–8.

Schmechel, D.E. and Rakic, P. (1979). A Golgi study of radial glial cells in developing monkey telencephalon. *Anat. Embryol.* **156**, 115–52.

Simeone, A., Gulisano, M., Stornaiulo, A., Rambalki, M., and Boncinelli, E. (1992). Two vertebrate homeobox genes related to *Drosophila empty spiracles* gene are expressed in the embryonic cerebral cortex. *EMBO J.* **11**, 2541–50.

Steindler, D.A., Cooper, N.G.F., Faissner, A., and Schachner, M. (1989). Boundaries defined by adhesion molecules during development of the cerebral cortex. The J1/tenascin glycoprotein in the mouse somatosensory cortical barrel field. *Devel. Biol.* **131**, 407–18.

Szenthagothai, J. (1978). The neuron network of the cerebral cortex. A functional interpretation. *Proc. Roy. Soc. (London), B* **201**, 219–48.

Tan, S.-S. and Breen, S. (1993). Radial mosaicism and tangential cell dispersion both contribute to mouse neocortical development. *Nature* **362**, 638–40.

Walsh, C. and Cepko, C.L. (1992). Widespread dispersion of neuronal clones across functional regions of the cerebral cortex. *Science* **255**, 434–40.

Walther, C. and Grauss, P. (1991) *Pax-6*, a maurine paired gene is expressed in the developing CNS. *Development* **113**, 1433–49.

Wikler, K. and Rakic, P. (1991). Emergence of the photoreceptor mosaic from a protomap of early-differentiating cones in the primate retain. *Nature* **351**, 397–400.

DISCUSSION

Participants: J.-P. Changeux, M. Mishkin, P. Rakic, A. Roch Lecours, M.P. Stryker, P.V. Tobias

The effects of enucleation on the cytology of cortical areas and on their connections

Responding to questions by **Lecours** and **Changeux, Rakic** detailed the cytological characteristics of the experimentally induced cortical area X which shows some features of area 17, others of area 18, and additional novel characteristics. In particular, area X differs from either adjacent area by the relative thickness and cytological composition of cortical layers. For example, the size and cytoarchitectonic composition of layers II and III resemble area 18, but the distribution of cytochrome oxidase within these two layers has a pattern characteristic of area 17 rather than the stripes characteristic of area 18. It appears to be a 'hybrid' cortex formed in response to a reduced number of incoming fibres from the lateral geniculate nucleus, but the density of synapses in it seems to be normal. The connectivity of this area with other cortical and subcortical regions is not known.

Defects in cortical organization of lissencephalic and gyrencephalic brains

Rakic replied to **Tobias**: There are two broad groups of congenital cortical malformations. (1) The lissencephalic or 'smooth' cortex have fewer radial columns and the entire cortical mantle is smaller in surface area. This defect apparently occurs before embryonic day 40 when proliferative units are being formed in the human cerebrum. After that age, each unit apparently produces a normal number of postmitotic cells and, as a result, the cortex has a normal or even higher thickness. Recently, a defective gene was identified by Reiner and her colleagues that may, in at least one form of lissencephaly, cause an absence of convolutions, presumably by disrupting neuronal migration. (2) The other major group of malformations is the polymicrogiria, in which the cerebrum has a normal size of the surface, but the cortex itself is much thinner than in the controls. The radial unit hypothesis predicts that this deficit occurs later in foetal life and that mutant genes may act on the stage of unit production in the fetal cerebral vesicles, which, in human, begins only after E40. So, in this group of congenital malformations, one usually can find a normal number of radial units in the neocortex, but each radial unit produces a much smaller number of cells. Of course, in reality, many malformations of the cerebral cortex are a mixture of the two basic types.

Geniculo-cortical connections

Mishkin: Suppose that if you could go into the geniculate and remove it completely, would you still have a striate cortex?

Rakic: We do not know the answer to that question, but I venture to predict that the striate cortex, as defined by anatomists, probably would not develop because thalamic input is essential for its normal cytological differentiation. The problem is that, technically, we cannot ablate the retina before embryonic day 60. I tried to perform intrauterine surgery at day 50 and lost several embryos. However, we had a very high success rate (80 per cent) in prenatal surgery performed after that gestational age. Before embryonic day 50, the uterine wall in macaque is thick, the placenta is covering most of its inner surface, and an abortion results if one touches it with the surgical knife or cause traumatic bleeding, so experiments are not practical.

Functions of area X

Several people asked what is the function of the novel area induced by the deprivation of visual input.

Rakic: What is the function of this part of the cortex in anophthalmics and is it useful to adult animals? Well, we do not know much about it at this time, but we will try to examine this question in operated monkeys. However, the technical problems are enormous. I would like to emphasize that one might address this same question in blind humans with the new non-invasive imaging methods. In Belgium, Dr Veraat and his colleagues investigated neuronal activity of the occipital lobe in humans with congenital blindness, called secondary anophthalmia. The occipital lobe in such individuals is metabolically highly active, even though the visual cortex never had any input from the retina *via* the lateral geniculate or had any light stimulation and visual experience. When, as a control, they examined human beings who became blind due to retinal degeneration or carcinoma occurring during adulthood, the occipital cortex in such individuals, under the very same conditions, is significantly less metabolically active. So, the possibility exists that the congenitally blind have the equivalent of an area X and that they may use it for some non-visual function. This hypothesis is, however still in the realm of speculation and, hopefully, experiments in animals can provide an idea of what is going on in the deprived occipital cortex.

Cortical ontogenetic columns

The origin and composition of ontogenetic columns was questioned by **Changeux** and **Stryker**.

Rakic: The ontogenetic columns are clearly visible in histological preparations of the fetal primate cortex. Unlike adult cortex, which is organized horizontally into layers, fetal cortex is composed of radial columns. Initial autoradiographic studies, using ^{3}H-thymidine to label postmitotic cells, indicated that these columns are polyclones. Furthermore, the original light and electron microscopic examination indicated that the majority of cells within a given column in the cortex originate from the same spot in the proliferative ventricular zone. More recent studies using retrovirus gene transfer to study the cell lineage relationship confirmed that more than one cell clone contributes to each radial column. In the macaque monkey, our assessment indicates that there must be between four and ten stem cells that contribute to a single column.

The validity of the radial unit hypothesis of cortical development has been challenged by the finding that, in rodents, some postmitotic cells do not migrate along radial glial fibres and do not obey radial constraints imposed by glial scaffolding. However, recent studies in transgenic and in mosaic mice show clearly and directly that a majority of postmitotic neurones are moving radially to the cortex following radial glial fascicles and eventually form radial columns. Several investigators suggested that tangentially moving cells may be a separate class and their number may be larger in some rodents than in primates. The radial organization is particularly prominent in fetal primates, including human.

6

Brain, locomotion, diet, and culture: how a primate, by chance, became a man

YVES COPPENS

First step

About 8 million years ago, in an East African region which was becoming dry, lived a primate of the size of a pygmy chimpanzee, unusually intelligent, imaginative, active, expressive, and social. Its name was *Australopithecus* the first representative of the family Hominidae.

What could have been the scenario for the emergence of *Australopithecus*? Let us imagine the common ancestor of *Australopithecus* and of the pre-chimpanzee, around 10 million years ago, in the forest or the mosaic of forest and woody savannah that existed at that time across Equatorial Africa, from the Atlantic Ocean to the Indian Ocean. We know that the process of rifting had a period of reactivation around 8 million years ago — intense volcanic activity as well as uplifting of the rift shores (especially the western shoulder of the Western African Rift). Also, there was uplifting of the whole eastern region between the Rift and the Indian Ocean. Because of the new topography, rains were at the time affected by the escarpments of the western rift and by the altitude of the new East African plateau. The west remained moist and the east became increasingly dry (today 180 mm of rain a year on the western side, and less than 100 mm on the eastern side). The numbers of taxa of trees and their density decreased dramatically in the east and the forest retreated to the west.

As it was also the time of the uplifting of the Tibetan plateau, it was the reason for the beginning of what we call the seasonal phenomenon of monsoon in the north-western part of the Indian Ocean. The endemic East African fauna (Ethiopian), adapted to open landscapes, then appeared between the Rift Valley and the Indian Ocean, and included *Australopithecus*. It was obviously a particularly important local change. The population of Hominoids, which were to become the common ancestors of both Hominidae (*sensu stricto*, *Australopithecus* and *Homo*) and Panidae (*Pan* and *Gorilla*), was then divided by this tectonic accident into two parts. This separation became an ecological barrier, and hominoid descendants continued to evolve right where they were — the western in the west, in the area which maintained the humidity and woody environment, the eastern in the east, in the area which became increasingly drier providing an open environment savannah and grassland. Contrary to earlier hypotheses, there was no migration from savannah to forest, or vice versa. The western group became the Panidae, the eastern group, the Hominidae. I first proposed this model in 1982 during a symposium on the evolution of Primates (Coppens 1983) because during the course of the symposium, David Pilbeam, for palaeontological reasons, and Gerald Lowenstein, for biochemical reasons, suggested that *Ramapithecus* and *Sivapithecus* should no longer be classified as being Hominidae. Furthermore, in the same symposium, Russel Doolittle and Leonard Greenfield, for molecular reasons, and Jérôme Lejeune and Bernard Dutrillaux, for cytogenetical reasons, confirmed that African apes, and not Asian ones, were undoubtedly the closest relatives of Man. These new conclusions on *Ramapithecus* as well as the confirmation of the short genetic distances between chimpanzee and *Homo* were strong enough for me to reconsider the situation in Africa at the end of the Miocene. The molecular clock was at that time indicating 7.5 million years for the divergence of Hominidae and Panidae. Today, there is general agreement for a period of 5.5–7.7 million years, and geophysicists were indicating a period of 7.5–8 million years for the reactivation of rifting, and the vertebrate palaeontologists 7 to 8 million years for the emergence of the endemic Ethiopian fauna.

In East Africa, australopithecines could have existed as early as 7 million years ago, if we accept the maxilla of Samburu Hills in Kenya, formerly called *Motopithecus*, as belonging to the Hominidae, 6.5 million years if we consider the tooth of Lukeino in Kenya as belonging to this family or 5 to 5.5 million years if we consider the jaws of Lothagam and Tamarin, in Kenya, as being jaws of australopithecines (Hill and Ward 1988). On the other hand, following 20 years of intensive excavations in East Africa, from a total collection of about 200 000 fossil vertebrates, 2000 were specimens of Hominidae but none belonged to Panidae. Accepting the African origin of Hominidae and their close resemblance to Panidae, it becomes necessary to explain the contradiction that the Panidae and their Hominidae-cousins were never found together, which is why I proposed this model, because it does provide such an explanation — they were never together at any time. It is because of the geographical division of their ancestors that they are so derived. If this is the case, it is an example of a speciation by geographical peripatric isolation, which is very common in vertebrates' phylogeny (Fig. 6.1). The theory is original only in so far as the separation between the main population and the isolated one was tectonic. One further comment: this proposal was possible in 1982 but no earlier, because the most important excavations which began in East Africa in 1965 (1967 for the Omo where I was co-leader with Francis Clark Howell, 1968 for Eastern Turkana that Richard Leakey led, 1972 for Hadar where I was co-leader with Donald Johanson) were not completed until 1982 and only by then had the large number of vertebrate fossils collected been classified. A few years earlier we were not certain of what we had discovered.

The newborn first Hominidae *Australopithecus* had a new brain, a new

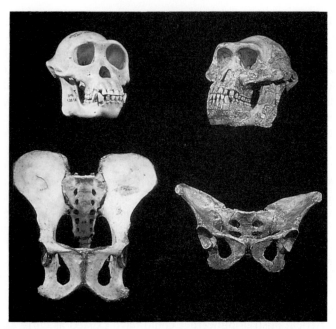

Fig. 6.1. The rifting process and the uplifting of East Africa can be the reasons for the emergence, 8 million years ago, of the Hominidae. From left (West) to right (East), skull and pelvis (arboreal) of *Pan* and skull and pelvis (bipedal) of *Australopithecus*; the rift valley can be visualized as cutting the picture vertically into two halves.

posture, a new locomotor repertoire, a new diet, and for the very first time, a culture. All these new features: brain, posture, locomotion, teeth, as well as tools, can be interpreted as adaptations of these primates to the new open environment. These new characteristics, except culture, which appeared by genetic mutations or heterochronies affecting life-history parameters, were determined by environment because they represented advantages.

Brain

The very detailed study carried out by Ralph Holloway on the endocast of the Hadar AL 162-68 skull fragment, arrived at the conclusion that the *lunate sulcus* would have had to be, according to the position of the interparietal *sulcus*, in a more posterior position than in *Pan* brains (Holloway and Kimbel 1986, and Holloway, this volume). The result of this interpretation is that *Australopithecus afarensis* — the *Australopithecus* species which we currently consider as being the earliest — already had an expansion of the parietal association cortex and consequently a brain reorganization. Some authors considered that the morphology of the *A. afarensis* endocast is fairly similar to that of the chimpanzee. However, if the body size of *A. afarensis* is taken into consideration in the interpretation of the brain size, the volume of the

Australopithecus brain appears to be 30 to 40 per cent greater than the brain of a chimpanzee of the same size (Martin 1983; McHenry 1988). And as there is a distance of 3 million years between the chimpanzee and *A. afarensis* it seems that, with the emergence of the Hominidae, a reorganization of the brain took place, reflected by a differential development of its parts, as well as by a general increase in its volume, along with possible allometric changes (Holloway, this volume).

Posture and locomotion

The new posture of the australopithecines and the new locomotor repertoire (Coppens 1991*a*) are other adaptations to the new habitat and are of particular significance. *Australopithecus afarensis* stood erect, walked bipedal, but still climbed trees. Bipedalism is easily documented, thanks to the shape of the pelvis, the curves of the rachis (a bit more elongated than ours), the morphology and orientation of the femur and, of course, the analysis of the Laetoli footprints. Climbing capacity appears in the opposability of the hallux, the amplitude of rotation of the knee, the medio-lateral enlargement of the distal epiphysis of the femur, the curvature of the phalanges, the development of the muscles of the forelimb, etc.

Diet

A new diet was obviously a necessity in this new environment. It was documented in *Australopithecus*, dentition — small size of the canines, tendency towards the molarization of the premolars, thickness of the enamel, etc. (Picq 1990).

Culture

Many surveys and seven important excavations carried out by Jean Chavaillon and Harry Merrick in the lower Omo valley provided thousands of artefacts and, in association with them, several remains of australopithecines, in levels dated by the potassium/argon method from less than 2 to more than 3 million years (Coppens *et al.* 1976). The earliest stone tools were collected on the surface of level B2, i.e. were 3.2 to 3.3 million years old, plus or minus 50 000 years. The majority of these artefacts are very small quartz flakes, and 5 to 6 per cent of them have retouches. They are genuine stone *tools*. It is obviously an unexpected assemblage in such early levels. Their age, and their association with australopithecine remains, indicates that *Australopithecus* was probably the very first tool-maker. I think that it is not difficult to accept that the use of hands, freed by bipedalism, is very ancient, and that it existed a long time before *Homo*. The Hominidae must have used pieces of wood, stone, or bone, as living chimpanzees do today, and one day may have knocked one of them against another to make the first one more efficient. In doing so, they were only one step, but a significant one, further on than living chimpanzees. In other words, it is not surprising that tool-making activity preceded Man (*Homo*). It would have been more surprising, even suspect, to see man and tools emerging together. I have maintained this opinion for about 20 years, and, for anatomical reasons, David Ricklan, in South Africa, a student of Phillip Tobias (Ricklan 1988), and Randy Susman, in Stony Brook, in the United States (Susman 1988) now confirm that *Australopithecus* was probably a tool-maker.

Second step

About 3 million years ago, in the same East African region, which was getting still drier, lived a hominid of the size of a chimpanzee, exceptionally intelligent, imaginative, inventive, creative, talkative, emotional, and social. His name was *Homo*.

What could have been the scenario for the emergence of the Homininae? Let us imagine the australopithecines, ancestors of *Homo*, around 3 million years ago, in the savannah and in the forest areas which existed at that time in East Africa, between the Rift Valley and the Indian Ocean. During that period there was a world-wide cooling, because of the development of ice at the poles. This was well-documented recently by oxygen 16 and 18 measurements made in the Atlantic Ocean foraminifera collected in deep-sea cores (Prentice and Denton 1988). When these results and those of my own Omo Valley surveys of 1967–68 were compared, I was able to demonstrate a clear correlation between this climatic crisis and the evolution of hominids. I published my findings (Coppens 1975) and described it as being the '(H)Omo event' (Fig. 6.2). There was an obvious correlation with the transformation of the faunal assemblages and pollen sequences, the subsequent transformation of the climate, and the evolution of hominids (more precisely, the emergence of the genus Homo and of the robust australopithecines). At a symposium in 1981 (Coppens 1985), the 19 participants came to the same conclusion that robust *Australopithecus* as well as Man emerged between 3 million years or a little more and 2.5 million years or a little less, during this global climatic change. The following results are those I obtained in the Omo Valley; they have been restated more recently (Turner and Wood 1993).

During the period from 3 to 2 million years, as examples, the number of folds or enamel in the upper cheek teeth of several species of Hipparion, indicative of a diet of leaves when it is high and of a diet of grasses when it is low, decreases from 24 to 6; the number of fossil monkeys remains collected on the exposures, indicative of trees and humidity when it is high, decreases from 367 per square kilometre to 39; the ratio of tree and grass pollens decreases from 0.4 to 0.01; the frequency of allochtonour pollens compared with autochtonous ones (indicative of humidity when it is high, because of a closer vicinity of the trees and because of a greater capacity of transportation by the Omo river) decreases from 21 per cent to 2 per cent. There are many

Fig. 6.2. The (H)Omo-event: 3 million years ago robust *Australopithecus* and *Homo* were born at the same time because of the climatic changes. From left to right, one of the latest robust *Australopithecus* (*A. boisei*) and one of the earliest *Homo* (*H. rudolfensis*).

more examples collected from vertebrate fauna, molluscs, sediments, fossilized wood, ostracodes, etc., which also indicate the effects of the change in the climate. This change and its influence on the evolution of many vertebrates, including Hominidae, is generally accepted. However, many questions still remain: why *two* adaptative answers to the same climatic change: robust *Australopithecus* and *Homo*? Why did the same answers occur in South Africa? Was it, for *Australopithecus* as well as for *Homo*, an evolution in East Africa and then a migration to South Africa?

The newborn Hominidae, *Homo*, had a larger brain, a more exclusive bipedalism, a new diet, and an improving culture. All these new features, brain, locomotion, teeth, better tools, can be interpreted again as adaptations of these primates to the new, still more open, environment. These new characteristics, except culture, which appeared by genetic mutations or heterochronies affecting life-history parameters, were determined by this environment because they represented advantages.

Brain

Homo, as early as *Homo habilis*, has a larger brain (Tobias 1991), with better drainage. The meningeal vascular system of the well-known 1470 skull, for instance, shows some interesting new developments when compared to the skull of *Australopithecus*: first appearance of clear asymmetry and anastomosis between the branches (Saban 1984, and this volume). The endocranial capacity, in absolute and relative size, increased dramatically with an additional acceleration at the time of *Homo erectus* — it doubled in 1 million years.

Locomotion

Homo locomotion was probably exclusively biped, or showed, anyway, more bipedalism and less climbing abilities.

Diet

The 3-million-year drought obviously changed the environment. Many species of trees and plants disappeared, and with them many types of fruits, seeds, and leaves. A new diet was a necessity in this new landscape. *Homo* became less vegetarian, still relying on fruits and roots, but also eating more meat. His front teeth were larger and his jugal teeth smaller.

Culture

The first tool-maker was probably *Australopithecus*, but tool-making was then somewhat anecdotic. With *Homo*, the tools became permanent, numerous, and diverse. A method of quantifying the development of this new cultural environment and the progress of technology is to measure the total length of the cutting edges of 1 kg of stone tools. This was carried out by André Leroi-Gourhan (1964) by weighing 1 kg of stone tools of the same material (flint) aged 2 million years, 500 000 years, 50 000 years, and 20 000 years. He obtained 10 cm of cutting edge for 1 kg of stone tools of 2 million years, 40 for the one of 500 000 years, 2 m for the one of 50 000 years, and 20 m for 1 kg of 20 000 years (Fig. 6.3). According to Arambourg (1962), for each level of brain evolution there is a level of technology, and he proposed *H. habilis*, 10 cm, *H. erectus*, 40 cm, *H. sapiens*, 2 m, *H. sapiens sapiens*, 20 m. Unfortunately, it does not work that way. Following Jean Chavaillon's 1976 excavations in Melka Kunture, Ethiopia, I found (Coppens 1982, 1991*b*): *H. habilis*, 10 cm of cutting edges for 1 kg of stone tools, but for early *H. erectus*, also 10 cm. Then, a later *Homo erectus* produced 40 cm of cutting edges for 1 kg of stone tools, but early *Homo sapiens*, also 40 cm. During this period there is a more rapid biological development, and a slower technological one. Then appeared *Homo sapiens* with 2 m, *Homo sapiens* with 20 m, etc. It was just the contrary. Around a few hundred thousand years, there is a sort of reverse order; biology is going slower, technology faster.

This progress, very impressive and also exponential, indicates that for the first 2½ to 3 million years of tool-making, biological development evolved more rapidly than technology, and for the past 100 000 years technology has developed more rapidly than biological evolution. It appears that 'instinct' was more important than knowledge during initial evolution, but that the volume of data to be learnt was becoming more important than 'instinct' 100 000 or 200 000 years ago. By then, the cultural environment was becoming more influential than the natural environment; it was beginning to solve the different demands imposed by the different natural environments. This situa-

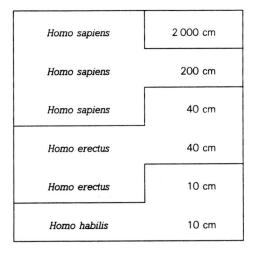

Homo sapiens	2 000 cm
Homo sapiens	200 cm
Homo sapiens	40 cm
Homo erectus	40 cm
Homo erectus	10 cm
Homo habilis	10 cm

Fig. 6.3. The acceleration of culture, 100 000 years ago, enabled *Homo sapiens* to be free.

tion reduced and probably even dramatically slowed down biological evolution — *Homo sapiens* was becoming free.

References

Arambourg, C. (1982). *La genèse de l'Humanité* (6th edn). Coll. Que sais-je? PUF. Paris, 1943.

Chavaillon, J. (1976). Evidence for the technical practices of Early Pleistocene hominids, Shungura Formation, Lower Omo Valley, Ethiopia. In *Earliest man and environments in the Lake Rudolf Basin* (ed. Y. Coppens, F.C. Howell, G. Ll. Isaac, and R.E.F. Leakey), pp. 563–73. Chicago University Press, Chicago, Ill.

Coppens, Y. (1975). Evolution des Hominidés et de leur environnement au cours du plio-pléistocène dans la basse vallée de l'Omo en Ethiopie. *C.R. Acad. Sci. Paris* **281**, 1693–6.

Coppens, Y. (1982). Qui fit quoi? Les plus anciennes industries préhistoriques et leurs artisans. *Bull. Soc. préh. fr. CRSM Paris* **79** (6), 163–5.

Coppens, Y. (1983). In *Recent advances in the evolution of Primates*. Pontificiae Academiae Scientarium Scripta Varai, 50. Vatican City, Rome.

Coppens, Y. (ed.) (1985). *L'environnement des hominidés au Plio-Pléistocène*. Fondation Singer-Polignac, Masson, Paris.

Coppens, Y. (1991*a*). L'originalité anatomique et fonctionnelle de la première bipédie. *Bull. Acad. Nat. Med.* **175** (7), 977–93.

Coppens, Y. (1991*b*). L'évolution des hominidés, de leur locomotion et des environnements. In *Origine(s) de la Bipédie chez les Hominidés* (ed. Y. Coppens and B. Senut), pp. 295–301. Cahiers de Paléoanthropologie. CNRS, Paris.

Coppens, Y., Howell, F.C., Isaac, G. Ll., and Leakey, R.E.F. (ed.) (1976). *Earliest man and environments in the Lake Rudolf Basin*. Chicago University Press, Chicago, Ill.

Doolittle, R. (1983), see Coppens, Y. (1983).

Dutrillaux, B. (1983), see Coppens, Y. (1983).

Greenfield, L. (1983), see Coppens, Y. (1983).

Hill, A. and Ward, S. (1988). Origin of the Hominidæ: the record of African large hominoid evolution between 14 My and 4 My. *Yearb. Phys. Anthropol.* **31**, 49–63.

Holloway, R. L. and Kimbel, W. H. (1986). Endocast morphology of Hadar hominid AL 162.68. *Nature* (321), 536.

Lejeune, J. (1983), see Coppens, Y. (1983).

Leroi-Gourhan, A. (1964). *Le geste et la parole.* Vol. 1, *Techniques et langage.* Albin Michel, Paris.

Lowenstein, G. (1983), see Coppens, Y. (1983).

Martin, R.D. (1983). *Human brain evolution in an ecological context.* 52nd James Arthur Lecture. American Museum of Natural History, New York.

McHenry, H.M. (1988). New estimates of body weight in early hominids and their significance to encephalization and megadontia in 'robust' australopithecines. In *The evolutionary history of 'robust' australopithecines* (ed. F. E. Grine), pp. 133–48. Aldine Press, New York.

Picq, P. (1990). Le régime alimentaire d'*Australopithecus afarensis*: un essai de reconstitution. *C.R. Acad. Sci. Paris* **311**, série II, 725–30.

Pilbeam, D. (1983), see Coppens, Y. (1983).

Prentice, M.L. and Dention, G.H. (1998). The deep sea oxygen isotone record, the global ice sheet system and hamimu evolution. In *The evolutionary history of 'robust' australopithecines* (ed. F.E. Grine). Aldine Press, New York.

Ricklan, D.E. (1988). A functional and morphological study of the hand bones of early and recent South African Hominids. D.Phil. thesis, University of Witwatersrand, Johannesburg.

Saban, R. (1984). Anatomie et évolution des veines méningées chez les Hommes fossiles. *ENSB-CTHS, Mem. Sec. Sci. Paris,* **11**, 289.

Susman, R.L. (1988). Hand of *Paranthropus robustus* from Member 1, Swartkrans: fossil evidence for tool behavior. *Science* **240**, 781–4.

Tobias, P.V. (1991). *Olduvai Gorge. The skulls, endocasts and teeth of* Homo habilis, Vol. 4. Cambridge University Press, Cambridge.

DISCUSSION

Participants: L. de Bonis, J. Chavaillon, Y. Coppens, B. Huberman, P. Tobias

L. de Bonis: I am absolutely in agreement with the hypothesis placing orang-utans on one side and humans and African great apes on the other. However, I think that the divergence between the lineage represented by extant chimpanzee and gorilla and the lineage leading to modern humans was more ancient. My first point concerns the time of divergence as given by studies in molecular biology. We should recall that the molecular clock was calibrated on the fossil evidence, i.e. the separation between New World monkeys (Platyrrhini) and Old World monkeys (Catarrhini), which was supposed to occur about 30 million years ago. But new fossil discoveries already indicate the existence of catarrhine primates 40 million years ago. If these dates are confirmed, then we have to revise all the estimates yielded by this molecular clock. My second point is based on the fossil evidence. There is the fossil from Samburu Hills, the well-known *Motopithecus* maxilla dated from 7 million years ago. The morphology of the maxilla is similar to extant gorilla, except for the thick enamel. The other set of evidence is offered by the numerous hominoid fossils of the genus *Ouranopithecus* discovered by my team in Greece and dated from 9 million years ago. These fossils display many characters present in more recent hominids, especially *Australopithecus afarensis*. My interpretation is that, according to this fossil evidence, one can consider an older separation between the African great apes lineage and the hominid lineage.

P. Tobias: The latest molecular biological research has shown that there were two distinct divergences within Africa. One was somewhere between 9 and 7 million years ago, when gorilla ancestors separated from the other African hominoids, leaving proto-chimpanzee and proto-hominid still sharing a common ancestry for another few million years. The other divergence between proto-chimpanzee and proto-hominid occurred between 6.5 and 4.9 million years ago. So, your model has to allow for a succession of at least two divergences.

Y. Coppens: I was discussing here only the divergence between the proto-chimpanzees and the proto-hominids. For the dates, as usual in molecular biology, they have a tendency to be a little too recent. I agree with de Bonis about the numerous resemblances between *Ouranopithecus* and the australopithecines. However, our respective hypotheses are very different. He places the divergence between the African great apes and human lineages before *Ouranopithecus*, perhaps about 10 to 12 million years ago, whereas I think that the divergence took place more recently, about 7 to 8 million years ago, as a consequence of the tectonic events and the associated ecological changes, that occurred in East Africa at this time. All these events took place in Africa. All the fossil primates preceding the divergence between the Panidae and the Hominidae, the African Proconsul for example, are found in Africa and, after the divergence, the australopithecines are African and so are all the first hominids. Finally, the fossil from Samburu Hills (*"Motopithecus"*) is very interesting to me because it presents many characteristics which could have very well been the ones of the common ancestors of chimpanzees and humans.

Palaeo-environment and pattern of divergence

L. de Bonis: I agree that local tectonic events modelling the Rift Valley in East Africa may have influenced the pattern of early hominid evolution. However, several more global events were also likely to affect the emergence of the first austrolopithecines a little earlier. By the end of the Miocene, during the Messinian, a dramatic drought had the effect of drying out the Mediterranean basin. It was the result of an increasing aridity, beginning about the end of the Middle Miocene, and reaching its peak about 5 million years ago. I think that these global environmental events played an important role for the emergence of bipedally walking primates.

P. Tobias: I am interested in your clear-cut division of the east from the west side of the Rift Valley for the emergence of the Panidae and the Hominidae. I would like to know how rigid the separation was between the two ecozones. The rifting was a very long process that went on for millions of years, and we know that animals can and do get across the Rift Valley; on the slopes of both sides of the Rift Valley there are abundant deposits of animal, fossils, such as bovines and antelopes. We know also that apes are forest-bound to a very large extent, and forest conditions are unsuitable for fossilization and preservation. Following Alan Walker's ideas, it is possible that the retreating great wet forest was reduced down to gallery forest along the streambeds, and australopithecines with inter-digitation with the spreading savannah. Could the apes living in the gallery forest have strolled across the savannah from one strip of shade to another?

Y. Coppens: The process of rifting (Oligocene) took several tens of million years and an intense process of reactivation took place between 7 and 8 million years ago. The separation and the isolation of small populations may have occurred at that time and for a sufficient length of time to allow a process of allopatric speciation. Then, a few million years later, there was a great variety of species but no further interbreeding. This is a common way of speciation in vertebrate zoology. Again, the most convincing data supporting my hypothesis are that, between 8 and 3 million years ago in East Africa, there are thousands of fossil specimens, including hominidae, but not a fragment of a chimpanzee or a gorilla.

Biological, cultural, and behavioural evolution

B. Huberman: According to your observations, hominid evolution was first characterized by biology evolving more rapidly than technology and, more recently, the process was reversed. Is there any evidence that this change is associated with an increased level of social interaction?

Y. Coppens: Initially, these observations were made on several archaeological localities in the same site and, also, between sites in East and South Africa. According to Jean Chavaillon, the level of organization of the remains of occupied sites, seems to evolve less rapidly than technology and, in turn, technology changes less rapidly than biology at this time.

J. Chavaillon: This pattern of human evolution is based on the organization of archaelogical deposits. We have noted in the same site, with the same research method, and probably for the same general environmental conditions, some progressive transformations with time. It seems that technical advances were ahead relative to the way of life, not to say traditions. For example, during the early Acheulean, which is characterized by the appearance of the first handaxes, the organization of archaeological sites was similar to those existing during the pre-Acheulcan (Oldowan). We were able to observe genuine Middle Palaeolithic tools with, apparently, a level of social organization already present in the Palaeolithic. Then, our tentative conclusions were that the main characteristics of human evolution did not evolve simultaneously. Biology evolved first, then technology, and finally behaviour which is held back by the inertia of tradition.

Part II

Genetics

7

The human genome
JEAN-LOUIS MANDEL

The human genome contained in the 22 autosomes and the X and Y sex chromosomes corresponds to a DNA sequence of 3×10^9 base pairs (bp), and codes for an estimated number of 50 000–100 000 genes. A majority of these genes may be expressed in brain. Almost all protein-coding genes have a mosaic structure, consisting of exons which contain the protein-coding sequences, and introns which are left out of the messenger RNA and have no protein-coding capacity. Introns are usually much larger than exons. The ratio of size of gene to size of the protein-coding sequence ranges from 2.5 for the β globin gene to 200 for the dystrophin gene (defective in Duchenne muscular dystrophy), which has 79-exons for a coding sequence of 11 055 bp and a total size of 2.3 million bp. The average ratio is probably in the order of 10. It is estimated that only 3–7 per cent of the human genome is protein coding. Another few per cents may correspond to regulatory DNA sequences (that allow control of gene expression) and to sequences important for chromosome structure (telomeres, centromeres) or DNA replication. Thus 90 per cent of the genome may have little or no specific function as evidenced by the very rapid changes in most intron or intergenic sequences during evolution (compared to conservation of protein coding or regulatory sequences). An important proportion of the non-coding DNA is made of repetitive sequences dispersed throughout the genome (sometimes described as 'selfish' or 'junk' DNA).

Protein-coding genes often belong to gene families: proteins with similar functions are coded by benes which share significant sequence similarity. This is extremely useful in practice since it allows the fishing out of new members of a gene family either in different organisms, or within the same organism. One striking example is the very large family of receptors for neuromediators, peptide hormones, light (opsins), and odoriferous substances, the G protein-coupled receptors, which include at least nine adrenergic receptors, five dopamine receptors, eight serotonin receptors, and recently identified receptors for cannabinoids and opioids, etc. (Kieffer *et al.* 1992; Parmentier *et al.* 1992; Matthes *et al.* 1993; Sokoloff *et al.* 1993).

Gene families arise by successive gene duplications throughout evolution and they can be clustered and/or dispersed on different chromosomes. They are extremely important in the evolution of new functions or regulatory processes and allow also for some functional redundancy. The multiple receptors for dopamine or serotonin differ by their affinity for ligands, their use of signal transduction mechanisms, and their distribution within the brain, which can be broad for some receptors or restricted to specific structures. An added level of diversity is generated, for some genes, by a plasticity in the choice or definition of particular exons. This mechanism, called alternative splicing, can generate two or more protein isoforms from a single gene, which may differ by their localizations or by some functional properties (as in the case of the dopamine D2 receptor, Montmayeur et al. 1993).

Fine tuning of gene expression is necessary during development and differentiation, or in response to environmental signals. This is mediated in large part by control of transcription through specific interactions between regulatory DNA sequences in or near the gene, and proteins called transcription factors. Several families of transcription factors have been uncovered, which share some structural motifs (leucine zipper, helix-turn-helix or homeo domains, zinc fingers of the C2H2 or Cx type) Mitchell and Tjian 1989). The complexity of these families is gradually uncovered. For instance, it has been estimated that 100–300 genes might code for finger proteins of the C2H2 type, although it remains to be seen whether they are all transcriptional regulators (Bellefroid et al. 1989). The Hox (homeobox-containing) gene family comprises 38 identified genes in mammals, which play a major role in embryogenesis, and especially in the segmental patterning of the brain (Krumlauf 1993). Thus, a significant part of protein-coding genes may be involved in the control of expression of other genes, through intricate functional networks.

Variability and evolution of the human genome

About one base pair in 400–500 is polymorphic in the (nuclear) human genome. Thus two copies of the genome will show $1-2\times10^6$ sequence differences. The vast majority of these differences are selectively neutral (and found at higher frequency in repetitive sequences, especially within tandem repeats of small motifs, which are probably functionless). Some of them might alter in a subtle way the function or regulation of a gene. For instance a single nucleotide difference can lead to increased expression of a fetal globin in adult life. Genes involved in antigen presentation to T-lymphocytes, that are located in the major histocompatibility complex (MHC) on chromosome 6, are characterized by an extremely high level of polymorphism that affects functionally important aminoacid residues in the corresponding proteins. It is generally thought that different alleles of these genes may modulate protection against different infectious diseases, and have thus been maintained by selective pressure in human populations (Trowsdale 1993). Much of this polymorphism was already present before separation of the human, chimpanzee, and gorilla lineages (Gyllensten and Erlich 1989). Extensive polymorphism of the dopamine D4 receptor gene in the human population creates a large number of variant forms of this receptor (localized in the frontal cortex), which differ by the number and exact sequence of 16 amino-acids repeat units (Van Tol *et al.* 1992; Lichter *et al.* 1993). Other rarer changes have a more drastic effect and correspond to mutations causing genetic diseases. About 3000 genetic diseases are known in man, many of them affecting brain function in some way (McKusick 1992).

Comparison of homologous sequences in human and primate genomes have shown about 1.6 per cent sequence differences in non-coding regions between human and chimpanzee, and about 7.5 per cent between human and a macaque (Savatier *et al.* 1987; Miyamoto *et al.* 1988). Coding regions show less variation. The vast majority of the 5×10^7 sequence differences between human and chimpanzee have probably little or no functional impact. However, especially in clustered gene families, one can demonstrate recent gain or losses of functional genes, which are of greater significance. A relatively recent duplication of a colour pigment (opsin) gene that occurred less than 30 million years ago on the X chromosome, allowed the evolution of trichromatic colour vision in higher primates (Yokoyama and Yokoyama 1989). Mutations (by unequal crossing-over) which led to the loss of one of the two adjacent alpha globin genes on chromosome 16 are present at very high frequency in some populations, as they gave selective advantage in regions of malarial endemy. A rare example of loss of gene function by point mutation is documented for the gene coding for urate oxydase, which was inactivated by a single mutational event in a common ancestor of hominoids, after divergence from the Old World monkeys (Yeldandi *et al.* 1991). The silenced gene is still present in the human genome. Absence of urate oxydase activity is responsible for the elevated plasma levels of uric acid in man, and this has been proposed to be an important factor in lengthening life span during primate evolution, as uric acid may protect from lesions caused by oxygen radicals (Ames *et al.* 1981). It may be of anecdotal interest to note that significant correlations have been reported in population studies between higher serum levels of uric acid and 'academic achievement, drive, range of activities and leadership' (Wyngaarden and Kelley 1983).

Repetitive sequences can be classified in several families, and play an important role in genome evolution. The most numerous are the Alu sequences (SINEs for short interspersed elements) that constitute, with $\sim 10^6$ copies, 5 per unit of the human genome, and the L1 sequences (or LINEs for long interspersed elements). They have accumulated in the genome by transposition mechanisms, from a relatively small pool of active elements. They are useful markers for evolutionary studies (Ryan and Dugaiczyk 1989). For instance it has been estimated that a subclass of 'young' Alu elements has expanded to about one thousand members during the radiation of the human population following divergence of the gorilla and chimpanzee lineages. Ten per cent of these are dimorphic at a particular locus in the human population (i.e. present or absent), and represent recent events in the evolution of man (Schmid and Maria 1992). Repetitive sequences may cause deleterious mutations, very rarely by insertion in gene, and more often by promoting deletions between two repeats (by a homologous recombination mechanism).

Short tandem repeats of di-, tri-, or tetranucleotide units are highly polymorphic, showing stably inherited variations in the number of repeat units. These are most useful as markers for genetic mapping (Weissenbach *et al.* 1992, see below). It has been recently shown that unstable expansion of trinucleotide repeats present within a gene may lead to genetic diseases with peculiar modes of transmission, such as the fragile X mental retardation syndrome, myotonic dystrophy, or Huntington's disease (Mandel 1993).

Genomes evolve not only by accumulation of point mutations or insertions or deletions of small DNA segments but also by much rarer events of chromosome rearrangements (translocations, inversions). The latter are, however, most important for speciation. Many studies of chromosomal evolution in primates have been conducted using morphological analysis of chromosomes (banding patterns). This can be now done much more precisely by studying the chromosomal localization of cloned genes, for instance by fluorescent *in situ* hybridization (O'Brien and Marshall Graves 1991).

The genome project

The ultimate goal of the genome project is to obtain a complete sequence of the human genome, and interpret it in terms of gene function. However, most of the sequencing is expected to take place in 10–15 years from now. The immediate goals are to develop a genetic map and physical map. The genetic map is constructed by analysing segregation of genetic markets (DNA polymorphisms) in families, and allows the localization of disease genes, especially those of unknown biochemical function that are defined only by a clinical phenotype. The physical map is the ordering of cloned DNA segments covering the whole genome, which is a prerequisite for total sequencing, but which meanwhile will be extremely useful to gain rapid access to disease-related genes, once they have been genetically mapped. Major progress has been reported very recently concerning the construction of dense and complete genetic and physical maps of the human genome (Bellané-Chantelot *et al.* 1992; Chumakov *et al.* 1992; Weissenbach *et al.* 1992).

Total genome sequencing has been criticized on the basis that most of it will be of little functional interest (and not interpretable). As a short cut to obtain biologically significant information, several projects have been initiated for the specific sequencing of protein-coding regions (cDNA sequencing). In particular, partial sequencing of brain cDNAs (i.e. that correspond to genes expressed in brain) is being actively pursued (Adams *et al.* 1992, 1993), and sequence tags have been obtained for 8600 cDNA clones. This allowed the identification of over 700 different proteins present in brain based on sequence similarities with genes of known function characterized in other tissues or organisms. However, many sequences do not show a significant match with known genes. A very hard task will be to ascribe a precise function to such sequenced genes. Their expression in various regions of the brain or during ontogeny can be studied by *in situ* hybridization (in mouse, but also in human fetal brain). In some cases, it may be necessary to study the equivalent mouse gene, by constructing transgenic mice where the gene is inactivated or abnormally expressed, and analysing the resulting phenotype.

As part of the human genome project, it was rapidly understood that sequencing of smaller genomes would provide both a testing ground for methodologies and a tool for functional analysis through sequence comparison. The complete sequencing of a yeast chromosome has revealed that only one gene in three shows homology to presently known genes (Oliver *et al.* 1992). Two organisms likely to be sequenced are of interest for understanding brain development and function. The sequencing of the genome of the worm *Caenorhabditis elegans* (10^8 bp) has just started (Sulston *et al.* 1992). The nervous system of this organism is made of 302 neurones, derived from 407 neural precursors and is known in exquisite details, allowing for instance the analysis of genetic mechanisms of neuronal differentiation or programmed cell death (Chalfie and Au 1989; Hengartner *et al.* 1992). The knowledge of genes involved in neuronal development in drosophila is advancing at a rapid pace (Thomas and Crews 1990). It is likely that sequencing of this 1.6×10^8 bp genome (10–15 000 genes), will provide important functional information.

Mapping of disease genes

Very efficient tools are now available for the mapping and isolation of monogenic genes that affect the nervous system, based on the analysis of patients and their families (Ballabio 1993). Although in most cases the effect on brain function may be rather trivial (lack of a structural protein, or toxic effect due to the accumulation of a metabolite), the understanding of the mechanisms of some of these diseases might prove biologically important; for instance the late onset selective loss of neurones in Huntington's chorea, or diseases which affect the development of specific brain structures.

Three recent examples illustrate this point well. Kallmann's syndrome (hypogonadotropic hypogonadism and inability to smell) is caused by a defect in the migration of olfactory neurones, and of neurones producing gonadotropin-releasing hormone. The corresponding gene was isolated by Franco *et al.* (1991) and Legouis *et al.* (1991), and shares homology with neural cell adhesion molecules. A gene for lissencephaly, a disease that also involves abnormal neuronal migration, was identified by Reiner *et al.* (1993) and encodes a protein resembling a subunit of G proteins, that may be implicated in a signal transduction pathway crucial for cerebral development. Finally, genetic linkage analysis in a family where several males in different generations demonstrated abnormal and often violent behaviour, associated with mild or borderline mental retardation, allowed the identification of a causative mutation in the X-linked gene monoamine oxydase B (a gene involved in catecholamine catabolism) (Brunner *et al.* 1993).

Multifactorial diseases are caused by interaction between a genetic predisposition (which may involve several genes) and environmental factors. The mapping of genes involved in predisposition to psychiatric disease (manic depressive illness, schizophrenia, even alcoholism), is the object of great current interest, but initial claims of success could not be replicated (Pauls 1993). The methodological issues are complex (problems of diagnosis, gathering of informative families, data analysis etc. . . .). Success can be guaranteed only if a small number of major genes are involved in the genetic predisposition. However, the cloning of 'candidate' genes (genes involved in synthesis, degradation, or action of neuromodulators) is providing important tools for this research. Similar approaches might also be used in the future to localize and identify genes involved in traits such as left-handedness or perfect pitch.

Conclusion

In conclusion, one can expect that the identification of genes involved in brain development will advance rapidly as a result of systematic analysis of the genomes of man, mouse, and other model organisms, coupled to detailed study of the function of such genes at the biochemical, cellular or organismic level. However, the identification of genetic changes that underlie the evolution of the human brain remain a formidable challenge for the decades to come, as the genetic make-up of our hominoid ancestors cannot be determined, and the sequence differences with our closest relative, the chimpanzee, are so numerous. Furthermore, genetic studies in man are hampered (compared to mouse) by the small size of families and lack of possibilities of experimental manipulation. Detailed analysis of the genetic factors involved in normal or pathological variations in brain development and higher functions within the human population, should, however, provide important information.

Note: Because of the very rapid progress in analysis of the human genome, this text has been modified with respect to the original presentation to include recent examples and references, without changing its general focus and structure.

References

Adams, M.D. *et al.* (1992). Sequence identification of 2,375 human brain genes. *Nature* **355**, 632–4.

Adams, M.D., Kerlavage, A.R., Fields, C., and Venter, J.C. (1993). 3,400 new expressed sequence tags identify diversity of transcripts in human brain. *Nature Genet.* **4**, 256–67.

Ames, B.N., Catheart, R., Schweirs, E., and Hochstein, P. (1981). Uric acid provides an antioxidant defense in humans against oxidant and radical caused aging and cancer: a hypothesis. *Proc. Natl Acad. Sci. USA* **78**, 6859–62.

Ballabia, A. (1993). The rise and fall of positional cloning? *Nature Genet.* **3**, 277–9.

Bellanné-Chantelot, C. *et al.* (1992). Mapping the whole human genome by finger-printing yeast artificial chromosomes. *Cell* **70**, 1059–68.

Bellefroid, E.J., Lecocq, P.J., Benhida, A., Poncelet, D.A., Belayem, A., and Martial, J.A. (1989). The human genome contains hundreds of genes coding for finger proteins of the Kruppel type. *DNA* **8**, 377–87.

Brunner, H.G. *et al.* (1993). X-linked borderline mental retardation with prominent behavioral disturbance: phenotype, genetic localization, and evidence for disturbed monoamine metabolism. *Am. J. Human Genet.* **52**, 1032–9.

Chalfie, M. and Au, M. (1989). Genetic control of differentiation of the caenorhabditis elegans touch receptor neurons. *Science* **243**, 1027.

Chumakov, I. *et al.* (1992). Continuum of overlapping clones spanning the entire human chromosome 21q. *Nature* **359**, 380–7.

Franco, B., Guioli, S., Pragliola, A., Incerti, B., Bardoni, B., Tonlorenzi, R. *et al.* (1991). A gene deleted in Kallmann's syndrome shares homology with neural cell adhesion and axonal path-finding molecules. *Nature* **353**, 529–35.

Gyllensten, U.B. and Erlich, H.A. (1989). Ancient roots for polymorphism at the HLA-DQalpha locus in primates. *Proc. Natl Acad. Sci. USA* **86**, 9986–90.

Hengartner, M.O., Ellis, R.E., and Horvitz, H.R. (1992). Caenorhabditis elegans gene ced-9 protects cells from programmed cell death. *Nature* **356**, 494–9.

Kieffer, B.L., Befort, K., Gaveriaux-Ruff, C., and Hirth, C.G. (1992). The delta-opioid receptor: isolation of a cDNA by expression cloning and pharmacological characterization. *Proc. Natl Acad. Sci. USA* **89**, 12 048–52.

Krumlauf, R. (1993). Hox genes and pattern formation in the branchial region of the vertebrate head. *Trends Genet.* **9**, 106–12.

Legouis, R. *et al.* (1991). The candidate gene for the X-linked Kallmann syndrome encodes a protein related to adhesion molecules. *Cell* **67**, 423–5.

Lichter, J.B., Barr, C.L., Kennedy, J.L., Van Tol, H.H.M., Kidd, K.K., and Livak, K.J. (1993). A hypervariable segment in the human dopamine receptor D4 (DRD4) gene. *Human Mol. Genet.* **2**, 767–73.

Mandel, J.L. (1993). Questions of expansion. *Nature Genet.* **4**, 8–9.

Matthes, H. *et al.* (1993). Mouse 5-hydroxytryptamine 5A and 5-hydroxytryptamine 5b receptors define a new family of serotonin receptors: cloning, functional expression, and chromosomal localization. *Mol. Pharmacol.* **43**, 313–19.

McKusick, V.A. (1992). *Mendelian inheritance in man*. The Johns Hopkins University Press, Baltimore, MD.

Mitchell, P.J. and Tjian, R. (1989). Transcriptional regulation in mammalian cells by sequence-specific DNA binding proteins. *Science* **245**, 371–8.

Miyamoto, M.M., Koop, B.F., Slightom, J.L., Goodman, M., and Tennant, M.R. (1988). Molecular systematics of higher primates: genealogical relations and classifications. *Proc. Natl Acad. Sci. USA* **85**, 7627–31.

Montmayeur, J.P., Guiramand, J., and Borrelli, E. (1993). Preferential coupling between dopamine D2 receptors and G-proteins. *Mol. Endocrinol.* **7**, 161–70.

O'Brien, S.J. and Graves, J.A.M. (1991). Report of the committee on comparative gene mapping. *Cytogenet. Cell Genet.* **58**, 1124–51.

Oliver, S.G. *et al.* (1992). The complete DNA sequence of yeast chromosome III. *Nature* **357**, 38–46.

Parmentier, M.D. *et al.* (1992). Expression of members of the putative olfactory receptor gene family in mammalian germ cells. *Nature* **355**, 453–5.

Pauls, D.L. (1993). Behavioural disorders: lessons in linkage. *Nature Genet.* **3,** 4–5.

Probst, W.C., Snyder, L.A., Schuster, D.I., Brosius, J., and Sealfon, S.C. (1992). Sequence alignment of the G-protein coupled receptor superfamily. *DNA Cell Biol.* **11,** 1–20.

Reiner, O. *et al.* (1993). Isolation of a Miller-Dieker lissencephaly gene containing G protein beta-subunit-like repeats. *Nature* **364,** 717–21.

Ryan, S.C. and Dugaiczyk, A. (1989). Newly arisen DNA repeats in primate phylogeny. *Proc. Natl Acad. Sci. USA* **86,** 9360–4.

Savatier, P., Trabuchet, G., Chebloune, Y., Faure, C., Verdier, G., and Nigon, V.M. (1987). Nucleotide sequence of the beta-globin genes in gorilla and macaque: the origin of nucleotide polymorphisms in human. *J. Mol. Evol.* **24,** 309–18.

Schmid, C. and Maraia, R. (1992). Transcriptional regulation and transpositional selection of active SINE sequences. *Curr. Opin, Genet. Dev.* **2,** 874–82.

Sokoloff, P., Martres, M.P., and Schwartz, J.C. (1993). La famille des récepteurs de la dopamine. *Médecine/Sciences* **9,** 12–20.

Sulston, J. *et al.* (1992). The *C. elegans* genome sequencing project: a beginning. *Nature* **356,** 37–41.

Thomas, J.B. and Crews, S.T. (1990). Molecular genetics of neuronal development in the Drosophila embryo. *FASEB Journal* **4,** 2476–82.

Trowsdale, J. (1993). Genomic structure and function in the MHC. *Trends Genet.* **9,** 117–22.

Van Tol, H.H.M. *et al.* (1992). Multiple dopamine D4 receptor variants in the human population. *Nature* **358,** 149–52.

Weissenbach, J. *et al.* (1992). A second-generation linkage map of the human genome. *Nature* **359,** 794–801.

Wyngaarden, J.B. and Kelley, W.N. (1983). Gout. In *The metabolic basis of inherited disease* (ed. J.B. Stanbury, J.B. Wyngaarden, D.S. Fredrickson, J.L. Goldstein, and M.S. Brown), pp. 1043–114. McGraw-Hill, New York.

Yeldandi, A.V. *et al.* (1991). Molecular evolution of the urate oxidase-encoding gene in hominoid primates: nonsense mutations. *Gene* **109,** 281–4.

Yokoyama, S. and Yokoyama, R. (1989). *Mol. Biol. Evol.* **6,** 186–97.

8

Mitochondrial DNA and human evolution
REBECCA L. CANN

Introduction

Recent attempts to describe the structure and function of the human genome are summarized in Mandel's chapter. The 50 000–100 000 genes in the cell's nucleus represent the bulk of DNA in the human genome, which is usually biparentally transmitted from parent to offspring. This DNA is organized in linear fragments, 22 autosomes, an X chromosome, and the Y chromosome. Any discussion of human DNA would be incomplete, however, if we did not also consider the 37 genes of the mitochondria, some 16 569 base pairs, which have been fully described for one human (Anderson *et al*. 1981). In contrast to nuclear genes mitochondrial genes are maternally transmitted, organized in a double-stranded circle, and all function in energy production and electron transport for the generation of ATP to be used by the cell. Mitochondrial DNA contains its own ribosomal subunits, its own suite of tRNAs, two independent origins of DNA replication, and 13 polypeptide-coding regions.

Mitochondrial DNA (mtDNA) studies of modern human populations have made unique contributions to our understanding of human evolution. This fact stems from two important features of mitochondrial genes, their lack of recombination and their rate of mutation.

Rapid evolution and maternal transmission affect phylogenetic reconstruction

The lack of recombination between genomes of males and females, or between mutated genomes within an individual has a unique consequence for phylogenetic studies. A human female passes on her mitochondrial genes whereas a human male does not. Because human males are sometimes unsure of paternity, but females are almost always sure of maternity, mtDNA is preferred for tracing genealogies in many populations with a different cultural norms for mate choice and long-term pair bonds.

A high rate of mutation for mtDNA, usually an average of 10 times that of a nuclear gene in a similar functional class, implies that the scale of resolution is very fine for population studies using mtDNA sequences. In fact, if we use the most rapid changing regions, the control regions on either side of the origin of DNA replication for the heavy strand, we will see the fixation of mutations along maternal genetic lineages which shared a common ancestor only 1000 years ago. This very high level of polymorphism makes mtDNA excellent for providing genetic markers to trace migration and fissioning in human societies. For example, two native Hawaiians, who last shared a common mother only 2000 years ago, may differ at 3 to 4 nucleotides in a 100 nucleotide stretch.

The high rate of evolution in mtDNA is thought to be due to two factors. First, there is little to no DNA repair in the mitochondria, and there are many copies of this DNA per cell. The paradox is that although these genes are absolutely essential for life, there are so many copies of them that mutations are tolerated. Some are eventually fixed, and these are usually single nucleotide substitutions that appear in the third positions of codons. Because there are only a few gene products translated, there may also be relaxed functional constraints on the rRNAs and tRNAs.

A possible human bottle-neck?

When I began my study of mtDNA in the late 1970s with Dr Allan C. Wilson, one of his postdoctoral fellows, Dr Wesley Brown, was writing up his work on a study of 21 human mtDNAs. Dr Brown had discovered that using restriction fragment length polymorphisms (RFLPs), humans as a species looked 'different' to other mammals. He found that in comparison to two chimpanzees, or two gorillas, or two orang-utans, or two gibbons, or even two pocket gophers, humans had only one-half to one-fifth of the intraspecific variability seen in our closest primate relatives and other genetically well-characterized mammals. In 1980, Brown proposed that the level of variability sampled in his study was consistent with the derivation of the human mito-chondrial sequence from a single female about 200 000 years ago. This was the origin of the bottle-neck hypothesis and mitochondrial 'Eve' (Brown 1980).

A genetic bottle-neck can be described as a successive loss of genetic diversity due to local extinction events which prevent the transmission of all but a few alleles in the case of nuclear loci, or maternal lineages in the case of mtDNA. If a population undergoes a reduction even to a single mating pair due to disease, climatic change, or predation, nuclear genetic diversity may not be affected too greatly if the population expands quickly. This is because the surviving nuclear alleles can undergo genetic recombination, as well as mutation, to generate new diversity in the population. A bottle-neck event is much more severe for mtDNA, and reduces the diversity to n alleles, instead of 4n alleles as would be the case for nuclear genes, if we compare any one locus.

Random lineage loss over time can also generate a single ancestor

Mitochondrial DNA is a very sensitive indicator of population history, but population bottle-necks are not necessarily the only reason that humans may show reduced mitochondrial variability, compared to other species. The other reason is a simple consequence of what has been called by John Avise and his students in Georgia, United States, 'lineage sorting' (Avise 1991). Consider first that if we think about an individual today, we see the 32 distinct ancestors that individual had, going back five generations. If we try to trace the person's nuclear alleles, we have to consider the probability of transmission in each generation, and the probability that any one ancestor might be polymorphic at the locus of interest. We must also account for the possibility of genetic recombination. The result is some complicated mathematics for nuclear gene transmission. Yet, if we consider the person's mtDNA only, there is one and only one ancestor in this family pedigree. That is the maternal great-great-grandmother. There may be a mitochondrial 'Eve' in this family, but she is not the only woman contributing genetic loci to the proband, and she is not the only woman alive in her generation. So, the sense in which 'Eve' matches her biblical model is rather remote. I think this helps us to understand some of the confusion generated in the popular press about the mitochondrial mother of us all and the playful misnomer 'Eve'. The biological truth is that she may represent the only distinct common ancestor that we all share.

Another way to think of this can be illustrated in the following example. Imagine a population where we start with 15 females. Over time, some of these females have sons only. Some other females have daughters, but eventually their daughters leave no daughters. So, the end-product in 10 generations can be only a single surviving mitochondrial lineage, that which my colleagues and I have called the 'Lucky Mother'. Even though in the past there were many females and many other mtDNA lineages, in the slice in time that we are examining, all those other lineages today are absent from the population.

The mitochondrial tree for our species

Mitochondrial studies of many humans were summarized in a phylogenetic tree, that showed how each of us tested were related to each other. For those of us who learn genetics by first learning about the genes for eye colour or pea wrinkling, we have to unlearn some of our biases. What we illustrate in these diagrams are not gene frequencies in different populations, or similarity measures, but are actual representations of lineal evolution of alleles due to successive fixation of mutations along diverging matrilines.

One way to illustrate this point is to consider what the tree is saying, starting from today. If we were to do this, we would draw you at the tips, and attach you by a line to your siblings, if you have any. Further out would be your cousins, on your mother's side. Your most distant maternal cousins would be further out, and you would all be connected by a node in the tree which stems from your maternal grandmother many generations back. For some of us, we are not even aware of these family connections, because during 200 years, many branches of the tree are lost to memory.

Imagine now that you are looking at the mtDNA tree of our species. Past migrations, changing rules of kinship, and new dialects of spoken languages have altered the memory of various cultures as to their closest genetic relatives. However, we each carry within us this record of maternal genealogy, and it was discovered by extensive analysis both with fine scale restriction fragment length polymorphism (RFLP) mapping and direct sequencing of the control regions around the heavy strand origin of DNA replication.

Human mtDNA trees have two features in common, whether discovered using RFLPs or direct sequencing. The first is a pattern of two branches uniting the deepest or most divergent branches in the tree. The tree has one branch which contains only Africans, and a second branch which contains Africans and other ethnic and racial groups. The most genetically diverse humans today, from the mitochondrial perspective, are Africans. Two Africans, taken at random, show a larger number of mutations than two Chinese, or two Italians.

We infer that this has something to do with time, and perhaps also, with the geographical origin of anatomically modern humans. It looks as if the centre of modern human genetic diversity is in Africa. Human lineages were present in Africa, starting at least 2 million years ago. Some nuclear loci began to diverge in sequence from this time point. We know now that archaic humans left Africa at least 1 million years ago, but Africa has remained a refugia for the specific set of maternal lineages which eventually led to modern people. We might also infer from this mtDNA tree that some people who once left Africa returned.

Such great genetic diversity has a number of consequences for the study of the human genome, and for understanding why some DNA sequences described now only from Caucasian populations may be just the tip of the iceberg, as far as polymorphisms. The medical community is now acknowledging this problem for proper matching of tissue transplantation and organ donors. Hypervariable regions in HLA loci (the major histocompatability genes), identified first in people of northern European ancestry, often fail to cross react at the serological level when tested in the American Black population. So much genetic divergence has occurred in these regions that the Black population appears monomorphic in its lack of identity to some arbitrary standards. The correct interpretation of this state of events is rather that northern Europeans contain only a limited number of the possible HLA genotypes that characterize our species.

A second feature of the tree is a confirmation of fact known already from studies of protein polymorphisms using electrophoretic methods. That is, human races do not adequately describe how genetic diversity is apportioned in our species. The variability of human ancestral groups, defined broadly as Asians, Africans, Europeans (including Europe, North Africa, and the Middle East), and Australian/New Guinea populations, is quite high. On the mitochondrial tree, this is reflected in the branching pattern that may show a European with his closest mitochondrial relative being an Indonesian, or Japanese.

Intra-racial diversity accounts for something like 92 to 94 per cent of human nuclear polymorphism, and mitochondrial diversity is in accord with these estimates. Simply put, this means that it is often possible to find the same clusters of maternal genetic lineages in all human groups, even those now geographically separated. Of course, there may be private polymorphisms specific only to one geographical area, but these polymorphisms are a small subset of all the genetic diversity present in our species.

We estimate that at least 20 000 years is necessary for isolation alone to produce region-specific lineages by mutation. If a population bottle-neck is combined with isolation, genetic drift may also help to produce private poly-morphisms in even shorter periods of time. These estimates come from comparisons of geographically isolated populations where the time of isolation can be estimated geologically (New Guinea from Australia, or Polynesians from Melanesians) (Stoneking, Bhatia, and Wilson 1986). The internal calib-ration points for humans reduce our reliance on estimates of the rate of human mtDNA sequence evolution derived from comparisons of cloned and sequenced homologous genes. Such studies estimated a 2 per cent/per million years of genetic divergence average rate of change. Given the two most divergent human lineages, Allan Wilson, Mark Stoneking, and I stated in 1987 that modern human mitochondrial lineages derived from a common ancestor sometime in the last 280 000–140 000 years, although to be statistic-ally rigorous, the actual errors of the estimate could make the true time 500 000–50 000 years ago (Cann *et al.* 1987).

Reactions

In the past, some anthropologists have chosen to attack these ideas as being nonsensical, pointing to the small numbers of people examined, the lack of precision in estimating the rate of mitochondrial mutation, and the use of American Blacks as examples of average 'Africans'. Another criticism pro-posed that the pattern might be correct, but that the time scale was wrong. By that, critics suggested that because the mitochondrial clock was mis-calibrated, we were looking at genetic divergence times stemming from the migration of *Homo erectus* from Africa, and not the divergence within *Homo sapiens sapiens*.

The hypothesis of an African origin for *Homo sapiens sapiens* was even demoted to a myth in the one human genetics journal. In the last three years, my colleagues at Berkeley and Japanese researchers have attempted to answer these criticisms, refine the controversies, and examine greater num-bers of donors. The invention of the polymerase chain reaction (PCR), or cell-free cloning, has improved our ability to investigate human population genetics. Our conclusions have not changed drastically since 1982, but only strengthened due to the accumulation of additional evidence.

The mitochondrial genome

The polymerase chain reaction (PCR) has now given us a highly expanded view of human mitochondrial polymorphisms, and allowed us to reconstruct a much more sophisticated hypothesis about human mitochondrial evolution. Because we can amplify, in three hours, mtDNA from plucked or cut hairs in a test tube we can easily study people from everywhere on this planet. Mitochondrial DNA has also been extracted from 3000-year-old Egyptian mummies, and 8000-year-old palaeo-Indian brains preserved in peat in Florida (Paabo 1989). We can also easily sequence hundreds of nucleotides in a single day. However, while the precision of our hypothesis about human population expansion has increased, the pattern is basically that which was first obtained using high-resolution RFLP analysis.

The mitochondrial genome, as seen with PCR, now reveals the true level of interaction with the cell's nuclear genome, and new classes of mutations previously undetected have been described by Douglas Wallace and co-workers that affect large portions of the central nervous system (Wallace *et al*. 1988). Some families in the human population are segregating large deletions of mtDNA which affect optic neurones and large portions of the cortex, in addition to peripheral nerves. One mitochondrial gene is known to code for a cell surface antigen, thus polypeptide communicates with many chemical messengers in the body. Note the fact that mtDNA encodes three subunits of cytochrome oxidase (CO), CO-1, CO-2, and CO-3. Dr Rakic has already shown us antibodies to these proteins studding areas of the visual cortex that have greatly expanded in primate evolution, and natural selection for improved function of these enzyme subunits in the brain may alter their rate of evolution over same background average rate.

Our post-PCR view of mtDNA evolving over time in the human population still shows that single base substitutions are the predominant mutation. Mitochondrial DNA deletions, while dramatic and linked to a variety of neuro-pathies, are still quite rare. I think that many are probably so deleterious that they can only exist in cells where most of the mtDNA is not very heavily damaged. With PCR, many patients in the maternal myopathy families can be shown to be genetic mosaics, or are heteroplasmic for two types of mtDNA, the mutated and the normal-sized molecules. The penetrance of the conditions affecting these patients points to modifying loci in the nuclear genome that limit the percentage of deleted mtDNAs in any one cell.

In phenotypically normal people, it appears that over 99.9 per cent of the changes scored are functionally silent substitutions. These are mutations that hit the third position of a three-letter codon for a particular amino acid, but do not cause a substitution in the amino acid sequence for the protein coded by this region, because of the degeneracy of the genetic code (ex, the amino acid glycine can be coded for by GG*G*, GG*A*, GG*C*, or GG*T*). We also know that the most common mutation is a transition mutation, or a purine for another purine, and a pyrimidine for another pyrimidine (ex, GG*G* to GG*A*, or GG*C* to GG*T*). Mutagenesis studies like this can reveal the presence of sensitive areas that can affect our estimates of time of divergence for molecular clock calculations. They also refine our understanding of the forces which act on mtDNA over time, and our modelling of the importance of dietary mutagens or increased lifespan.

Clock estimates

The molecular clock is a hypothesis that states that many mutations in the cell appear to accumulate in a time-dependent manner. It does not state why this should be (natural selection or drift?), or the types of genes affected (regulatory or structural?), or any parameters of the population impacted (large or small?). It simply states that if the DNA sequences of two species are examined, the amount of change between them will be proportional to the time when they last shared a common ancestor.

These estimates can be quite crude, depending on the ultimate calibration of a particular clock with fossils known to be part of any one particular lineage. Fossils are real, but our conclusions about which specific lineage they belong to change over time. For primate genes, the molecular clock estimates are now quite accurate because of general agreement among palaeontologists that a divergence point of human from chimpanzee and gorilla is somewhere between 5 and 8 million years ago, not between 20 and 25 million years ago as was thought in 1967 (Andrews 1992). Clock calibration and accuracy discussions are a standard part of any publication on primate molecular evolution.

For the region of mtDNA which changes the fastest, the control region, a comparison of DNA sequences done by Tom Kocher and Linda Vigilant shows that humans have a divergence time proportional to 180 000–220 000 years (Kocher and Wilson 1991; Vigilant *et al.* 1991). There is no way this can be inaccurate by a factor of 5, which would please the *Homo erectus* migration champions. This would place the divergence of humans and chimpanzees at 25 million years ago, instead of 5 million years ago. There is also no good evidence to suggest that the rate of DNA sequence evolution for mitochondrial loci is different between humans and chimpanzees, which might invalidate the use of a clock over the time scale in which we use it.

We are unable to support the conclusions of other anthropologists who suggest that Asian populations today are the direct descendants of archaic Asian *Homo* species isolated from Africans for over 1 million years. We suspect that the features used by these palaeontologists to unite ancient and modern peoples represent morphological convergences, low statistical accuracy due to very small numbers of fossils unrepresentative of ancient populations, and the artefact of selective sampling in modern people. Other palaeontologists support this criticism, and it is important to recognize that there is a significant disagreement among them about the evidence for regional continuity in human evolution (Stringer and Andrews 1988). We hope that modern biotechnology will improve DNA identification and extraction techniques to the point where recovery from fossils in the plus-100 000 year time range will solve this problem.

The spread of *Homo sapiens sapiens*

It is impossible for us to say with certainty that the 'Lucky Mother' was an anatomically modern human female. There exists the possibility that she was more archaic, and a representative of a maternal lineage that eventually gave rise to our direct ancestors. Our model then requires that her descendants leave Africa. We cannot state when this migration took place, but are cheered by the dating of relatively ancient (around 100 000-year-old) modern people in the Middle East, which is a logical place to be if they are also found in East and South Africa at approximately the same time (Valladas *et al.* 1988).

Did the descendants of our 'Lucky Mother' replace all other archaic or modern lineages? The mitochondrial evidence is clear that high genetic diversity is found among donors in Africa. Parsimony and distance based methods suggest the spread of our ancestors was rapid, with little mixing. (Saitou and Nei 1987; Swofford 1993). Such a perspective leads us to ask if natural selection was a part of this replacement process, because huge populations would be required to minimize the danger of the loss of certain lineages due to random extinction events. What kept these populations of Africans so large? We often wonder if language played a part of this process, and that our ancestors all had some new mutations which allowed them to spread, at the expense of other indigenous peoples. For these reasons, the discovery of extensive mitochondrial DNA involvement in neurological disorders is a productive and exciting avenue of current research.

References

Anderson, S. *et al.* (1981). Sequence and organization of the human mitochondrial genome. *Nature* **290**, 457–65.

Andrews, P. (1992). Evolution and environment in the Hominoidea. *Nature* **360**, 641–6.

Avise, J. (1991). Ten unorthodox perspectives on evolution prompted by comparative population genetic findings on mitochondrial DNA. *Ann. Rev. Genet.* **25**, 21–44.

Brown, W. (1980). Polymorphism in mitochondrial DNA of humans as revealed by restriction endonuclease analysis. *Proc. Natl Acad. Sci. USA*, **77**, 3605–9.

Cann, R.L., Stoneking, M., and Wilson, A.C. (1987). Mitochondrial DNA and human evolution. *Nature* **325**, 31–6.

Kocher, T.D. and Wilson, A.C. (1991). Sequence evolution of mitochondrial DNA in humans and chimpanzees: control region and a protein coding region. In *Evolution of life* (ed. S. Osawa and T. Honjo), pp. 391–413. Springer-Verlag, Tokyo.

Paabo, S. (1989). Ancient DNA: extraction, characterization, molecular cloning, and enzymatic amplification. *Proc. Natl Acad. Sci. USA* **86,** 1939–43.

Saitou, N. and Nei, M. (1987). The neighbor-joining method: a new method of reconstructing phylogenetic trees. *Mol. Biol. Evol.* **4,** 406–25.

Stoneking, M., Bhatia, K., and Wilson, A.C. (1986). Rate of sequence divergence estimated from restriction maps of mitochondrial DNAs from Papua New Guinea. *Cold Spring Harbor Laboratory Symp. Quant. Biol.* **51,** 433–9.

Stringer, C.B. and Andrews, P. (1988). Genetic and fossil evidence for the origin of modern humans. *Science* **239,** 1263–8.

Swofford, D. (1993). *PAUP, phylogenetic analysis using parsimony*, version 3.1.1, computer program. Laboratory of Molecular Systematics, NMNH.

Valladas, H., Reyss, J.L., Joron, J.L., Valladas, G., Bar-Yosef, O., and Vandermeersch, B. (1988). Thermoluminescence dating of Mousterian 'Proto-Cro-Magnon' remains from Israel and the origin of modern man. *Nature* **331,** 614–16.

Vigilant, L., Stoneking, M., Harpending, H., Hawkes, K., and Wilson, A.C. (1991). African populations and the evolution of human mitochondrial DNA. *Science*, **253,** 1503–7.

Wallace, D.C. *et al.* (1988). Familial mitochondrial encephalomyopathy (MERRF): genetic, pathophysiological, and biochemical characterization of a mitochondrial disease. *Cell* **55,** 601–10.

DISCUSSION

Participants: R. L. Cann, M. P. Stryker

Stryker: What is the prospect of getting mitochondrial DNA sequences from anatomically characterized early man?

Cann: So far, the oldest DNA sequences that have been extracted from human bone are around 7000 years old (Pääbo 1989). There have been sequences taken from palaeo-Indian brains which are about 8000 years old (...). What increases the chances that the DNA will survive? If an individual dies in an environment where the soil is salty or very basic and the skeleton is covered rather quickly, the probability of survival for 10000 years of a DNA sequence is very high. Right now, our level of detection is so good that we need very much to guard against contamination. I am fortunate in that my ethnic group is so different from the group I am currently working on that I can tell when I contaminate a sample. This would be more of a problem if I were to work with Europeans aged 30000 years. I would not necessarily be able to differentiate with certainty from that material, even though I could sequence it.

Mitochondrial African divergence vs. regional continuity

This subject opposing geneticists to palaeoanthropologists was extensively discussed by R. L. Cann and B. Vandermeersh (cf. their respective chapters). The question is still highly controversial and difficult to clarify in the absence of additional information on mitochondrial DNA and fossil records.

9

Mammalian homeo box genes: evolutionary and regulatory aspects of a network gene system

FRANK H. RUDDLE AND CLAUDIA KAPPEN

Homeotic genes in the fruit fly *Drosophila* are involved in pattern formation during embryonic development. Based on sequence similarities more than 30 homeo box-containing genes of the class Antennapaedia have been identified in mouse and man. The genomic organization of this multi-gene family in gene clusters located on different chromosomes supports the hypothesis that the homeo box gene complexes evolved by duplications. We present evidence that these cluster duplications happened early during vertebrate evolution. Based on the comparison of the arthropod and vertebrate homeo box gene systems we consider the function of multi-gene regulatory networks for the specification of regional and cellular identity during development and in the adult organism. We argue that during evolution the homeo box gene network may have been crucially involved in the development of new body plans and behavioural attributes.

Introduction

The homeo box was first discovered as a common DNA sequence element in homeotic genes of the fruit fly *Drosophila melanogaster* (McGinnis *et al.* 1984; Scott and Weiner 1984). These genes regulate the development and morphogenesis of the fly, and mutations in such genes lead to characteristic phenotypes, so-called homeotic transformations, e.g. the transformation of antennae into legs. The resulting malformations change the identity of a given body segment without affecting the total number of segments. The identity of a segment with regard to its position along the anterior posterior axis appears to be changed. In concordance with their effects in certain regions of the body, homeotic genes are expressed in respective regions of the embryo that later give rise to the affected structures (for a review see Gehring 1987). Homeotic genes are organized in two gene clusters in the *Drosophila* genome, the Antennapaedia (*Ant-C*) and Bithorax (*BX-C*) clusters. It is intriguing to note that the linear order of the genes on the chromosome in these complexes is co-linear with the gradient of expression of homeotic gene products along the anterior-posterior axis in the embryo.

Homeotic gene products are transcription factors that regulate the activity of their target genes during developmental and differentiation processes. They may also regulate the activities of other homeotic genes and control their own activities in an autoregulatory fashion. Thus, homeotic genes of the fly encode for a network of regulatory interactions controlling gene expression during development. Accordingly, it has been shown recently that homeo domain proteins bind DNA. The homeo domain consists of 61 amino acids and forms a helix–turn–helix structure similar to the DNA-binding domain of prokaryotic repressor molecules (reviewed in Scott *et al.* 1989). Thus, on the molecular level, the action of homeotic genes is mediated by their role as DNA-binding proteins modulating the transcriptional activity within the homeotic gene network as well as the expression of target genes in determining segment identity during fly development.

Mammalian homeo box-containing genes

Homeo box genes have subsequently been isolated from a variety of organisms including frogs, chicken, mice, and humans. They can be grouped into different categories based on their sequences as well as their biological features. Homeo domains have been identified in ubiquitous transcription-factors, such as Oct-1, as well as in tissue-specific transcription factors. Examples for the latter are Pit-1, a pituitary-specific transcriptional activator of the gene for growth hormone, and Oct-2 which regulates the transcriptional activity of immunoglobulin genes. Genes of the so-called class Antennapaedia are the most similar to the *Drosophila* homeotic genes, and this similarity suggests that they may also play important roles in development in vertebrates (Fienberg *et al.* 1987).

Structure and expression of mammalian homeo box genes

Homeo box genes in vertebrates are typically about 10 kb (kilobases) long and consist of at least two exons. The genes encode proteins of the average size of 240 amino acids. The homeo domain resides in the carboxy-terminal part of these proteins. Additionally, the amino-terminus as well as a hexa-peptide sequence preceding the homeo domain is conserved between several proteins. The hexapeptide sequence has been implicated in mediating dimerization of homeo domain proteins.

The expression patterns of homeo box genes are quite complex and have been recently covered extensively (for a review see Shashikant *et al.* 1991). The most striking feature of the expression patterns of class Antennapaedia homeo box genes in developing embryos is the region-specific accumulation of their respective mRNAs. Each homeo box gene has a characteristic anterior boundary of expression in the spinal cord, and while the posterior boundary is less well defined, a gradient can be observed in the succession of anterior expression levels. This graded expression is also found in the meso-derm. These patterns are reminiscent of those observed in *Drosophila* in that homeo box gene expression is detected in structures originating from differ-ent germ layers, such as neural tissue, which is of ectodermal origin, and somitic mesoderm. Moreover, as in the fly, the region of expression is re-flected by the chromosomal position of the respective gene (Duboule and Dolle 1989; Graham *et al.* 1989).

Genomic organization of mammalian homeo box genes

The genomic organization of mammalian class Antennapaedia homeo box genes is described in Fig. 9.1. Homeo box-containing genes reside in four clusters on different chromosomes, *HOXA* on chromosome 7 in human (6 in the mouse), *HOXB* on chromosome 11 (17), *HOXC* on chromosome 12 (15), and *HOXD* on chromosome 2 (2) (Ruddle 1989). The transcriptional orienta-tion of all genes on one cluster is identical. They are spaced approximately every 10 kb, and no pseudogenes have been identified.

Each of the genes in a given cluster has closely related cognate genes located on at least one of the other clusters. Thus, the highest similarity between genes is found across clusters rather than along one cluster. These similar genes form subfamilies which are termed cognate groups. Moreover, the physical order of the corresponding genes is identical along the four clusters. This parallel arrangement has served as evidence that the four clusters of homeo box genes have arisen through duplication events that involved entire clusters. A virtually identical organization is found in mice and humans (Kappen *et al.* 1989a). The arrangement as well as the overall spacing of the genes is extremely conserved (see Fig. 6.1). This indicates that the duplication events leading to four clusters must have taken place at least before the divergence of these two species. Analysis of all available sequence information of vertebrate homeo boxes supports this hypothesis.

Furthermore, homeo box sequences representing genes from at least three clusters have been isolated from non-mammalian vertebrate species, such as frog, chicken, and zebrafish. Information on the genomic organization of those sequences is limited but in all cases compatible to the mammalian situation. These data strongly suggest that lower vertebrates may also have four clusters of homeo box genes similar to those in mammals.

Duplication of homeo box gene clusters during vertebrate evolution

A more detailed calculation of the timing of the duplication events with respect to vertebrate evolution has been derived from analysis of nucleotide

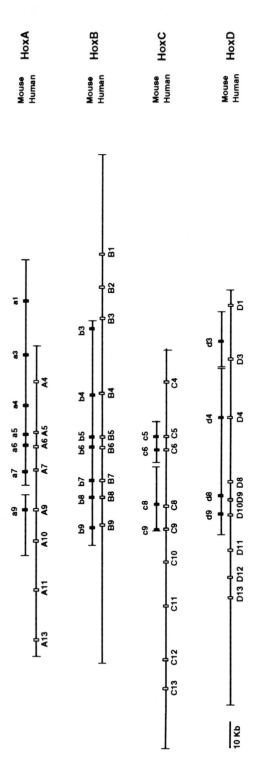

Fig. 9.1. Organization of mammalian homeo box-containing genes. ■, murine homeo box genes; □, human homeo box genes. (From Kappen et al. 1989; Acampora et al. 1990; Boncinelli et al. personal communication.)

sequences of mammalian homeo boxes (Kappen *et al.* 1989*b*). When the human and mouse sequences for each of the homeo boxes are compared in pairs, they differ by an average of 11.5 nucleotides. More than 90 per cent of these differences represent silent substitutions as they do not result in amino acid differences. These data show that human and mouse homeo box sequences have been very strongly conserved since the divergence of the two species. Comparisons of cognate homeo boxes located on different clusters in the mouse show that each pair of sequences differs by an average of 32.4 nucleotides, and 84.6 per cent of these changes account for silent substitutions. Such a comparison for human sequences yields an average of 37 differences (87.3 per cent silent) between cognate pairs (Kappen and Ruddle 1993). These figures represent the distance between homeo box clusters as measured by the number of nucleotide differences. If one of the clusters arose from a very recent duplication it should differ by a smaller number of nucleotides from its predecessor than from the remaining clusters. However, the four clusters in the mouse and human appear to be equally distant by several different criteria (Kappen and Ruddle 1993). Therefore it is likely that the four clusters arose in a rapid succession of duplications rather than over a long time period.

Timing of the cluster duplications

We have attempted to calculate the evolutionary time scale of the cluster duplications by taking the mouse–human divergence time as a reference point. Mouse and human species are thought to have diverged at least 70 million years ago (a figure which recently has been corrected to up to 100 million years ago; Li *et al.* 1990). During this period an average of 10 silent nucleotide differences has accumulated between mouse and human homeo box sequences. Based on the distance between individual mouse clusters of 27.4 silent nucleotide differences (and 32.3 silent nucleotide differences for the distances between human clusters) the minimum time progressed since the cluster duplications can be calculated to be around 300 million years (Table 9.1). This figure represents a minimum estimate since it is based on silent nucleotide differences only. However, if the occurrence of multiple mutations at the same site and of 15.4 per cent (in the mouse; 12.7 per cent in humans) replacement changes are taken into account, the presumed time point for the duplications might be assumed to be even earlier at around 350–400 million years ago. This estimate coincides well with the evidence on homeo box genes from lower vertebrates suggesting that the duplication events happened early during or possibly preceding vertebrate evolution (Kappen and Ruddle 1990). For whatever reason a four cluster arrangement arose, we posit that such a system was essential for the emergence of the vertebrates as we know them today, characterized by their unique body plan, size, and complexity.

Table 9.1. *Timing of Hox-cluster duplications*

Sequence comparisons	Silent nucleotide differences	Species divergence time
Human–mouse 23 cognate pairs	10.9	70–100 million years ago
Human–human 34 cognate pairs	32.3	*Estimated time after duplications*
Mouse–mouse 23 cognate pairs	27.4	300 million years ago

The number of silent nucleotide differences obtained from sequence comparisons (Kappen, unpublished) was correlated with evolutionary time. The divergence time for the mouse–human divergence was taken from Li *et al.* 1990.

Comparison of mammalian and *Drosophila* homeo box genes

In this context it is interesting to compare mammalian homeo box clusters to the homeotic gene complexes of *Drosophila* where homeo boxes were first detected. As described above, both in the fly as well as in mammals, the domains of expression in the embryo along the anterior-posterior axis are reflected in the relative chromosomal position of the genes. The more anterior a gene is expressed, the more 3′ it is located in the mammalian *Hox*-cluster, and as shown in Fig. 9.2, this situation is similar to that in the fly. However, in *Drosophila* the homeotic genes are located in two physically separated complexes on the same chromosome. Furthermore, the Ant-C complex contains a number of additional genes that are not homeotic, and in some instances do not contain homeo boxes. For example, *bicoid* is a homeo box gene that is non-homeotic, and *amalgam* does not contain a homeo box and is not a homeotic gene. Such 'extra' genes are absent from the mammalian *Hox*-clusters. Additionally, along each cluster, the transcriptional orientation of mammalian *Hox*-genes is the same. In *Drosophila*, however, several genes are transcribed in directions opposite to each other, for example *Antp* and *ftz* as well as *Scr* and *Dfd*.

When the amino acid and nucleotide sequences of the fly homeo box-containing genes of these complexes are compared to vertebrate homeo box sequences, it becomes apparent that not all sequences have direct counterparts. Good matches are found for the following groups: *Abd-B* and the *Hox-9* cognate group, *Dfd* and the *Hox-4* cognate group, *labial* and the *Hox-1* cognate group, and *pb* has *Hox-2* as a counterpart. The strong conservation of these homeo box sequences between fly and mammals has been taken as indication that they represent homologous genes (Lobe and Gruss 1989). However, the mammalian homeo box genes in the middle of the clusters cannot easily be correlated to the *Drosophila* sequences, and for more 5′ genes such as human in cognate groups Hox-10, Hox-11, Hox-12, Hox-13, (Duboule and Dolle 1989; Acampora *et al.* 1990), no direct counterparts have been described in the fly.

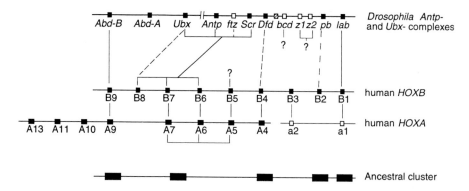

Fig. 9.2. Comparison of the genomic arrangement of human homeo box genes to the *Drosophila* homeotic gene clusters. In the fly clusters: ■, homeotic genes; □, homeo box-containing genes; the hatchbox represents the filled box represents the non-homeotic gene amalgam. In the mammalian clusters: ■, human homeo box genes; □, information additionally available from the mouse. The bottom line shows the minimal number of ancestor genes contained in the proposed primordial cluster.

These differences can be taken as an indication for separate lines of evolution in the arthropod and vertebrate lineages starting from an ancestral cluster of a few genes. In arthropods this cluster acquired more genes by gene duplications, followed by inversions or transpositions of genes into the Ant-C complex. The Ant-C and BX-C are split into separate complexes in *Drosophila*, whereas genetic evidence suggest that they are still closely linked in the red flour beetle *Tribolium* (Beeman *et al.* 1989). In the lineage leading to vertebrates, an independent increase in the number of homeo box genes was achieved by gene duplications along a primordial cluster. These local duplications involved genes for the more posteriorly expressed group (*Hox-9* and further 5′ cognate groups), genes that are located in the middle of the clusters, and only to a limited degree the more anteriorly expressed genes (creation of the *Hox-3* cognate group) (Kappen and Ruddle 1993). Alternatively, it may be possible that ancestor genes for the more posterior genes were present before the arthropod–vertebrate divergence but were subsequently lost in the insect lineage (Acampora *et al.* 1990). Finally, in the vertebrate lineage a further increase in the number of homeo box genes was brought about by duplications of the primordial complex to form at least four clusters of genes.

Implications of conserved cluster linkage domains

One of the striking properties of the mammalian homeo box gene clusters is the relationship between the clusters and genes in linkage. If one compares linkage relationships between clusters within a species, such as humans, one finds many instances of paralogous gene relationships between the homeo box cluster chromosomes 2, 7, 12, and 17. These paralogues include, for example, many of the collagen genes, retinoic acid receptor genes, epidermal growth factor genes, to name but a few (Ruddle *et al.* 1987; Ruddle 1989). The presence of these paralogous patterns provides strong evidence that cluster duplication involved large chromosomal domains (Schughart *et al.* 1989). More significantly, as stated above, cluster duplication most likely preceded the evolution of the vertebrates, and thus there is reason to suspect that paralogous relationships of genes in the vicinity of the clusters may also date back to the origin of the vertebrates. The long time period in which these linkage relationships would have been maintained is remarkable and suggests a functional relationship between the homeo box gene clusters and the genes in linkage.

In order to obtain evidence for the retention of linkage relationship structure around the homeo box gene clusters, it is necessary to develop detailed linkage maps for these domains within and between species. Such data is beginning to appear as a consequence of recent interest in the organization and structure of large and complex genomes. Evidence for the retention of structural similarities is provided by recent studies in our laboratory on the relationships between the *HOXB* cluster and 3′ linked genes encoding nerve growth factor receptor (*NGFR*) and erythrocyte band 3 (*EB3*). Using pulse field electrophoresis analysis, we have been able to show that *NGFR* maps 500 kb to the 3′ side of the homeo box gene cluster, and that *EB3* maps an additional 500 kb to the 3′ side of *NGFR* in both human and mouse (K. Bentley, unpublished). Recent reports suggest that man and mouse diverged some 100 million years BP, and this together with the above findings would suggest that linkage relationships extending over 1500 kb *HOXB*/NGFR/ EB3) have been maintained with little change over a relatively long period of time. It will be of interest to expand the linkage data set, and especially to include organisms that show a greater divergence time to humans.

We currently have no explanation for the striking similarities in linkage patterns in the homeo box cluster domains. One possibility is that there are *cis*-regulatory elements that co-ordinate gene expression over long distances in these regions. Evidence has been presented that long transcripts may be generated in *HOX*-clusters by transcriptional read-through (Simeone *et al.* 1988). Such a transcriptional mechanism could be consistent with conserved linkage patterns in the vicinity of the clusters.

A second possible explanation is that genes in the vicinity of the *Hox*-clusters have a functional relationship with the *Hox*-genes as target or input genes to the homeo box gene network. Recent evidence suggests that *wingless*, the fly counterpart of the *Wnt-1* gene (in linkage with the *HoxC* cluster) may be regulated by homeo box genes (Immerglück *et al.* 1990). In such a functional relationship, potential changes in one part (for instance, a homeo box gene) will affect the other part (target or input genes) at the same time. This would imply that a co-evolution of polymorphisms within these genes has been and continues to be important to the evolution of vertebrates. Thus, close linkage of genes, as exemplified in the chromosomal regions involving the *Hox*-cluster, appears to be adaptive to this process.

Implications of expansion of the homeo box gene regulatory network by cluster duplication

All invertebrates studied to date have a relatively small number of homeo box genes. *Drosophila*, representing evolutionarily advanced arthropod forms, has two closely linked clusters of homeotic genes, which map to a single chromosome, and which consist of a total of 10 genes of the Antennapaedia type. As discussed above, the vertebrates as represented by human and mouse have four clusters of genes of a related type which map to separate chromosomes. These are postulated to have arisen by a large chromosomal duplication event, most probably by a polyploid amplification of the entire genome (Schughart *et al.* 1989). The total number of mammalian genes of the class Antennapaedia has been estimated to be no less than 38. Thus, between advanced invertebrates and vertebrates, there is a four-fold difference in gene number. If, as we postulate, the genes are networked in a combinatorial fashion, then the vertebrate network could provide significantly more epigenetic states. Assuming the simplest situation with each gene capable of existing in two conditions, that is to say, on and off, then the *Drosophila* system permits only 1000 (2^{10}) states, whereas the mammalian system provides approximately 10^{12} (2^{40}), a difference of about 10^9 times. This example is, of course, a gross simplification, since each gene most probably has many more conditions than the minimum of two.

The *Drosophila* genes can be very large compared to mammalian genes, with *Ubx* having a complementation unit estimated to be 100 kb. The mammalian genes, interestingly, are roughly one-tenth as large. This relationship suggests that the conditions per gene as determined by the *cis*-flanking, non-coding sequences may be greater in insects than in mammals. In fact, one might postulate that the invertebrate strategy, as exemplified by *Drosophila*, has been to maintain a small genome size, but to maximize the epigenetic power of the homeo box network by increasing the number of conditions possible for the individual genes.

Taking these arguments into account, let us recalculate the epigenetic potential in the two species by making the generous assumption that each *Drosophila* gene has 50 conditions for each of 10 genes, whereas in the case of the mouse, each gene has five conditions (most probably a gross under-estimate) for each of 40 genes. Under these assumptions, the number of epigenetic states possible in the insects is approximately 10^{17} (50^{10}), and in the mammals about 10^{27} (5^{40}). If one makes the not irrational assumption that the complexity of the developmental programme and the resulting body plan are dependent on the combinatorial power of the epigenetic regulatory gene network, then one sees the overwhelming superiority of a multi-cluster system. States of 10^{27} would permit the unique identification of each of the 10^{15} cells in a mammalian organism, plus 10^{10} states reserved for each cell. For most cells of the organism, this degree of identification and specification is most probably not needed, however, the critical case may be that of the nervous system, where in fact, individual neurones require a highly refined degree of identification and specification in order to make specific synaptic and end-organ connections.

It has been pointed out previously that for particular organisms or groups of organisms, the number of neurones is fairly rigidly fixed early in development. Increases in body size between closely related groups affects the number of neuronal elements minimally. This is also true when one manipulates body size experimentally, that is to say, there is a minimal influence on the number of nerve cells, although the somatic cells at large may be increased significantly. These observations are consistent with the view that the design of the nervous system is under tight genetic/epigenetic control.

Implications of expansion of the homeo box gene family by gene duplication

Strong evidence as cited above supports the idea that the homeo box gene family has arisen by gene duplication and gene cluster duplication. Evidence also supports the notion that homeo box genes are networked in the sense that a transcription factor encoded by one gene may interact with response elements located within that same gene and other genes within the network (Odenwald *et al.* 1989). Such interactions are termed autoregulatory for those operating within a gene, and transregulatory for those acting between genes (Fig. 9.3A). It should be noted that an autoregulatory circuit, if copied at the time of gene duplication, becomes at least potentially a transregulatory circuit between the newly duplicated genes (Fig. 9.3B). In this manner, it is easy to imagine how a network of homeo box gene interactions may have arisen as the gene number increased by the process of duplication.

One might further postulate that when a duplication event occurs, the initial developmental effect is minimal, since only a dosage effect is produced, and no or only a few qualitative changes in gene expression take place. It might also be postulated that in the period of time following the duplication event, one of the duplicated genes will be conserved in function, and that the other gene will diverge and take on new adaptive functions. It would seem logical for the conserved gene to be located proximal to the cluster, since its behaviour would be influenced by the *cis*-regulatory elements in the cluster from which it had arisen at an earlier time. Thus, according to this scheme, the gene network is to a large extent conservative in nature, since the aggre-

A

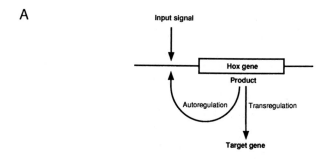

B

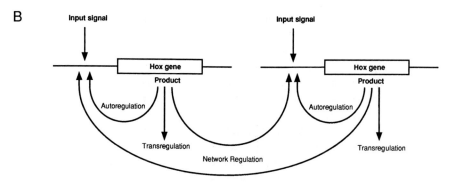

Fig. 9.3. Auto- and transregulation by homeo box gene products. A, simple regulatory circuit of one Hox-gene; B, regulatory circuit containing two Hox-genes constituting a network regulatory system.

gate gene family is controlling, and thus, integrated into the developmental plan of the organism. One imagines that mutations and polymorphisms would only rarely be tolerated, and indeed, we have previously shown that this is the case (for a review see Ruddle 1989).

Under strong functional conservation, the addition of new regulatory elements into the network system may only be possible in close physical vicinity that would place the new copy under the regulatory constraints by which the whole network is guided. The new gene may then take on new functions by mutation. The available evidence suggests that such a process of local gene duplications may have been operative during the evolution of the homeo box gene network (Kappen *et al.* 1989*b*; Kappen and Ruddle 1993). Over time, one would expect the 'innovative' gene to evolve relatively rapidly, and for fewer changes of a reflexive nature to occur in the homeo box gene network, the respective input genes to the network and target genes would have to evolve similarly. Thus, the birth of a new gene into the network would open up new possibilities of change for the entire ensemble of genes regulating the developmental programme, and hence the body plan and the behavioural attributes of the organism. The direction of adaptation would depend to a great degree on the environmental conditions existing at the time of gene duplication.

It is easy to see how the process of gene duplication within an epigenetic network of regulatory genes might serve as the basis of saltation in the evolutionary process. The duplication of a homeo box gene can be regarded as a mutation with little intermediate effect on the organism, but with wide-ranging implications for radical innovations over a period of time depending on selective forces. The existence of an additional gene at once significantly increases the number of epigenetic states available to the homeo box gene network. Moreover, the new gene, as it undergoes modification, opens possibilities for adaptive change throughout the regulatory gene network, including the homeo box genes themselves, and more peripheral members of the network, such as the immediate target and input genes. Once a homeo box gene duplicates and then becomes established within a population, many possibilities arise with respect to novel pathways for evolutionary change and radiation of variant forms. Given the conservative nature of the homeo box gene network, it may be possible to obtain definitive data which supports this hypothesis. The hypothesis would predict that the addition of genes during evolution would be unique to particular clades, but absent in others. These candidate genes could then be tested in terms of their expression patterns during development to determine if they are expressed predominantly in the primordia of structures novel and common to the clade in question. Additional tests, such as directed mutation and transfer into related, but different clades to produce predicted loss of function and gain of function developmental mutations, could also be carried out.

Conclusions

In summary, the study of the evolution of the homeo box gene system has provided evidence for discernible phases of variation and conservation within this multi-gene family. Strong conservation in the arthropod lineage is evident from comparison of homeotic genes of *Drosophila* and the red flour beetle *Tribolium*, species thought to have diverged 300 million years ago. The genomic organizations and homeotic mutations in both species appear very similar (R. Dernell, personal communication). In the vertebrate lineage conservation was favoured after the multiplication of the number of elements in the homeo box gene system by duplications of large clusters. Before the divergence of arthropods and vertebrates, variability may have led to the development of new forms as would be consistent with the radiation of forms observed for the pre-Cambrian period. We speculate that the homeo box gene system may have been crucially involved in these events since it is able to provide a large number of regulatory states and sufficient complexity to allow for the rapid acquisition of new functions. The function of the homeo box gene regulatory network in development and the adult organism may provide experimental tools to address the phenomena of radiation and saltatory events during evolution.

Acknowledgements

We are grateful to Dr E. Boncinelli (Naples) and Dr M. Frohman for making data available to us before publication. We thank Dr J. Michael Salbaum for discussion and a critical review of the manuscript, and Marie Siniscalchi for secretarial help in preparing it. This work was supported in part by a post-doctoral fellowship of the Deutsche Forschungsgemeinschaft to Claudia Kappen; and NIH-grant GM09966 to Frank H. Ruddle.

References

Acampora, D. *et al*. (1990). The human Hox gene family. *Nucleic Acids Res.* **17**, 10385–402.

Beeman, R.W., Stuart, J.J., Haas, M.S., and Denell, R.E. (1989). Genetic analysis of the homeotic gene complex (HOM-C) in the beetle *Tribolium castaneum*. *Devel. Biol.* **133**, 196–209.

Duboule, D. and Dolle, P. (1989). The structural and functional organization of the murine HOX gene family resembles that of *Drosophila* homeotic genes. *EMBO J.* **8**, 1497–505.

Fienberg, A.A. *et al*. (1987). Homeo box genes in murine development. *Curr. Top. Dev. Biol.* **23**, 233–56.

Gehring, W.J. (1987). Homeo boxes in the study of development. *Science* **236**, 1245–52.

Graham, A., Papalopulu, N., and Krumlauf, R. (1989). The murine and *Drosophila* homeobox genes complexes have common features of organization and expression. *Cell* **57**, 367–78.

Immerglück, K., Lawrence, P.A., and Bienz, M. (1990). Induction across germ layers in *Drosophila* mediated by a genetic cascade. *Cell* **62**, 261–8.

Kappen, C. and Ruddle, F.H. (1990). *Duplication of homeobox gene clusters in early vertebrate evolution*. Abstract and poster presented at: Evolution of molecules and developmental systems. Indiana University, Bloomington.

Kappen, C. and Ruddle, F. (1993). Evolution of a regulatory gene family: *HOM/ HOX* genes. *Curr. Opin. Genet. Devel.* (In press.)

Kappen, C., Schughart, K., and Ruddle, F.H. (1989*a*). Mammalian Antennapedia class homeo box genes: organization, expression, and evolution. In *Molecular genetics of early* Drosophila *and mouse development* (ed. M.R. Capecchi), 67 pp. Current communications in molecular biology. Cold Spring Harbor Laboratory Press.

Kappen, C., Schughart, K., and Ruddle, F.H. (1989*b*). Two steps in the evolution of Antennapedia-class vertebrate homeobox genes. *Proc. Natl Acad. Sci. USA* **86**, 5459–63.

Li, W.-H, Gouy, M., Sharp, P.M., O'hUigin, C., and Yang, Y.-W. (1990). Molecular phylogeny of Rodentia, Lagomorpha, Primates, Artiodactyla, and Carnivora and molecular clocks *Proc. Natl Acad. Sci. USA* **87**, 6703–7.

Lobe, C.G. and Gruss, P. (1989). Mouse versions of fly developmental control genes: legitimate or illegitimate relatives? *The New Biologist* **1**, 9–18.

McGinnis, W., Garber, R.L., Wirz, J., Kuroiwa, A., and Gehring, W.J. (1984). A homologous protein-coding sequence in *Drosophila* homeotic genes and its conservation in other metazoans. *Cell* **37**, 403–8.

Odenwald, W.F., Garbern, J., Arnheiter, H., Tournier-Lasserve, E., and Lazzarini, R.A. (1989). The *Hox-1.3* homeobox protein is a sequence specific DNA-binding phosphoprotein. *Genes Devel.* **3**, 158.

Ruddle, F.H. (1989). Genomics and evolution of murine homeobox genes. *The physiology of growth* (ed. J.M. Tanner and M.A. Priest), pp. 47–66. Cambridge University Press.

Ruddle, F.H., Hart, C.P., Rabin, M., Ferguson-Smith, A.C., and Pravtscheva, D. (1987). Comparative genetic analysis of human homeo-box genes in mouse and man. In *New frontiers in the study of gene functions* (ed. G. Poste, and S.T. Crooke), pp. 73–86. Plenum, New York.

Scott, M.P. and Weiner, A.J. (1984). Structural relationships among genes that control development: Sequence homology between the *Antennapedia*, *Ultrabithorax* and *fushi tarazu* loci of *Drosophila*. *Proc. Natl Acad. Sci. USA* **81**, 4115.

Scott, M.P., Tamkun, J.W., and Hartzell, G.W. 3d. (1989). The structure and function of the homeodomain. *Biochim. Biophys. Acta* **989**, 25–48.

Schughart, K., Kappen, C., and Ruddle, F.H. (1989). Duplication of large genomic regions during the evolution of vertebrate homeobox genes. *Proc. Natl Acad. Sci. USA* **86**, 7067–71.

Shashikant, C.S. *et al.* (1991). Homeobox genes in mouse development. *Crit. Rev. Eukaryotic Gene. Express.* **1**, 207–45.

Simeone, A., Pannese, M., Acampora, D., D'Esposito, M., and Boncinelli, E. (1988). At least three homeobox genes belong to the same transcription unit. *Nucleic Acids Res.* **16**, 5379.

DISCUSSION

Participants: R.L. Cann, J.-P. Changeux, P. Rakic, F.H. Ruddle

The relevance of homeo box genes to evolution and morphogenesis of the brain was the major topic of many questions asked by **Changeux, Rakic, Ruddle,** *and* **Cann**

Ruddle: There is growing evidence from transgenic mouse experiments that homeo box genes are involved in development in mammals as they are in *Drosophila*. Also, in numerous experiments, when one introduces these genes, there are profound developmental problems which result in lethality at earlier time in development. In mammals, so far we have not found any naturally occurring mutants in these genes (...). It is possible, however, that spontaneous abortion in humans is somehow connected to these changes and for that reason they have not been detected.

Homeo box genes could be involved in the development of the neocortex or other parts of the brain. It would be most interesting to make some very careful comparisons between the lower and higher primates, including ourselves, in order to detect the presence or absence of a particular homeo box gene which could be important in the different morphologies (...). One point that ought to be kept in mind is that the homeo box clusters seem to have a relationship to the anterio-posterior axis in development: their expression affects the nervous system but also the mesoderm, the ectodermal derivatives, and to some extent the endodermal derivatives as well. You can look at this system as a kind of a morphogene transducer providing information to cells with regard to the gradient of a particular morphogene across the anterior-posterior axis. The system seems not only to be involved in pattern formation, but may play a role in targeting axons to particular organs. All of that is part of the puzzle of the rich combinatorial nature of the system which is adapted to a number of different functions in the course of the life cycle.

Part III

Culture

10

Life in the fast lane: rapid cultural change and the human evolutionary process
ROBERT BOYD AND PETER J. RICHERSON

Introduction

Modern humans are uniquely reliant on culture as a means of adaptation. Like all organisms, human populations evolve genetically, and, like most other animals, we also adapt using individual learning. Humans, however, acquire a greater proportion of their adaptive information by imitation, teaching, and other forms of cultural transmission than do other creatures. The human reliance on cultural adaptation probably changed the human evolutionary process in many important ways. Here, we will focus on just one of them: the speed of cultural adaptation.

Culture clearly allows humans to adapt much more rapidly than non-cultural species. The historical record provides many striking examples of the speed with which cultural change may occur, even in simple societies. For example, the sweet potato had become a staple in the New Guinea highlands by the late seventeenth century, perhaps only two centuries after its introduction to the Old World (Yen 1974). Other examples come from the archaeological record: the shift from nomadic, egalitarian, food-foraging societies to sedentary, stratified agricultural ones occurred in many areas over the period of a few thousand years. While these changes occurred slowly compared to the diffusion of technical innovations in the modern world, they were much faster than any analogous major adaptive change in non-human organisms.

In this chapter we argue that an increase in the rate at which a population adapts can cause qualitative changes in the nature of the adaptations that characterize that population. We begin by briefly reviewing population-based models of cultural evolution developed by ourselves (Boyd and Richerson 1985) and others (Pulliam and Dunford 1980; Cavalli-Sforza and Feldman 1981; Lumsden and Wilson 1981; Rogers 1989). Then we use this approach to argue that rapid cultural change leads to more accurate adaptation to local ecological conditions, greater variation among groups in social behaviour, the evolution of group-beneficial behaviours, and the subdivision of human populations into stylistically marked groups, and, therefore, cultural adaptation can lead to different long-run evolutionary outcomes compared to genetic adaptation.

Cultural change as a Darwinian process

The population approach to understanding cultural change begins with the premise that culture constitutes a system of inheritance. People acquire skills, beliefs, attitudes, and values from others by imitation and teaching, and these 'cultural variants' together with their genotypes and environments, determine their behaviour. Since determinants of behaviour are communicated from one person to another, individuals sample from and contribute to a collective pool of ideas that changes over time.

Because cultural change is a population process, it can be studied using Darwinian methods. To understand why people behave as they do in a particular environment, we must know the nature of the skills, beliefs, attitudes, and values that they have acquired from others by cultural inheritance. To do this we must account for the processes that affect cultural variation as individuals acquire cultural traits, use the acquired information to guide behaviour, and act as models for others. What processes increase or decrease the proportion of people in a society who hold particular ideas about how to behave? We thus seek to understand the cultural analogues of the forces of natural selection, mutation, and drift that drive genetic evolution. We divide these forces into three classes: random forces, natural selection, and the decision-making forces. In this chapter we will focus on the decision-making forces.

Decision-making forces result when individuals evaluate alternative behavioural variants and preferentially adopt some variants relative to others. If many of the individuals in a population make similar decisions, especially if similar decisions are made for a number of generations, the decision-making forces can transform the pool of cultural variants. Naïve individuals may be exposed to a variety of models and preferentially imitate some rather than others. Alternatively, individuals may modify existing behaviours or invent new ones by individual learning. If the modified behaviour is then transmitted, the resulting force is much like the guided, non-random variation often attributed to Lamarck.

Let us illustrate these ideas with a very simple example. Suppose that there is a culturally transmitted character with two variants. These variants might be simple beliefs; for example, taro is superior to sweet potato and vice versa. They could also be large integrated complexes of beliefs, for example, a belief in Christianity versus Islam. We describe the state of the population by the frequency of individuals who are characterized by each of the variants. Individuals acquire their initial beliefs from their parents. As adults, if they encounter another person who has different beliefs they attempt to evaluate the relative merit of their beliefs and those of the person they have encountered. If they decide the other person's beliefs are better, they switch. If a new, preferred variant is introduced into a population, this process will cause it to increase through time in the way that innovations have been observed it to spread. We refer to this kind of process as 'biased transmission'.

The decision-making forces are derived forces (Campbell 1965). Decisions require rules for making them; that is, if individual decisions are not to be random, there must be some sense of psychological reward or similar process that causes individual decisions to be predictable. For example, if sweet potatoes taste better or yield more than taro, and if sweet potato cultivation is introduced from another island, then the belief that sweet potatoes are superior to taro is liable to spread. The decision-making rules that drive biases may be acquired during an earlier episode of cultural transmission, or they may be genetically transmitted traits that control the neurological machinery guiding the acquisition and retention of cultural traits.

Like natural selection, decision-making forces allow populations to adapt to changing environments. This is easiest to see when the goals of the decision rules are closely correlated with genetic fitness. If human foraging practices are adopted or rejected according to their energy pay off per unit time (as is typically assumed in optimal foraging theory), then the foraging practices used in the population will adapt to changing environments much as if natural selection were responsible. If the adoption of foraging practices is strongly affected by consideration of prestige, such as that associated with male success in hunting dangerous prey, then the resulting pattern of behaviour may be different. However, there will still be a pattern of adaptation to different environments, but now in the sense of increasing prestige rather than calories.

The decision-making forces allow rapid adaptation compared to genetic adaptation resulting from natural selection. There are several reasons for this. First, learning allows a population to adapt without any selective load; an entire population may be transformed without a single death. Secondly, the 'generation time' of cultural change may be much shorter than a biological generation because cultural variants can spread horizontally among age mates. Individuals may modify their behaviour many times during their lifetimes. Thirdly, because individuals can acquire adaptive information from many individuals, not just their parents, beneficial variants can spread more rapidly through a population.

The evolutionary effects of rapid change

Increased variation

Over the last several million years the hominid lineage evolved the capacity for cultural inheritance. Here we argue that the resulting increase in the speed of adaptation may lead to striking changes in the *kinds* of adaptations that characterize humans compared to non-cultural species.

At first blush, there is no obvious reason why more rapid evolution should lead to different evolutionary outcomes in the long run. There is good reason to suspect that the goals which drive decision-making forces are closely correlated with fitness. Learning mechanisms embedded in the human psyche have been shaped by natural selection, and, thus, pleasurable, reinforcing events are usually fitness enhancing, and unpleasant, aversive events are usually deleterious (or, at least, were under Pleistocene food-foraging conditions). A number of authors (e.g. Dawkins 1980; Flinn and Alexander 1982; Maynard Smith 1982) have argued that when this is the case, one can ignore the details of evolutionary processes and simply assume that behaviour will maximize genetic fitness. Cultural evolution will yield the same adaptations as genetic adaptation, just more quickly.

In other work (Boyd and Richerson 1985), we have argued that this view is too simple — cultural evolution may involve novel dynamic processes even when the capacity for culture is shaped by natural selection. Here, we ignore such effects, and assume that decision-making forces on culture work just like natural selection would, only much faster. We propose that even this seemingly innocent change can lead to qualitatively different outcomes.

Faster adaptation leads to more accurate tracking of environmental gradients which produces greater within-species phenotypic variation, and a larger species range. These effects can be illustrated by the following very simple model: an 'island' of one habitat adjoins a much larger 'continent' with a differing habitat. At each time period some fraction of the population on the island are replaced by immigrants from the mainland. There are two phenotypic variants — an island variant which has higher fitness on the island, and a continental variant with higher fitness on the continent. Adaptive processes like decision-making forces cause the favoured variant to increase on the island, but immigration brings in individuals with the continental variant, and thus decreases the frequency of the island variant. As the speed of the adaptive process is decreased, the frequency of the favoured variant on the island decreases, and once the adaptive process becomes sufficiently slow, migration swamps adaptation and the island becomes just like the continent. Now consider a large area made up of many different habitat islands. If adaptive processes are slow compared to the effects of migration, then all the islands will be the same — only the phenotypic variant with the highest fitness averaged (in a complicated way) over all habitats will survive. If adaptive processes are fast enough, individuals in each island will tend to exhibit the phenotypic variant that is best for that habitat. Similarly, more rapid adaptation will allow populations to adapt to conditions at the edge of a species' range. Rather than being limited to a range where an average adaptation is successful, fringe populations will adapt to local conditions, allowing the population to have a wider range.

Even more interesting, rapid adaptation can lead to greater within-species variation in the absence of underlying environmental variation. Many types of social interactions have the property that once a behavioural strategy becomes sufficiently common, individuals using that strategy have higher pay off than individuals using alternative strategies. Such a strategy will increase under the influence of adaptive processes once its frequency exceeds a threshold. The simplest examples are co-ordination games in which fitness is frequency-dependent but there is no conflict of interest among individuals (Sugden 1986). Driving on the left versus right side of the road is an example. It does not matter which side we use, but it is critical that we agree on one side or the other. This property is shared by many symbolic and communication systems.

Many other kinds of social interactions mix elements of co-ordination and conflict. In such interactions, all individuals are better off if they use the same strategy, even though the relative advantages of using the strategy differ greatly from individual to individual, and some, or even all individuals would be much better off if another strategy were common. As long as the co-ordination aspect of such interactions is strong enough, multiple stable equilibria will exist. Reciprocity provides a good example. Models (Boyd 1992) suggest that there are a large number of different strategies which can capture at least some of the potential benefits of long-run co-operation. In order to persist when common, reciprocating strategies must retaliate against individuals who do not co-operate when it is appropriate. When a strategy is rare, it will interact mostly with other strategies which co-operate and expect co-operation in a different set of circumstances. Inevitably, a rare strategy will retaliate or suffer retaliation and co-operation will collapse. Thus, a common reciprocating strategy has an advantage relative to rare reciprocating strategies, even if the rare strategy would lead to greater long-run benefit once it became common. We believe that interactions of this kind are omnipresent in social life.

Processes leading to rapid adaptation can cause a population to be subdivided into subpopulations characterized by different behavioural strategies. Consider a population occupying a number of 'islands'. Initially, on a few islands most people drive on the right, while on most islands most people drive on the left. On each island the identity of the favoured behaviour depends on the local frequency. If adaptive processes are strong, they can maintain this situation indefinitely. New immigrants who drive on the wrong side rapidly learn that this is a bad idea, so that when yet more wrong-headed immigrants arrive they find themselves to be in the minority, rapidly learn to drive on the other side, and so maintain the strategy that originally arose merely by chance. In contrast, suppose that adaptive processes are weak compared to migration, as would commonly be the case in genetic evolution. On an island on which most people drove on the right, most of the immigrants would drive on the left. If selection is weak, it will have eliminated only a few immigrants by the time that more immigrants arrive. This reduces the benefit of driving on the right, and thus even fewer of these immigrants are eliminated, and eventually driving on the left becomes advantageous, and the subpopulation comes to resemble the global population. If one behaviour is common in some subpopulations, but not others, rapid adaptive processes can act to counteract migration and preserve variation among subpopulations, much as in the case where there are ordinary adaptive differences between islands. In contrast, weak adaptive processes cannot maintain an island at a different frequency than the continent in the face of migration, and all islands will be alike. Once either behaviour becomes sufficiently common in the metapopulation, it is favoured by adaptive processes, and one or the other behaviour comes to dominate the metapopulation; no one island will be able to resist the effects of migration from the other islands.

Thus a shift from genetic to cultural adaptation should lead to an increase in the amount of behavioural diversity among subpopulations within a species. Because a very wide range of social behaviours are characterized by some element of co-ordination, in any habitat there are a very large number of social arrangements that are stable once established in the absence of migration. When there is migration, however, natural selection can maintain a local population only if the selective advantage of the local type is greater than the migration rate. Thus, in non-cultural populations, different social systems can be preserved only among nearly isolated groups. Rapid cultural change can maintain social differences between groups with much greater levels of migration.

Group beneficial traits

Most contemporary evolutionists believe that behaviours should be explained in terms of their benefit to individuals or their kin, not to large social groups as a whole. This belief is a corollary of the conviction that selection among groups is rarely an important evolutionary process. By definition, altruistic individuals have lower fitness relative to non-altruists in their group. Thus, selection reduces the frequency of altruistic behaviour within each group. Selection among groups increases the frequency of altruism in the population as a whole because groups with more altruists have higher average fitness or lower extinction rates. Selection among groups is weak because there is typically little variation between groups. Variation between groups is destroyed both by migration and selection. Only genetic drift acts to create and maintain differences between groups. Unless groups are quite small, selection within groups weak, and migration rates low, the processes destroying variation between groups overwhelm the process generating it.

Group selection can be an important process when social behaviour leads to many alternative stable equilibria. As we have seen there are a variety of important kinds of social behaviour that have this property, and when adaptive processes are strong compared to migration this can lead to stable between-group variation — the basic raw material necessary for group selection. For many types of social behaviour, alternative stable equilibria have quite different levels of average fitness. For example, in models of the evolution of reciprocity, populations in which reciprocating strategies like tit-for-tat are common may have much higher average fitness than stable equilibria mainly composed of non-co-operators (Axelrod and Hamilton 1981). It is plausible that differences in average fitness might lead to differences in group persistence. Larger groups are less prone to extinction due to random demographic fluctuations or competition with other groups. Groups with harmonious social relations are more likely to persist than those with conflictual social relations. Either differences in average fitness or extinction rate make possible selection among groups. We have shown (Boyd and Richerson 1990) that this form of group selection can be important even if groups are very large and migration rates are substantial.

This form of group selection should be most important when the adaptive processes acting within groups are strong compared to migration. When social interaction creates multiple stable social arrangements, within-group adaptive processes act to *increase* between-group variations. Given that per capita migration rates between neighbouring groups will tend to increase as group size decreases (as the ratio of boundary length to group territory area increases), stronger selection (or learning) within groups will allow variation to be preserved between smaller groups. Since smaller groups are likely to have lower persistence, and higher rates of drift, stronger within-group processes should lead to more rapid group selection.

Stylistic variation

More rapid adaptation can also give rise to 'stylistic' variation among groups. By stylistic variation we mean differences in characters that have no direct effect on fitness — dialect, artefact style, ritual — but which come to serve as emblems of group membership. There are at least two reasons this might occur.

First, when the environment varies, different beliefs may be beneficial in different environments. Then individuals living in one environment will be better off if they can avoid imitating immigrants from other environments who bring with them locally inappropriate beliefs. One way to accomplish this would be to only imitate individuals who are similar to one's self with respect to some marker character which shows variation among groups, say dialect. Interestingly, individuals who use such a learning rule are more likely to be successful, which causes the local variant of the marker character to increase. This process causes neighbouring groups to become more different with respect to the marker, further increasing the utility of using the marker to bias imitation, and so on. We have shown (Boyd and Richerson 1987) that this process can cause initially similar groups to diverge dramatically for stylistic characters.

Even when the environment does not vary, stable behavioural variation between groups can generate stylistic variation by a similar mechanism. Recall that stable between-group variation can result when individuals are better off if they interact socially with others like themselves. This means that individuals living in a group in which one behavioural strategy is common will be better off if they can avoid interacting with immigrants from other environments who may use different behavioural strategies. Again, this can be accomplished by selectively interacting with others who are similar for some stylistic character which differs among groups, and, as in the previous case, the use of the character for this purpose will cause further stylistic differences among groups.

Conclusion

Cultural organisms are able to adapt more rapidly than non-cultural organisms. Such rapid adaptation can lead to:

- more within-species variation in characters that affect adaptation to local ecological conditions, and wider ecological range;
- greater within-species variation in social structure;
- more group beneficial social behaviour; and
- stylistic differentiation of local groups.

These results are roughly consistent with the observed variation between humans and other animals. Even if one restricts attention to hunter–gatherers, humans exhibit much greater variation in adaptive strategy than any other species—think of Inuit seal hunters, the sedentary gatherers of California, the nomadic foragers of Amazonia, and so on. The range of the human species is also much greater than that of any other species, excepting perhaps, human commensals such as the dog. In non-cultural organisms most of the variation in social structure appears to occur among species. In contrast, there is a great deal of variation in social organization among groups within the human species. Moreover, much of this variation cannot be related to ecology in any simple way. Culturally related societies occupying different environments often have similar social systems, and culturally unrelated societies in similar environments often have very different social systems (see Hallpike 1986 for a review). Humans exhibit larger, more complex, and more co-operative societies than most other social organisms. Social complexity and degree of co-operation are suggestive that human social arrangements are more group-beneficial than those of other animals. Finally, the human species is subdivided into symbolically marked groups, such as ethnic groups, that are indicated by dialect, clothing, decoration on tools, and many other stylistic characters. Moreover, ethnic and other symbolically marked groups are also often distinguished by economic and ecological differences and variation in social structure.

References

Axelrod, R. and Hamilton, W.D. (1981). The evolution of cooperation. *Science* **211**, 1390–6.
Boyd, R. (1992). The evolution of reciprocity when conditions vary. In *Coalitions and*

competition in humans and other animals (ed. A. Harcourt and F.D. Waal). Oxford University Press, Oxford.

Boyd, R. and Richerson, P.J. (1985). *Culture and the evolutionary process*, pp. 473–92. Chicago University Press, Chicago.

Boyd, R. and Richerson, P.J. (1987). The evolution of ethnic markers. *Cult. Anthropol.* **2**, 65–79.

Boyd, R. and Richerson, P.J. (1990). Group selection among alternative evolutionarily stable strategies. *J. of Theoret. Biol.* **145**, 331–42.

Campbell, D.T. (1965). Variation and selective retention in sociocultural evolution. In *Social change in developing areas: a reinterpretation of evolutionary theory* (ed. H.R. Barringer, G.I. Blanksten, and R.W. Mack), pp. 14–99. Schenkman, Cambridge, Mass.

Cavalli-Sforza, L.L. and Feldman, M.W. (1981). *Cultural transmission and evolution: a quantitative approach*. Princeton University Press, Princeton, NJ.

Dawkins, R. (1980). Good strategy or evolutionarily stable strategy? In *Sociobiology: beyond nature/nurture* (ed. G.W. Barlow and J. Silverberg), pp. 331–67. Westview, Boulder, Colo.

Flinn, M. and Alexander, R.D. (1982). Culture theory: the developing synthesis from biology. *Human Ecol.* **10**, 383–99.

Hallpike, C.R. (1986). *The principles of social evolution*. Clarendon Press, Oxford.

Lumsden, C.J. and Wilson, E.O. (1981). *Genes, mind, and culture: the coevolutionary process*. Harvard University Press, Cambridge, MA.

Maynard Smith, J. (1982). *Evolution and the theory of games*. Cambridge University Press, London.

Pulliam, H.R. and Dunford, C. (1980). *Programmed to learn: an essay on the evolution of culture*. Columbia University Press, New York.

Rogers, A.R. (1989). Does biology constrain culture? *Am. Anthropol.* **90**, 819–31.

Sugden, R. (1986). *The economics of rights, cooperation, and welfare*. Blackwell, Oxford.

Yen, D. (1974). The sweet potato and Oceania. *Bernice P. Bishop Mus. Bull.* **236.**

DISCUSSION:

Participants: R. Boyd, J.-P. Changeux, R. Hinde, B. Huberman, M. Mishkin, D. Premack, P. Rakic, P. Tobias

Hinde: All that you have said seems to me to be concerned with changes within a group as a whole, but, in fact, most cultural changes concern groups within groups. I think it is correct to say that religious beliefs are much more conservative than anything else in a society, and it is certainly true of religious buildings which are much more conservative in any society compared to palaces, for example. There is a psychological reason for this, in that it profits religious specialists dealing with unverifiable beliefs to keep everything constant because conservatism helps people to maintain their unverifiable beliefs. I do not see how you handle that sort of differential in your model.

Boyd: I think that your generalization might be a fair one for stratified agricultural societies and societies that are much more complex politically and socially than that. When you think about pre-agricultural and simple agricultural societies, I am not sure that it is fair to say that most of the cultural evolution takes place in subunits of the societies. To provide an example, let us consider the evolution of religious institutions in the New Guinea Highlands. In the Highlands, everybody does exactly the same thing to make a living. Their political and economic institutions are very similar, but according to Fredrik Barth their rituals and religious beliefs vary from valley to valley. Having said that, I think that your criticism is valuable.

Hinde: I do not think that the Barth case is quite relevant because, of course, it is true that they differ between valleys. But what I was talking about was the difference between the conservatism of religious beliefs and anything else in the society.

Boyd: Still, in the Barth case, they have very different religious beliefs, but they have exactly the same agricultural practices and very similar political systems. So, presumably at that time, their religious beliefs have changed fairly extensively, but their other institutions have been conserved. The reason they were conserved is that it is the only way to make a living in the Highlands. There is some kind of selective process acting more strongly on foraging or technology than on cosmology.

Premack: I would like to make three comments. First, I agree with your conjecture that the human alone teaches. We do not find pedagogy teaching on individuals by any other species, even if we find some approximations. One rapid way to capture that distinction is simply to observe that any time you see an adult chimpanzee, you can realize that there is absolutely nothing in various technologies that the young chimpanzee cannot acquire simply by some combination of learning or imitation. This is certainly not true of the human child. The unassisted child will not turn into an adult, whereas the unassisted young chimpanzee definitely does turn into an adult. The second thing related to that is the matter of innovation and combination that you can obtain for rapid cultural change through the fact of pedagogy or teaching in humans. There is a great variability of intelligence among humans, but also great deal of modularity, because human intelligence consists of specialized components. Unlike animals, in humans we have individuals capable of innovation. This is not acquired through unassisted learning. Innovation would die or not diffuse in the group if there is no pedagogy. Given that, we agree on the variability and intelligence that we find in humans, and the modularity of human intelligence. It is clear enough that there are portions of our population which are capable of dramatic innovation, but not even capable of acquiring existing knowledge in certain other spheres because of the modularity of their intelligence, except as we train one another, these innovations will die. Now, given that you have innovations preserved, through pedagogy, by that combination of the creative intelligence and pedagogy, you get diversity. That alone will give you a notable measure of diversity among humans than will appear in any group in which you have neither the pedagogy nor modularity and intelligence. But the combination of the two is a guarantee of diversity. Finally, about co-operation among humans, it is not merely a matter of co-operation. Co-operation has two sides: the self-evident one, which has to do with working together towards the common good; and a not so self-evident one, which is an agreement to share, which, I can show you, is not language dependent. It is the latter, in particular, which distinguishes all human co-operation from animal co-operation. Even in the chimpanzee where we find hunting and sharing in the sense of physical transfer of goods, we have absolutely no evidence for agreement to share of reciprocation. That is what distinguishes the human development ladder, for example, the expectation of reciprocation in human infants that is not acquired, and I think this needs to be seen to appreciate what is distinctive about human co-operation.

Boyd: The division of tasks is something you see in social insects. You do not see it in most other animals, except on the basis of sex or age. Co-operation in humans provides an enormous advantage. I think it is the basis of a huge amount of additional ecological compatibility of the species.

Rakic: It seems to me that all your examples are based on the earlier way of living on the earth, that is, when people were separated by valleys and geographical constraints, and also when they were dependent on local conditions for survival. I wondered why you did not mention modern societies that are changing very rapidly. For example, in the last two decades the green revolution has transformed agriculture all the way from India to South America or North America. Therefore, people no longer depend on local conditions for food. So, would these rapid migration and rapid transformation in electronics make pressure for conformity so that diversity, which was historically developed, would lose its evolutionary selective value, and therefore we will have a very uniform society in the future?

Boyd: First of all, I made the choice of discussing simple societies because this conference is about the evolution of the brain and human evolution. It seems to me that this is the social and ecological context in which human evolution of humans has occurred. Secondly, once you have writing, electronic transmission and hierarchically organized society, all the things that we take for granted as part of our own society, it just gets much more complicated to try to think about was going on and it is easier to study them in the simplest context.

Changeux: I feel interested in the way you can test experimentally your model. Economics is certainly one of the simplest schemes. So, could you test your model, which is Darwinian, with others which are not Darwinian, for example Lamarckian, and decide for one or the other?

Boyd: There are a lot of data from disciplines other than economics, and the best come from studies on the diffusion of innovation in developing countries. I think that the Darwinian-like model fits the time pattern of diffusion much better than a Lamarckian model.

Huberman: I totally agree with you, but I think economics provides a very good example of an area in which to test some of these ideas. On the other hand, I can tell you that this kind of model will never be able to account for economic behaviour because agents in economics, for example and perhaps also in culture, usually choose among strategies on the basis of expectations about the utilities of these strategies in the future. This means that the dynamics are essentially controlled by equations in which the present depends on the future, if you were able to write them down. That is not so in those simplistic equations.

Boyd: Expectations are based on some past data. There is no way that the present can be caused by the future. Any economic actor has to make forecasts about the future behaviour based on past data. You can model this process. Usually, the actors in such models come up with their expectations in response to the particular histories to which they are exposed.

Changeux: Why do you reject this?

Boyd: Because it seems to me that such equations do not model the actual situation. One way to model the generation of expectations is to suppose that people acquire beliefs from other members of their social group.

Changeux: But you can generate beliefs from what is stored in the long-term memory. I think that an important component is the entry in the long-term memory, and this is the place where a selection is taking place in human beings. The storing in long-term memory is something which is highly select-ive. What I want to say is that you can build expectation and anticipation from long-term memory. A given human community may share the same long-term memory store and make anticipations, and selections on these anticipations in a similar way.

Mishkin: I would be interested in any speculation or hypothesis that you have about the interaction between cultural evolution and genetic evolution.

Boyd: There are at least two issues that are involved: the first is how should natural selection shape the psychological mechanisms that cause people to differentially adopt some cultural variants rather than others? A second question is for particular characteristics, say foraging behaviour, how does cultural transmission affect genetic variation. I will deal with the second question first. It seems to me that a lot of people have said that genetic evolution has been slowed down by cultural evolution. For example, we acquire our subsistence behaviours by social learning, rather than acquiring such knowledge genetically. Monkeys in different parts of the world may have different diets because they differ genetically, while people in different parts of the world have different diets because they learnt them socially. To the extent that this is true, selection will no longer cause much genetic change directly affecting diet because most of the variation in diet is not genetically heritable. On the other hand, cultural adaptation can give rise to rapid change in aspects of phenotype which serve as an environment of the genes and, as a result, culture can give rise to very rapid genetic change. There are a couple of famous examples. The first is lactose intolerance. Europeans maintain the ability to digest lactose into adulthood, while most people in the rest of the world, like other mammals, do not have the necessary enzyme as adults. The most plausible story is that dairy farming is a culturally evolved trait that changed the environment in which genes affecting lactose absorption were selected. In places where it was possible to store fresh milk, the ability to digest lactose was useful, which, in turn, led to selection for the ability to digest lactose and rapid genetic change. The second example concerns malaria. Some people argue that malaria was not prevalent in Africa until forest clearing for agriculture began 2000 or 3000 years ago. If this story is correct, then the evolution of the genetic disease, sickle cell anaemia, in West Africa was caused by a cultural change, namely, the knowledge of steel tools which made it economic to clear large areas of tropical forest.

Now let me turn to the first question: How should natural selection shape the genes that affect social learning. In what kinds of environment is social learning favoured? Under what circumstances are you better off acquiring things socially, rather than genetically? What our model says, which I think makes sense, is that when environments change very rapidly, culture is no longer any good, because it is not possible to track the environment; the best that you can do is stay at the long run mean. Genetic evolution accomplishes this because it responds so slowly. When the environment is changing very slowly, genetic evolution also does well, because natural selection allows the population to adapt to environmental changes. However, when the rate of environmental change is intermediate, natural selection favours cultural adaptation because it can track environmental changes that are fast, but not too fast. A second example is illustrated by the diffusion of innovations. Empirically, there are two things that determine whether an innovation is adopted: one is the utility of the innovation and the second is the prestige of the demonstrator. If you want something to diffuse into a particular area, you have got to convince the local elite to adopt those variants. We have modelled this process and both these rules make sense from an adaptive point of view. Natural selection will favour forms of social learning that are structured so that people are more likely to acquire traits that are adaptive in the local environment. A final example is that our calculations predict that people who are exposed to a number of different cultural variants should not imitate at random, but should preferentially imitate what the majority of people are doing. Learning and other mechanisms often cause adaptive beliefs to be more common and, thus, imitating the majority is, on the average, a cheap mechanism of increasing the chance of acquiring adaptive beliefs. It is interesting that adaptive mechanisms of cultural inheritance such as imitate the majority or be more influenced by successful individuals give rise to processes of cultural evolution that lead to outcomes that you would never predict if you just ask what maximizes fitness. Each mechanism has positive effects, but they also lead to dynamic outcomes that have no analogues in genetic evolution.

Tobias: I wonder what you think about the idea which occurred to me some time ago when I was investigating the Kalahari San Bushmen during the 1950s and 1960s. It seems to me that there were certain features where natural selection clearly would have a role, and there were features which also had cultural selection operating on them. I am not using the Darwinian term 'sexual selection', but cultural selection deliberately. And I suggested then that there would be a potential for a great acceleration of evolution if natural selection and cultural selection pulled in the same direction. For instance, the advantage of steatopygia. Steatopygia presents certain selective advantages it seems, in relation to females and child-bearing, and during dry periods and drought and so on. But also, the men in that society said 'we like our women that way'. They do not like flat-bottomed women, and in a society where there is a high rate of divorce, and some women make weak contribution to the next generation and others not so, that factor maybe quite marks in the production in such features. Another one was less convincing but also a possibility. It is the small size of the Bushmen. The fact that they are small gives a great advantage in hunting in the Kalahari environment which had stunted semi-desert shrubs. But also, if there was a cultural selective advantage among the people, we like little men, we like little people, we like little women, it could well be that under those circumstances where natural selection and cultural selection were pulling in the same direction, you would have a potential for great expertising of evolutionary changes.

Boyd: I think that your hypothesis is correct. As I understand it, the human species is an extremely polytypic species, especially with regard to morphology, and yet we have only been spread across the globe for 1 or 2 hundred thousand years. The conventional dogma is that Darwin's argument was that he could not think of any reasons for these phenomena and so he proposed sexual selection. It seems to me that you could make the same argument for culture that rapidly gave rise to phenotypic differences.

11

The origins of cultural diversity
JANUSZ K. KOZŁOWSKI

Introduction

The aim of this chapter is to present hypotheses concerning the significance of the differentiation of various types of archaeological records and their interpretation in terms of the processes taking place in the anthropological sphere, i.e. in the 'living' culture of prehistoric man. In particular, we shall focus on the reasons for the diversity of cultural artefacts both of those independent of humans (associated with broadly understood ecological factors) as well as of man-dependent diversities which may have an arbitrary or symbolic character. The identification of the latter may indicate a relationship between the diversity of archaeological remains and the emergence of ethnic groups (bound by a common language), cultural groups (linked by a common tradition of behaviours and implements), or socio-political groups (bound by social organization).

Obviously the earlier the stage of human culture, the more limited is the range of this culture which is reflected in archaeological remains. This results not only from selective conservation of remains but also from the gradual expansion of the material base of human culture, namely, utilization of new raw materials, broadening of the basis of subsistence economy, a richer gamut of activities creating man's artificial environment (fire, dwellings, etc.), and the appearance and development of symbolic behaviours. In the outcome of the impact of all these factors the nature of fossil relics, as these evolved, became increasingly complex. At the same time, each cultural sphere shows differentiation determined either by man or resulting from ecological conditions, which require separate models of interpretation in anthropological terms.

Physical differentiation of man vs. diversity of archaeological records

A number of early archaeological syntheses assumed that particular stages of man's philogenesis corresponded to definite types of cultures demonstrated as units of differentiation of lithic tool kits. Such a view seemed especially well justified by the unilinear conception of the evolution of man. It held that the major stages of man's evolution (*Australopithecus, Homo erectus, Homo neanderthalensis, Homo sapiens*) coincided with arbitrarily isolated stages of the evolution of man's culture, i.e. the Lower Palaeolithic was synchronous with the first two types of man, and the Middle and Upper Palaeolithic with the last two.

The relinquishment of the concept of a unilineal evolution in favour of poliphyletic development questioned the possibility as well as the need for distinguishing cultural stages, in particular, the division into the Lower and the Upper Palaeolithic. The poliphyletic concept shows, in fact, that later forms of australopithecines were contemporaneous with early *Homo erectus*. In addition, the evolution of archaic modern man was synchronous with the evolution of Neanderthal man who persisted and coexisted with *Homo sapiens sapiens* proper. Similarly, it has been shown that assemblages regarded as typical of the Middle Palaeolithic appeared, in reality, very early on and for several hundred thousand years had been contemporaneous with the assemblages regarded as typically Lower Palaeolithic.

The question is, therefore: Can a certain physical type of man's ancestors be identified with a specific type of archaeological artefact? The present state of investigations into this problem compels us to answer: No, it cannot. Let us try to corroborate this negative answer.

During the period of the earliest development of man's culture (2.6–2 million years BP) there were predominantly small inventories, which were exclusively lithic. These assemblages contained flakes, mostly fine, sometimes fairly irregular in shape, which came from the removal of cores, and choppers occurring only sporadically. The characteristics of these earliest stone inventories is the result of the preservation of individualized, separate occupation levels and the fact that flakes were poly-functional tools in contrast to choppers, which were more specialized and were used primarily for working with wood (Beyries 1986). The creators of the earliest inventories could well have been *Australopithecus africanus* as well as *Homo habilis*. In any case, it does not seem plausible to link tool-making activity with *Homo habilis* only.

During the period between 2 and 1.5 million years BP (myr BP) there was further growth in the production of choppers and chopping tools. In various proportions, these tools accompany flake implements and retouched flake tools. Some forms emerge which are the initial stages of bifaces and cleavers. The differentiation of lithic inventories does in no way correspond to the differences between the last australopithecines (*A. robustus* and *A. boisei*), and *Homo erectus*.

In the period between 1.5 and 0.3 myr BP, when the distribution of *Homo erectus* extends over the entire territory of the Old World (Fig. 11.1) the dissimilarities are still maintained between lithic inventories in respect of the frequency of choppers and chopping tools as compared to bifaces and cleavers, and both these groups in comparison to retouched flakes. We can then conclude that *Homo erectus* coincides with a variety of expressions of material culture. The determinants guiding these cultural manifestations are discussed next.

During the period between 0.3 and 0.05 myr BP a marked differentiation of anthropological types takes place. First, in Europe the evolution of the pre-Neanderthals begins, and then subsequently the Neanderthals (0.3–0.15 myr BP), still later (about 0.2–0.15 myr BP) archaic forms of modern man emerge in South Africa. So far, none of these anthropological types is associated with specific types of assemblages of stone artefacts or a definite technology. Both the pre-Neanderthals and Neanderthals may be equally connected with the Levallois and non-Levallois technologies and also with assemblages with bifaces or choppers accompanied by various proportions of flake tools. Archaic forms of *Homo sapiens*, as the examples from South Africa and the Near East have shown, may, in turn, occur with inventories representing the Mousterian or the Levalloisian technologies, just as do later forms of *Homo erectus* and European pre-Neanderthals and Neanderthals. An exceptional position belongs to the discoveries in South Africa from the period of about 120 000 BP which identified archaic modern man with a specific blade culture. Thus, the culture known as the Howiesons Poort, is also recorded at Klasies river mouth, had no predecessor. This assemblage is accompanied by fine geometrical forms (segments, trapezes), and the burin technique is also used. It should be stressed, however, that early blade cultures like these, although with different tools (possibly earlier than the isotope stage 5), are found in the Near East (level E, Tabun), where, at the moment, it is difficult to identify them with the archaic *Homo sapiens*.

It is only with the appearance in Europe of *Homo sapiens sapiens* between 50 000 and 30 000 BP that a distinct cultural milieu is identified, which is different from that created by *H. neanderthalensis* or archaic *H. sapiens*. So far, Cro-Magnon man can be identified only with the Aurignacian culture in the period between 50 000 and 30 000 BP, whereas all other synchronous lithic variants (the so-called eastern Micoquian, Mousterian, leaf-point industries, the Châtelperronian, etc.) were created by Neanderthals. Unfortunately, we do not know to what extent the situation found in Europe corresponds to that in other parts of the Old World where *Homo sapiens sapiens* also emerges in the period between 40 000 and 30 000 BP, although possibly not always in a new lithic context.

The significance of the diversity of lithic inventories in the period before 120 000 BP

This is a period when stone artefacts constitute practically the sole testimony of material culture, although there is no evidence of a symbolic culture. From the point of view of the expansion of the oikumene two periods can be distinguished: the African world up to the late palaeomagnetic Matuyama phase, i.e. up to about 1.3 myr BP, and later the expansion of the oikumene into the Eurasian territory. Then, a technological boundary can be accepted between the production of core tools and non-standardized flakes and the manufacture of flakes with predetermined shapes (the beginnings of the Levalloisian technique about 0.3 myr BP).

The shift of the *Homo erectus* oikumene to high latitudes depended on the adaptational potential to cooler climates. In other words, this was a break-through for the further development of culture as a means of adaptation to ecological conditions. Similarly, one must recognize the great importance of the beginnings of predetermined flake production: the blanks production cycle became very complex (pre-core treatment, core reduction) and was separated for the first time into phases which took place in different locations (workshops, living sites). It can be assumed that this tool production in separate sites required a system of communication which in turn suggests the existence of a language using abstract concepts of time and space.

In the period preceding the expansion of the oikumene by *Homo erectus* we find a varying intensity of occurrence of particular categories of artefacts in assemblages, alongside the general tendency towards the evolution of pebble tools into bifaces and cleavers, and the differentiation of retouching techniques and flake-tool forms. These diversities cannot possibly be ex-plained as being an expression of different cultural traditions. The fact that within the distribution range of these units there are no essential geographical differences, that they have been distinguished on the basis of differing arte-fact frequencies rather than on a total exclusion of a certain category, disclaim their interpretation as being different cultural traditions. It is important, at the same time, to add that analysis of the correlation between the various intensities of occurrence of major stone artefact categories and elements of the natural environment, which was performed by Clark and Kurashina (1976), has not revealed statistically significant dependencies (Fig. 11.2). Such relationships appear only after 1.3 million years. On the other hand, the early sites frequently show spatial differentiation of the scatter pattern of major artefact classes. The use/wear of these artefacts indicates important differences between core and flake implements. Primarily, the difference has a functional character associated with the varying intensity of basic activities in time and space (temporarily occupied sites point to seasonal differentiation of these activities, whereas the differences within the sites themselves may mean that work has a sequential character, or even indicate the beginnings of the division of work).

After the oikumene spread into high latitudes a distinct line of demarcation can be recorded between the inventories with bifaces (Abbevillian–Acheulian) and the inventories with choppers and flake tools (Clactonian). The second demarcation line separates the inventories with large, mainly bifacial tools (the Abbevillian–Acheulian complex) and the microlithic inventories (in Europe: the 'Buda culture'). The types of inventories listed above are not distinctly territorially limited. At most, they can show some tendencies to group in climatic-ecological zones both synchronically as well as diachronically. J. Svoboda (1986) examined these tendencies in detail and quite rightly pointed out that assemblages of small tool-making industries group in milder climatic periods and in temperate climate zones with abundant trees, where-as more standardized and technologically more progressive biface industries

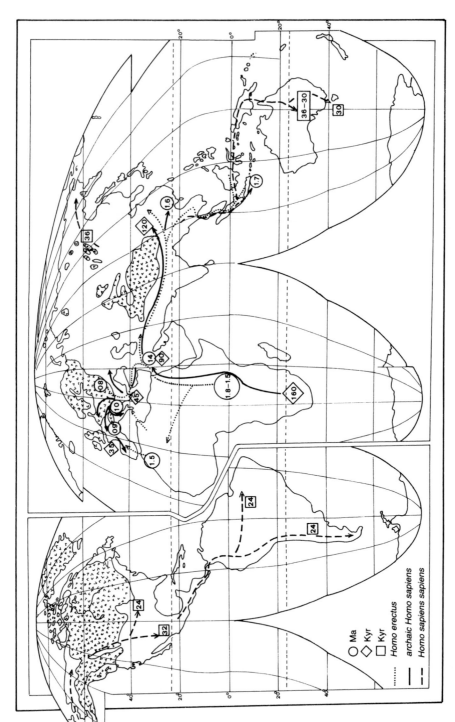

Fig. 11.1. The expansion of the oikumene of *Homo erectus*, archaic *Homo sapiens*, and *Homo sapiens sapiens*.

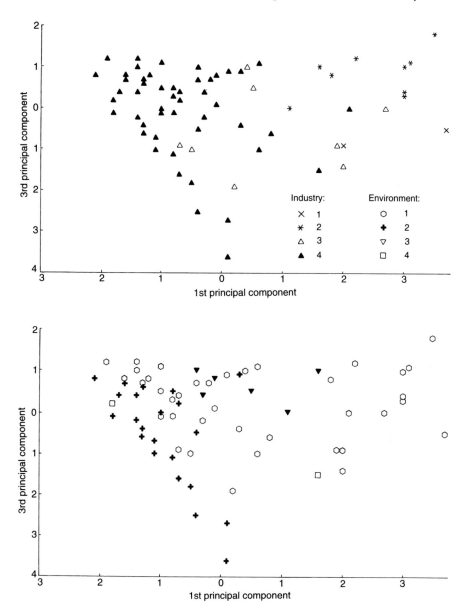

Fig. 11.2. Principal component analysis of the African Early Palaeolithic stone assemblages in relation to the natural environment: components 1 vs. 3. *Industry*: 1, Oldowan; 2, developed Oldowan; 3, lower Acheulian; 4, upper Acheulian. *Environment*: 1, open grassland; 2, deciduous woodland; 3, montane grassland; 4, semi-arid dry thorn-bush/steppe. (From Clark and Kurashina 1976.)

occur in cooler periods, mainly in a steppe environment, devoid of forest cover. These climatic-ecological diversities are increased by the differences related to the availability of raw material which must have influenced the size of artefacts and applied technologies.

Finally, it should be stressed that already during the period under discussion (between 1.3 and 0.3 myr BP) some of the technological differences may have resulted from the isolation of particular groups of *Homo erectus* caused by natural geographical boundaries or certain types of geographical environment (e.g. forest). Such isolation restricted the communication and exchange of information, therefore, possibly stimulating the emergence at the end of this period, of first time-space-limited events in chipped-stone technology and tool morphology.

The following period (0.3 to 0.12 myr BP) is characterized by the introduction of technology associated with core preparation and obtaining flakes whose shapes were predetermined and standardized. This innovation must have occurred simultaneously in a variety of regions, and certainly in Eastern Africa and Europe. In Europe, this was related to the progressive adaptation to cooler climatic technology conditions following the Holstein interglacial period. The diversity of assemblages at that time comprises either the occurrence or absence of the Levallois technique and a different intensity of the frequency of Lower Palaeolithic-type implements (initially, bifaces), and flake tools in that period already represent a typical Middle Palaeolithic tool-kit. If we assume that the expansion of the Levallois technique and its frequency were related to by the flow of information from hypothetical centres of this innovation and the availability of suitable high-quality lithic raw materials, then the occurrence of bifaces must have depended on some other factors whose nature still remains unknown. It is interesting that before 120 000 BP the greatest number of assemblages with bifaces is found in Western Europe and in the Caucasian Mountains, whereas after 120 000 BP a nearly homogeneous region of assemblages with bifaces (the Eastern Micoquian*) covers the entire north-eastern European Lowlands.

Assemblages with predominance of the Middle Palaeolithic flake tool-kit display, already in that period, variability similar to that in the following period between 120 000 and 50/40 000 BP

---

* The term 'Micoquian' has been used in two ways: firstly, to denote a specific facies of the western European Acheulian as defined by D. Peyrony (1932) on the basis of assemblage from the red clay layer from La Micoque Cave in the Dordogne. Secondly, the term was used in a different sense by G. Bosinski (1967) to describe the Middle Palaeolithic industries of central-eastern Europe characterized, besides Micoquian bifaces, by the presence of asymmetrical knives and some types of foliated implements. Recent investigations at La Micoque have questioned the existence of a genetic link between two units because of the very long chronological gap separating them. In this paper the term 'Micoquian' is used in the sense similar to that defined by Bosinski, but the adjective 'eastern' is added to distinguish it from the Micoquian *sensu* Peyroni. The Micoquian interpreted in this way is one of the few Middle Palaeolithic units with a homogeneous distribution zone described here as the 'Micoquian oecumene'.

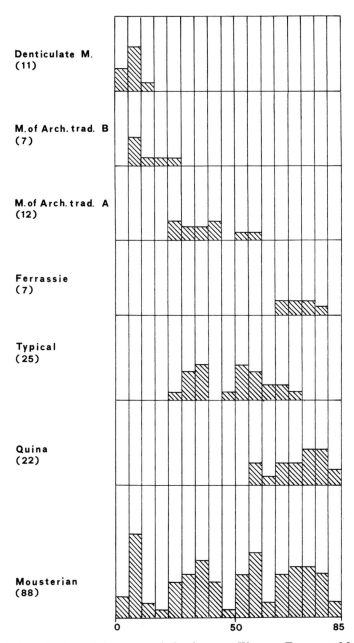

Fig. 11.3. Distribution of the scraper index in some Western European Mousterian assemblages: the main base for Bordes' hypothesis of the subdivision of the Mousterian complex into parallel 'cultures'. (From Bordes 1961.)

The significance of lithic inventory diversity in the period between 120 000 and 50/40 000 BP

The question of diversity in Middle Palaeolithic assemblages have frequently been discussed. F. Bordes (1961) established that within the 'Mousterian' assemblages the differentiation manifested itself as the discontinuity expressed in proportional percentages of major tool groups (side-scrapers, Mousterian points, denticulated tools, bifaces; Fig. 11.3). He interpreted this phenomenon as a development of parallel 'cultures' corresponding to ethnic-cultural groups. Bordes' paradigm has been compared to other interpretations of assemblage differences, notably with the functional interpretation by S. Binford and L. Binford (1966), which accounted for specific types of assemblages by the seasonal differences in activities of the same population groups, and secondly, with the diachronic model by Mellars (1969) who saw in the differentiation under discussion unilineal temporal succession. Subsequently, other feasible explanations have been added, such as the association of some facies with palaeoecological factors (Rolland 1977), or with the intensity of tool reshaping (Dibble 1986).

Recently, Otte (1989) and Rolland (1990) have finally resolved the issue of one-sided interpretations of 'Mousterian variability'. M. Otte has shown that a single human group was fully capable of producing diagnostic forms of 'distinct traditions', i.e. 'Quina' side-scrapers, Levallois *débitage*, and bifaces. N. Rolland and demonstrates, on the other hand, that in reality the statistical discontinuity and the polymorphism of the classification of the 'Mousterian complex' is illusory. Multivariate simulation using a computer reduces this complex to a continuing phenomenon.

In the light of the above discussion, it could be assumed that either the classification criteria used for the description of 'Mousterian' assemblages are inadequate (or simply incorrect), or that in the period under discussion the dissimilarities between stone assemblages corresponding to different ethnographic groups had not become isolated. In any case, the technological–morphological differentiation which can now be seen in assemblages of stone artefacts in the outcome of a complex range of factors, such as the type of lithic raw materials, the position of a given assemblage in the dynamic sequences of processing and use of tools, and the articulation of these sequences in time and space.

At the same time, investigations into certain types of tools in the period under discussion (e.g. leaf points, choppers, and some side-scraper knives) reveal their distinct regional distribution (Fig. 11.4). It is possible that these forms, together with other discrete 'stylistic' criteria, might be an expression of a specific 'iconological style of variability of lithics' (Sackett 1982). This style would constitute an initial manifestation of true diversity between human groups. Whether this view is correct or not can only be shown by further detailed investigations into the morphology of artefacts from this period.

It is important to add that in the period between 120 000 and 50/40 000 BP,

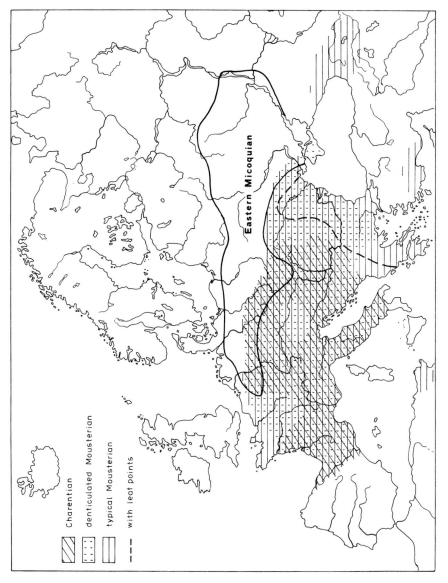

Eastern Micoquian

Charentian

denticulated Mousterian

typical Mousterian

with leaf points

Fig. 11.4. Distribution of the major industrial variants of the Middle Palaeolithic in Europe (80 000–45 000 years BP).

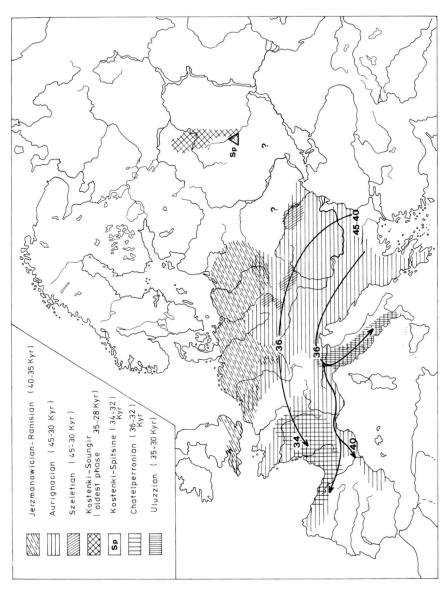

Fig. 11.5. Distribution of the major industrial variants of the Early Upper Palaeolithic in Europe (45 000–30 000 years BP).

Jerzmanowician–Ranisian (40-35 Kyr)

Aurignacian (45-30 Kyr)

Szeletian (45-30 Kyr)

Kostenki–Soungir (oldest phase 35-28 Kyr)

Kostenki–Spitsine (34-32 Kyr)

Chatelperronian (36-32 Kyr)

Uluzzian (35-30 Kyr)

Sp

other non-lithic records are extremely rare and cannot be taken into account in the discussion of assemblage variability.

Variability after 50/40 000 years BP

During this period (Upper Palaeolithic) and the appearance of *Homo sapiens sapiens*, two basic differences in tools are evident:

1. As a result of the common use of blade blanks, tools become conspicuously more standardized. Tools produced using all types of different retouching techniques show territorially limited distribution. Moreover, the morphology of these tools does not correspond to their functional differentiation.

2. With the mass appearance of artefacts which use other raw materials (e.g. bone) and the appearance of the first relics of symbolic behaviour (e.g. personal ornaments and figurative art) we can correlate particular types of objects. Occasionally, the extent of variability overlaps, in other cases, the diversities only cross each other, but generally they all have a limited regional character.

This evidence indicates that the Upper Palaeolithic variability may be, for the first time, the expression of ethnographically different human groups (Fig. 11.5).

Without doubt, the formation of this variability was a complex process. We can conjecture that its appearance, as mentioned earlier in this chapter, may have already occurred in the period just preceding 40 000 to 30 000 BP. This theory is supported by the fact that parts of assemblages still evident after 40 000 years had been made by the Neanderthals. These assemblages (e.g. the Châtelperronian, Bohunician, Uluzzian, Szeletian) show a spatio-temporal character similar to the contemporaneous objects made by the modern men. At the same time, they retained *fossile directeurs* known already from the Middle Palaeolithic milieu (e.g. backed points and leaf points).

A factor that also determined the Upper Palaeolithic variability was the natural environment (Fig. 11.6). Multivariate analysis of Upper Palaeolithic assemblages made by Dulukhanov *et al.* (1980) shows that the connection between lithic variability and palaeoecological factors was more evident during periods of extreme climatic change (the stadial periods), whereas it was weaker in more stable climates (interstadial periods). Additional factors influencing and shaping the Upper Palaeolithic variability were: a larger population density in some geographical micro-regions, the hunting of migrating animals which required either permanent or seasonal hunting teams, and other influences which increased the aggregational nature of the settlement network. Finally, certain types of individual communities were formed with the corresponding stylistic diversity of materials and symbolic culture. We could conjecture that the consolidation of these diversities was parallel to the diversities in communication, the first and foremost being language.

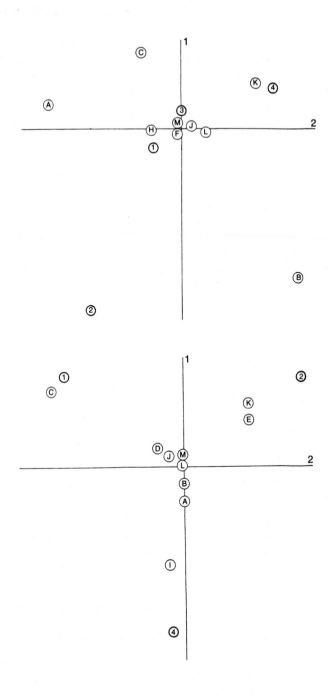

Fig. 11.6. Results of the factor analysis concerning the relationship between major tool classes and environments in the European Upper Palaeolithic. Upper diagram: Interpleniglacial; lower diagram: Upper Pleniglacial. Major tool groups: A, end-scrapers; B, other scrapers; C, burins; D, retouched truncations; E, retouched blades; F, perforators; G, composite tools; H, core-tools; I, leaf points; j, tanged points; K, backed blades and microliths; L, splintered pieces; M, others. Environments: 1, periglacial; 2, coniferous forest; 3, deciduous forest; 4, steppe. (From Dolukhanov *et al.* 1980.)

In contrast to earlier periods, the nature of Upper Palaeolithic variability requires the application of different models of interpretation. So much so that, in the Upper Palaeolithic where interpretation is much more straightforward, the concurrence of 'iconological style' may be explained as being the expression of phyletic ties in the diachronic sense or as migration. At the same time, the overlapping distribution range of various types of variability (economic, various kinds of material, and symbolic culture) strongly corroborate the interpretation of units distinguished in this way as not only ethnographic but also even socio-political.

References

Beyries, S. (1986). Evolution des comportements des premiers hominidés à travers l'étude des industries lithiques. *Ann. Fond. Fyssen* **2,** 63–73.

Binford, L.R. and Binford, S.R. (1966). A preliminary analysis of function variability in the Mousterian of Levallois facies. *Am. Anthropol.* **79,** 238–95.

Bordes, F. (1961). Mousterian cultures in France. *Science* **134,** 803–10.

Clark, D.J. and Kurashine, H. (1976). New Plio-Pleistocene archaeological occurrences from the Plain of Gadeb, Upper Webi Shabele Basin, Ethiopia, and a statistical comparison of the Gadeb sites with other Early Stone Age assemblages. In *Les plus anciennes industries en Afrique.* V. Congrès UISPP, Nice.

Dibble, H. (1986). Reduction sequences in the manufacture of Mousterian implements in France. In *Regional perspectives on Old World prehistory* (ed. O. Soffer), pp. 33–43. Plenum, New York.

Dolukhanov, P.M., Kozłowski, J.K., and Kozłowski, S.K. (1980). *Multivariate analysis of Upper Palaeolithic and Mesolithic stone assemblages.* PWN, Kraków.

Mellars, P. (1969). The chronology of Mousterian industries in the Perigord region. *Proc. Prehist. Soc.* **35,** 134–71.

Otte, M. (1989). The significance of the variability in the European Mousterian. In *The human revolution* (ed. P. Mellars), pp. 438–56 Edinburgh. University Press, Edinburgh.

Rolland, N. (1977). New aspects of the Middle Palaeolithic variability in Western Europe. *Man* **16.**

Rolland, N. (1990). La place de Charentien dans le cadre de la variabilité du Paléolithique moyen: nouvelle perspective. In *Les Moustériens charentiens* (ed. Comté National de l'INQUA), pp. 75–7. Brive.

Sackett, J. (1982). Approaches to style in lithic archaeology. *J. Anthropolog. Archaeol.* **1,** 59–112.

Svoboda, J. (1986). Early human adaptations in Central Europe. *Pamatký Archeologické* **77,** 466–86.

DISCUSSION

Participants: J. Chavaillon, Y. Coppens, H. Delporte, J.K. Kozlowski, P.V. Tobias

Coppens: There is general agreement among palaeoanthropologists that *Homo erectus* emerged about 1.6–1.7 million years ago. It seems that in the French Massif Central we have artefacts dated between 2 and 2.5 million years. So, I wonder why anthropologists keep on thinking that *H. erectus* expanded the Ocumène and not *H. habilis*? Following Phillip Tobias, it is clear that *H. habilis* underwent important anatomical changes and also behavioural changes (e.g. diet, meat-eating, tool-making, hunting, mobility, etc.). *Homo habilis* had all the biological abilities to exploit new environments.

Kozłowski: I do have any strong objection to the theory of the geographical expansion of *H. habilis*. However, there are several dubious elements about the European industries prior to 1.3–1.5 million years BP. It is the same for the Pakistan industries, which are supposed to be 2 million years old and, moreover, for the ones from China given as being older than 2 million years. However, we have now fairly reliable dates for several sites in southern Poland and Bohemia with an age of between 0.5 and 0.6 million years. The presence of industries in this region and in transcarpatian Ukraine (0.7–0.8 million years), in a cold steppe environment as indicated by fossil pollens, would indicate a significant period of adaptation which was necessary in order to live in Central Europe. So, as a prehistorian, I think that the present archaeological evidence remains uncertain but, in theory, the geographical expansion of *H. habilis* is quite possible.

Tobias: According to the work of Jean Chavaillon and Hélène Roche, the oldest tools in Africa go back 3 million years. But Jack Harris has told me that those dates are too late, there is nothing in Africa before 2.2 million years. This means that, if the dates in France are correct, the oldest African tools are contemporaneous with the oldest European tools.

Coppens: Jean Chavaillon and Harry Merrick have made excavations in levels which are 2.5 million years old or possibly more. We found tools on surface exposures in the same conditions, down to the very well-dated levels of 3 million years, including level B2 of the Omo Valley. We are not sure whether these artefacts come from inside, i.e. the level itself, or not. However, in the French Massif Central we have several sites 1.5 to 2 m.y. old that yielded artefacts (Chilhac, La Roche-Lambert, Les Etouaires, Le Coupet) but without absolute dating. The stratigraphic and absolute dating were made somewhere else, not in direct context of the finds. Anyway, I think that Africa was first.

Chavaillon: If there is no doubt about intentional character of the artefacts found at Chilhac, their stratigraphic position has to be reconsidered. They are associated with fauna aged 1.8 million years, but we do not know whether the tools were in place or descended from another level. Moreover, the dates were made in another site and stratigraphic work needs to be carried out in order to calibrate both sites. We encounter similar problems with other sites. So, I am ready to accept those very old ages, but we need more reliable studies. However, I would like to turn our attention to the successors of *H. erectus*. In Central Europe we have industries which are not quite Acheulian and I wonder if we can attribute them to the successor of *H. erectus*?

Coppens: We do not know where the transition is between *H. erectus* and *H. sapiens*; there are so many intermediary fossils. In fact, the sequence *H. habilis–H. erectus–H. sapiens* represents a clear case of gradual evolution. We can say that some fossils are *erectus*-like (Bilzingsleben) or *sapiens*-like (Vértesszöllös), but to put down limits does not make sense.

Delporte: I wonder whether the geological conditions of the different regions of the world have an effect on the preservation of the archaeological sites, i.e. that tectonic or seismic regions are more favourable (e.g. the French Massif Central, East and South Africa). I raise this question because the number of archaeological sites is the only way to estimate, even roughly, the palaeodemography. The population density was very low in lower palaeolithic. It started to increase very slowly during the Middle Palaeolithic and more steadily during the Upper Pleistocene (there were about 50 000 people in France). At the end of the Upper Palaeolithic we have a demographic explosion, and I think that this very important event had a strong influence on the processes described by Janusz Kozłowski.

Kozłowski: I agree completely. I am convinced that there was a sudden increase of population density between the Middle and Upper Palaeolithic. I do not think that these estimates are biased by any possibility of differential conservation of archaeological sites. In Central Europe we have numerous Plio-Pleistocene sites having a rich fauna in a very different geological context.

12

Individuals and culture
ROBERT A. HINDE

Introduction

This volume includes chapters on the evolution of the human brain, and on the genetic and environmental factors that contribute to its development in the individual. Other chapters have discussed how forces of natural selection could bring about the evolution of relatively elementary forms of culture and of cultural differences. This chapter is concerned with the relations between our biological heritage and the complexities of cultural beliefs in more complex societies. That, of course, is a tall order and my aim here is merely to sketch an approach to the problem (see Hinde 1991). This involves successively an emphasis on the need to understand social behaviour in terms of levels of social complexity, a brief reference to the nature of 'culture', a discussion of human universals, and then some examples of the dialectical relations between individuals, the successive levels of social complexity, and the 'socio-cultural structure'.

Levels of social complexity

Children grow up in a complex social environment, and development is constantly mediated by interactions with other individuals. It is largely through interactions with other individuals that the baby learns to act on objects, to regulate interpersonal behaviour and later, by internalization, his or her own behaviour (e.g. Vygotsky 1934). The development of social behaviour depends in large measure on the formation of relationships with others — a consequence of interacting with particular others on successive occasions so that each interaction is affected by earlier ones and by the expectation of future interactions (Hinde 1987). As a heuristic device, it can be said that the child forms 'working models' of relationships which guide its future behaviour (Bowlby 1969; Bretherton 1985; Stern 1985). These relationships are set within families or groups, such that each relationship influences and is influenced by other relationships (e.g. S. Minuchen 1974; P. Minuchen 1985; Hinde 1989).

Thus we must come to terms with a series of levels of social complexity — individuals, interactions, relationships, groups, and societies (Hinde 1987). Each of these levels has properties not relevant to the preceding one. For

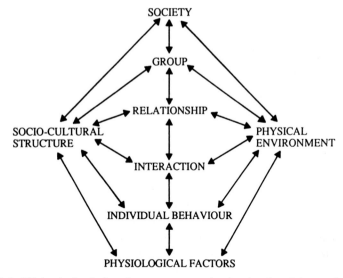

Fig. 12.1. Dialectical relations between successive levels of social complexity.

instance, the nature of a group varies with the patterning of the relationships within it — hierarchical, centripetal, etc., but such properties are irrelevant to particular relationships. And at each level, properties will depend on the adjacent levels: for instance, the course of an interaction depends both on the characteristics of the individuals involved and on the relationship in which it is embedded. Given such dialectical relations between levels, the levels must be thought of as processes continuously influenced through the agency of these dialectics (Fig. 12.1).

The socio-cultural structure

It is sometimes heuristically convenient to distinguish two processes in child development: (1) socialization (i.e. learning to go beyond a merely egocentric or ethnocentric view, and to live in co-ordination with others), and (2) acculturation (learning the specific values, beliefs, symbols, etc., of the culture in which the individual is living). The distinction, however, can never be absolute: the processes, nature, and degree of socialization depend on culturally constructed and learned forms, models, and conventions. The development of language, communicative skills, social cognition, and so on, are continuously subject to socio-cultural influences. These social and cultural influences on individuals are mediated primarily through relationships with particular others — parents, siblings, and so on. And the nature of the socio-cultural influences in any society depends in large measure on the natures of the individuals and relationships within the society. Thus, if we are to understand development we must come to terms not only with dialectical relations between successive levels of social complexity, but also between these and the 'socio-cultural structure' of the group or society in which the individual lives.

It may be helpful to say a few words about the way in which the terms 'culture' and 'socio-cultural structure' are used here. 'Culture' refers to those ways in which human social groups (or subgroups) differ that are communicated between individuals. For instance, the capacity for a spoken language is not a cultural phenomenon, but differences between languages are. It is acknowledged that there is a potential difficulty in putting the emphasis on differences because, by this definition, a characteristic possessed by all groups except one would be a cultural characteristic, whilst a characteristic possessed by every single group would not. However, this purely theoretical difficulty is of minor importance compared with those that arise from defining cultural characteristics as those acquired by a particular mechanism (such as learning). For instance, if there were common features of all languages that owed their existence to the fact that gravity acts 'downwards', or that causes precede effects (in so far as that is not a matter of definition), these features of languages would not be aspects of culture even though learning were involved in their acquisition.

'Culture' can be used in a descriptive sense to refer to the artefacts, values, customs, institutions, myths, etc., as described by an outsider (or indeed by a member of the society). Such a description, however, tends to imply a static whole external to individuals. In practice, culture is best viewed as existing in the minds (separately or collectively) of the individuals of a society and as in a continuous process of creation through the activities of individuals in their relationships. Of course, individuals differ somewhat in their perceptions of the customs, and so on, of their society and those differences can be an important source of creative change (cf. Goody 1987). Whilst 'culture' often refers to beliefs etc., common to all (or most) individuals, it must not be forgotten that aspects of the culture may be the special responsibility of particular categories of individuals. With this admittedly mentalistic view, the actual artefacts, institutions, myths, etc., are seen as expressions of the culture. Of course, they may in turn act back upon and influence culture in the minds of individuals.

The customs, institutions, myths, etc., of a society are interrelated in diverse ways. Thus, here the whole — the parts and relations between them — are referred to as the 'socio-cultural structure'. Again, the socio-cultural structure can be described, but it is not an entity and exists only in the minds of individuals (at least in non-literate societies: the advent of literacy is likely to be an important issue here, Goody 1968). Real-life relationships within the group may or may not have counterparts in the socio-cultural structure, and idealized relationships in the socio-cultural structure may be realized to only a limited extent in actuality. But the socio-cultural structure as perceived in the consciousness of individuals has a regulatory function. Thus the socio-cultural structure affects group structure, relationships, and individuals.

However, at the same time it is to be seen as continuously being created by the activities of individuals in their relationships.

Human universals

The conventional route to the analysis of behavioural development involves first the study of differences in order to implicate genetic or environmental factors in their genesis, and then the study of the mechanisms of action of the factors so isolated. In the absence of differences, naturally occurring or induced, development can be described but not understood. However, if we are to understand human diversity, we cannot rely solely on differences, because we must know something of the clay from which it is formed. We need some knowledge about what is common to all individuals as well as how they differ. In other words, we need to specify human universals.

The search for human universals, however, has a not entirely respectable history (Count 1973). There are several problems:

1. At what level of social complexity should we seek for universals? Some authorities focus on properties of individuals, others on properties of groups (e.g. hierarchical structures) and yet others on properties of the socio-cultural structure (e.g. ability to make fire, religious belief). However, universals at the more complex levels are related (albeit indirectly) to the psychological characteristics of individuals. In addition, it is difficult to imagine that a human individual brought up alone would make fire or elaborate a set of religious beliefs. What we must aim for, therefore, are aspects of human behaviour or psychology that may depend on features of the physical environment that all humans experience, but do not depend on interactions with other individuals less naïve than the individual in question. This implies nothing about the relative importance of genes and environment in their development.

2. A complete list of such characters is impracticable. Apart from the inherent difficulty in isolating 'characters', most of the items would be trivial.

3. In any case, adequate data for most items are lacking. Thus it is a reasonable supposition that all babies in all cultures show similar behaviour in searching for the nipple and sucking, but it has not been proven.

4. In practice, no character is absolutely constant across individuals: there is always some individual variation, and to that extent no universals. If variation is correlated with environmental or cultural differences, the character may be developmentally stable within the range of each environment or culture, but not across them.

In general, therefore, universals must be thought of as characters potentially present in all humans (or in all members of a particular age/sex class), actually present in most, but variable between individuals in some degree.

Given all these reservations, it may be said that human universals fall into the following categories.

1. Aspects of perception — for instance, differentiation between figure and ground, figure completion, and other Gestalt principles.
2. Motor patterns — for instance, walking, running, swallowing, the orientation reflex, and some expressive movements (e.g. smiling, laughing, crying) are basically similar in all cultures.
3. Responsiveness to some stimulus patterns — for instance, babies' responsiveness to loud noises, physical contact, loss of support, being left alone, and maternal responsiveness to some infant signals.
4. Motivation — for instance motivations to eat, drink, and behave sexually and parentally. Less obvious but potentially present ubiquitously, although developed to different extents in different cultures, are aggressive, assertive, acquisitive, and prosocial motivations, and a number of others. (The difficulties of defining motivation and differentiating between motivations are deliberately neglected here.)
5. Cognitive processes. Here the issues are more difficult, in part because cognitive processes also never develop in a social vacuum, and are never free from socio-cultural influence. Nevertheless, there is general agreement that there must be basic similarities in cognitive processes in all humans (Cole and Scribner 1974).
6. Predispositions to learn. Individuals are more prone to learn some associations than others; some skills are easier to acquire than others. In some cases, at least such predispositions are likely to be pan-cultural.
7. Certain aspects of language development, such as the sounds made in infantile babbling, the order of appearance of the sub-elements of phonemes, and the order of development of complex utterances.
8. Certain aspects of relationships. Thus the mother–infant relationship has common features across all cultures and, although they differ markedly in many respects, certain aspects of male–female relationships have considerable cultural generality.

The dialectics

We have seen that human development is profoundly influenced by aspects of the socio-cultural structure and that that structure must itself be created by the basic characteristics of individuals. We may now consider some examples of how this may work in practice.

Fear of snakes

A fear of snakes is extremely widespread, both amongst individuals within societies and across societies, but it is also variable between individuals (Hebb

1949; Marks 1987). A number of lines of evidence indicate that humans are predisposed to acquire a fear of snakes.

Children brought up in an institution, who have never seen a snake, show little fear if they first encounter one at one and a half or two and a half years, but avoid a snake crawling on the ground from about three years of age (Prechtl 1950; see also Jones and Jones 1928). Children also show spontaneous fears of other objects or situations which might have posed a real threat in our environment of evolutionary adaptedness, such as spiders, mammals, heights, darkness, falling, and being alone (Bowlby 1969). It is noteworthy that humans are much less prone to develop spontaneous fears of other situations that are genuinely lethal in modern societies, such as cars or bombs, and it is thus not unreasonable to suppose that a tendency to fear snakes, spiders, etc., is in part due to our biological heritage (Marks 1987). In harmony with this view, some early evidence suggests that fear of snakes shows a smaller tendency to decline with age than do other fears. Thus the percentage of 2-, 3-, 4- and 5-year-olds showing fear of a strange person were 31, 22, 7, and zero, and showing fear of dogs 69, 43, 43, and 12, but for snakes the percentages were 34, 55, 43, and 30 (Holmes 1936).

Anecdotal evidence suggests that the extent of the fear shown is much influenced by social referencing—the child looks at others, and especially at a trusted other, and models his or her response to the situation according to the response of that other (Emde 1980; Klinnert *et al.* 1983). Comparative data provides strong support for this view. For example, adult vervet monkeys give qualitatively different responses to three types of predator—leopards, snakes, and eagles. Young monkeys are more likely to give the correct response if they first look at an adult (Seyfarth and Cheney 1986). Even more interesting are data on rhesus monkeys (Mineka 1987) showing that:

1. Wild-reared rhesus monkeys tested in the laboratory nearly always show fear of snakes.

2. Lab-reared monkeys do not show fear of snakes.

3. Lab-reared monkeys shown a videotape of a wild-reared monkey showing fear of a snake become afraid of snakes thereafter.

4. Lab-reared monkeys shown a 'doctored' videotape of a wild-reared monkey apparently showing fear of a flower do not become afraid of either flowers or snakes.

There is thus clear evidence that rhesus monkeys have a propensity to fear snakes that depends for its full realization on the experience of seeing others respond fearfully to snakes. This in turn increases the plausibility of a similar explanation of snake fears in humans.

Some individuals develop snake phobias, showing a fear of snakes out of all proportion to the threat they present, a fear that is irrational and is beyond voluntary control. It is reasonable to suppose that the role of snakes as a symbol in our culture is related to these issues. Snakes play an important part, and have played an even more important part, in our mythology. In the myth of the Garden of Eden, in the Rubens paintings of snakes gnawing at the genitals of those cast down into Hell, snakes symbolize evil. If we are really to understand fear of snakes and the symbolic role of snakes, therefore, we must come to terms with a series of dialectical relations between the propensity to fear snakes, social referencing within relationships, snake myths, and the socio-cultural structure.

It is, of course, necessary not to push this sort of explanation too far. Just because snakes posed a real threat in our environment of evolutionary adaptedness (and in some societies still do), they may be a special case. While many symbols may have their origins in common human perceptions or in frequent or special experiences, they are not necessarily based on predispositions to learn. For example, lions have a significance common to many societies which presumably arises from common human perceptions of their qualities. And snakes have served as symbols not only of evil but also of prudence, wisdom, health and medicine, royalty, and so on — a diversity no doubt related to properties other than their danger — their characteristics of movement, stealth, the cobra's threat display, their ability to shed their skins, and so on (see e.g. Tervanent 1958–59).

Individualism vs. collectivism

Perhaps one of the most basic differences between cultural values concerns the relative emphasis placed on self-realization and self-advancement on the one hand, and the collective good on the other. To some extent the difference characterizes Eastern versus Western societies. Triandis (1991) describes the difference as one between 'collectivist' cultures, in which people are socialized to obey and conform to in-group norms and goals, and 'individualistic' cultures where individuals' goals are accorded priority and self-caring actions are accepted. Whilst every society has elements of each of these paradigmatic forms, at the present time Western nations are increasingly approximating the individualistic type, whereas Eastern nations are more collectivist. For instance, the Japanese term *ningen* means both humans and society: humans cannot be envisaged independently of society. And in discourse, the self is seldom referred to in the abstract: it cannot be without reference to a specific other, and changes depending on the specific other (Ohnuki-Tierney 1987).

Of course, the distinction between individualism and collectivism involves a difference between cultural values in the most general terms. We may ask, however, whether it is reflected in child-rearing practices. A number of authors have in fact suggested that North American mother–child interactions foster infant independence more than those of Japanese mothers, while Japanese mother–child interaction is oriented more towards cementing the mother–child relationship, and on inter-dependent relationships between the child and the members of a family or society (e.g. Caudill 1973; Stevenson

1991; Triandis 1991). Fogel *et al.* (1988) found that, in face-to-face interaction with their 3-month-old infants, Japanese mothers used more non-verbal communication and touching than American mothers. In a detailed study of mother–infant interaction in France, Japan, and the United States, Bornstein *et al.* (1992) found no differences in maternal responsiveness to infant vocalizing, but marked differences to infant looking. American mothers responded more to their 5-month-old infants looking at objects, and tended to direct their attention to objects, while the Japanese mothers responded more to infants looking at them, and directed their children's attention to themselves. They also found that, at 5 and 13 months, US mothers' speech and play was more information-oriented and dialectic, while Japanese mothers are more affect-oriented, empathic, and playful. These differences are consonant with the view that US mothers foster independence, while Japanese mothers tend to consolidate the mother–child relationship. The socio-cultural norms both affect and are affected by relationships in the society. (These data must be contrasted with the finding that, while mothers' dialectic interactions with their infants predict children's cognitive performance (Bornstein 1985; Tamis-Lemonda and Bornstein 1989), in adolescence the performance of Japanese children appears to be superior to that of US children and this appears to be related to a greater emphasis on academic matters by Japanese parents (Stevenson and Lee 1990).)

Gender differences

A third example concerns the genesis of gender differences in behaviour in close relationships. Although early anthropological studies claimed almost complete flexibility in the relative personalities and social roles of males and females (Mead 1935), in practice the directions of the differences (but *not* the extent or patterning) are similar over a far wider range of societies than could be accounted for by chance. Ubiquitously, or at least almost ubiquitously, women are more nurturant and obedient (see e.g. Williams and Best 1982; Peplau 1983). Other characteristics on which the direction of sex differences seems to be consistent across societies will be mentioned shortly, but it must be emphasized that, of course, we are speaking of central tendencies, and in every society there is enormous overlap between men and women: there is no suggestion that such characteristics are immutably fixed.

Three lines of evidence suggest that the direction of the mean differences reflects a biological universal.

1. There is strong evidence that comparable differences in rhesus monkeys stem from the influences of hormones prenatally (Goy 1978). There even appear to be male/female differences in the lateral asymmetry of androgen receptors in the foetal brain (Scholl and Kim 1990). Some evidence indicates that similar effects occur in humans (Money and Ehrhardt 1972; Meyer-Bahlburg *et al.* 1984). This of course in no way denies the importance of

postnatal experiential influences: indeed, in monkeys group composition (Goldfoot and Wallen 1978) and in humans assigned gender at birth (Money and Ehrhardt 1972) are known to be important.

2. Comparative evidence of the genital anatomy and physiology and the socio-sexual behaviour of the great apes indicates a close relation between these characteristics. Using similar principles for great apes and humans, human sex-size differences, genital anatomy, and physiology are in harmony with the scenario that in our environment of evolutionary adaptedness males competed for mates, mating was mildly polygynous, males had undisputed access to one or more females, male assistance was important in child-rearing and sex had a bond-maintaining as well as a reproductive function (Short 1979). (That there are other primate species which have monogamous bonds, but in which sex does not have this role, does not necessarily disprove the last point.)

3. Over a wide range of species, males' reproductive success is limited by the number of females they can fertilize, female reproductive success by the number of young they can rear. Since a male could have been cuckolded while a female could not, and since females must nurture the offspring at least until parturition, loss of an infant is more important to a female than to a male. This is in harmony with evidence indicating that male promiscuity is institutionalized in most societies, whereas female promiscuity rarely is; and with the fact that, at any rate in Western cultures, male–female relationships (as opposed to interactions) are more important to females than to males (Alexander 1980; Hinde 1984).

I have emphasized that these are differences only in central tendencies, and that there is, in practice, always enormous overlap. But in Western cultures at least, the difference is often *perceived* as absolute: the stereotypes portray men and women as differing absolutely. It can reasonably be suggested that this exaggeration of the real differences stems from basic psychological processes operating also in other contexts. Thus, when individuals perceive themselves as part of a group and dependent upon other group members, they immediately tend to exaggerate the difference between the in-group and the out-group, and usually to perceive the in-group as superior (Tajfel 1978; Rabbie 1991). As part of the development of the child's self-concept he or she comes to perceive him- or herself as belonging to male or female group, and to exaggerate the differences between the genders. Furthermore, pre-pubertally (at least) it is possible to rise in status amongst same-sex members by denigrating members of the other, and postpubertal individuals tend to exaggerate in themselves the characteristics that they deem to be attractive to the opposite sex (Hinde 1987).

Finally, we must ask why the degree of exaggeration, and its patterning amongst the various gender characteristics, varies so much between cultures. Presumably, influences from other aspects of the culture come into play. For instance, the Manus emphasize the differences between the sexes and also many other distinctions — between themselves and outsiders, between parts of their village (regarded as safe) and others (not safe). The Trobrianders, by contrast, have much greater sexual equality and emphasize distinctions in other aspects of their lives far less (Barnett 1969). Many other examples of the manner in which different aspects of the socio-cultural structure affect each other could be cited.

Conclusion

The thesis put forward, then, is that if we are to understand human social behaviour and human diversity, we must come to terms with successive levels of social complexity, strive to understand the dialectic relations between them, and between them and the socio-cultural structure, and not neglect the human universals which, in interaction, give rise to the diversity we perceive.

References

Alexander, R.D. (1980). *Darwinism and human affairs*. Pitman, London.

Barnett, L.E. (1969). Concepts of the person in some New Guinea societies. M.Phil. thesis, University of London.

Bornstein, R.H. (1985). How infant and mother jointly contribute to developing cognitive competence in the child. *Proc. Natl Acad. Sci. USA* **82**, 7470–3.

Bornstein, R.H. *et al.* (1992). Maternal responsiveness to infants in three societies: the United States, France and Japan. *Child Development,* **63**, 808–21.

Bowlby, J. (1969). *Attachment and loss*, Vol. 1. *Attachment*. Hogarth, London.

Bretherton, I. (1985). Attachment theory: retrospect and prospect. In *Growing points of attachment theory and research* (ed. I. Bretherton and E. Waters), pp. 1–2. Monographs of the Society for Research in Child Development, 209.

Caudill, W.A. (1973). The influence of social structure and culture on human behaviour in modern Japan. *J. Nerv. Ment. Dis.* **157**, 240–57.

Cole, M. and Scribner, S. (1974). *Culture and thought: a psychological introduction*. Wiley, New York.

Count, E.W. (1973). *Being and becoming human*. Van Nostrand, New York.

Emde, R.N. (1980). Levels of meaning for infant emotions. In *Minnesota Symposia on Child Psychology* (ed. W. A. Collins), Vol. 13, pp. 1–37. Erlbaum, Hillsdale, NJ.

Fogel, A., Toda, S., and Kawai, M. (1988). Mother–infant face-to-face interaction in Japan and the United States. *Dev. Psychol.* **24**, 398–406.

Goldfoot, D.A. and Wallen, K. (1978). Development of gender role behaviours in heterosexual and isosexual groups of infant rhesus monkeys. In *Recent advances in primatology* (ed. D.J. Chivers and J. Herbert), Vol. I. Academic Press, London.

Goody, J. (1968). *Literacy in traditional societies*. Cambridge University Press, Cambridge.

Goody, J. (1987). *Introduction to Barth. Cosmologies in the making: a generative approach to cultural variation in Inner New Guinea*. Cambridge University Press.

Goy, R.W. (1978). Development of play and mounting behaviour in female rhesus virilized prenatally. In *Recent advances in primatology* (ed. D.J. Chivers and J. Herbert), Vol. I. Academic Press, London.

Hebb, D.O. (1949). *The organization of behaviour*. Wiley, New York.

Hinde, R.A. (1984). Why do the sexes behave differently in close relationships? *J. Soc. Pers. Relation.* **1**, 471–501.

Hinde, R.A. (1987). *Individuals, relationships, and culture*. Cambridge University Press, Cambridge.

Hinde, R.A. (1989). Reconciling the family systems and the relationships approaches to child development. In *Family systems and life-span development* (ed. K. Kreppner and R.M. Lerner), pp. 149–64. Erlbaum, Hillsdale, NJ.

Hinde, R.A. (1991). A biologist looks at anthropology. *Man*, **26**, 583–608.

Holmes, F.B. (1936). An experimental investigation of a method of overcoming children's fears. *Child Dev.* **7**, 6–30.

Jones, H.E. and Jones, M.C. (1928). Motivation and emotion: fear of snakes. *Child. Educ.* **5**, 136–43.

Klinnert, M.D., Campos, J.J., Sorce, J.F., Emde, R.N., and Svedja, M. (1983). Emotions as behaviour regulators: social referencing in infancy. In *The emotions* (ed. R. Plutchek and H. Kellerman), Vol. 2. Academic Press, New York.

Marks, I.M. (1987). *Fears, phobias, and rituals*. Oxford University Press.

Mead, M. (1935). *Sex and temperament in three primitive societies*. Morrow, New York.

Meyer-Bahlburg, H.F.L., Feldman, J.F., Ehrhardt, A.A., and Cohen P. (1984). Effects of prenatal hormone exposure versus pregnancy complications on sex-dimorphic behaviour. *Arch. Sex. Behav.* **13**, 479–95.

Mineka, S. (1987). A primate model of phobic fears. In *Theoretical foundations of behaviour therapy* (ed. H. Eysenck and I. Martin). Plenum, New York.

Minuchin, P. (1985). Families and individual development: provocations from the field of family therapy. *Child Dev.* **56**, 289–302.

Minuchin, S. (1974). *Families and family therapy*. Harvard University Press, Cambridge, MA.

Money, J.W. and Ehrhardt, A.A. (1972). *Man and woman, boy and girl*. Johns Hopkins University Press, Baltimore.

Ohnuki-Tierney, E. (1987). *The monkey as mirror*. Princeton University Press, Princeton, NJ.

Peplau, L.A. (1983). Roles and gender. In *Close relationships* (ed. H.H. Kelley *et al.*). Freeman, New York.

Prechtl, H.F.R. (1950). Das Verhalten von Kleinkindern gegenüber Schlangen. *Wiener Zeits. f. Phil., Psych. and Paed.* **2**, 68–70.

Rabbie, J.M. (1991). Determinants of instrumental intra-group competition. In *Trust, co-operation, and commitment* (ed. J. Groebel and R.A. Hinde), pp. 238–62. Cambridge University Press, Cambridge.

Seyfarth, D.M. and Cheney, D.L. (1986). Vocal development in vervet monkeys. *Anim. Behav.* **34**, 1640–58.

Sholl, S.A. and Kim, K.L. (1990). Androgen receptors are differentially distributed between right and left cerebral hemispheres of the fetal male rhesus monkey. *Brain Res.* **516**, 122–6.

Short, R. (1979). Sexual selection and its component parts somatic and genital selection, as illustrated by man and the great apes. *Adv. Study Behav.* **9**, 131–58.

Stern, D. (1985). *The interpersonal world of the infant*. Basic Books, New York.

Stevenson, H.W. (1991). The development of prosocial behaviour in large-scale collective societies. In *Trust, co-operation, and commitment* (ed. J. Groebel and R.A. Hinde), pp. 89–105. Cambridge University Press, Cambridge.

Stevenson, H.W. and Lee, S.-Y. (1990). Contexts of achievement. *Monog. Soc. Res. Child Dev.* Serial No. 221, **55**, 1–2.

Tajfel, H. (1978). Contributions to *Differentiation between social groups* (ed. H. Tajfel). Academic Press, London.

Tamis-Le Monda, C.S. and Bornstein, M.H. (1989). Habituation and maternal encouragement of attention in infancy as predictors of toddler language, play and representational competence. *Child Dev.* **60**, 738–51.

Tervanent, G., de (1958–59). *Attributs et symboles dans l'art profane, 1450–1600.* Droz, Geneva.

Triandis, H.C. (1991). Cross-cultural differences in assertiveness/competition vs. group loyalty/co-operation. In *Trust, co-operation, and commitment* (ed. J. Groebel and R.A. Hinde), pp. 78–88. Cambridge University Press, Cambridge.

Vygotsky, L.S. (1934). *Thought and language.* MIT Press, Cambridge, MA.

Williams, J.E. and Best, D.L. (1982). *Measuring sex stereotypes.* Sage, Beverly Hills.

DISCUSSION

Participants: J.-P. Changeux, R.A. Hinde, R.L. Holloway, G.L. Lewis, M. Mishkin, L. Weiskrantz

Changeux: Why is there cultural diversity? Diversification of cultures, according to some anthropologists, originated before *Homo sapiens*, perhaps already with *Homo erectus*.

Hinde: Certainly, we get diversification of cultures with very small populations in non-human primates, which are imposed by local ecological conditions.

Changeux: You have mentioned that universals are not hard-wired, and then you mention motor patterns. There is some contradiction here.

Hinde: Some of them are, some of them are not. The motor patterns of smiling and crying are ubiquitous, but how much they are expressed in different societies is a cultural matter. They are ubiquitous but they vary between individuals and between societies.

Holloway: I enjoyed listening to that long list of universality of gender differences. In both humans and macaques, researchers found in the hypothalamus differences between males and females in the pre-optic nucleus and the size of the nucleus. Then, they found differences related to sexual dimorphism of the corpus callosum, which I think is the first demonstration of the possibility of anatomical hard-wired differences between both sexes, especially in the great body of fibres that joins to the cerebral hemispheres, and these might be involved in cognitive activities. There are many problems with these studies. What seems to be useful in these studies, however, is that even if these differences are hard-wired, they can be wired again or rewritten by cultural norms. Interestingly enough, we do not get these sexual dimorphic differences in monkeys at all within the corpus callosum, and it is not clear if they are in apes either. So, there is an interesting possibility that, at least in part, they might be a species-specific characteristic unique to *Homo*.

Hinde: There are, however, other brain differences between males and females in monkeys. For instance, there are differences in the distribution of androgen receptors between males and females in the fetus of rhesus monkeys.

Mishkin: I would like to make a statement about the gender differences beyond the hypothalamus in macaques. As a matter of fact, in the last several years there have been a few studies which have shown sex differences in learning ability, and we know some of the neural substrata in the developing monkey that are in place prenatally at about three or four months of age. Female infants learn visual discrimination habits more quickly than normal males. This can be reversed with appropriate hormonal treatment. Furthermore, because females seem to mature somewhat earlier (about two months), those females are impaired by removal of tissue that we know to be responsible for this developing learning capability. Male monkeys are not affected by the same lesion. There is more evidence from similar studies showing gender differences through reversal of hormonal treatment or differential effects of lesions. There are two or three examples in the literature on sex differences, apparently at the level of the cerebral cortex, even in the developing monkey.

Holloway: When we had completed our study on the structure of the corpus callosum in macaques, and in particular in the splenium, we could not find sexual dimorphic differences. In fact, we could not find it in any of the New World monkeys either.

Weiskrantz: The existence of structural differences does not necessarily lead to the kind of differences that are being discussed here. As far as the corpus callosum is concerned, there is a much bigger difference between left-handers and right-handers. Of course, left-handers and right-handers were often at war with each other and experienced other cultural difficulties, but they are different from the sexual ones.

Lewis: I am in accord with your basic position about propensities that are universal, and what interests me is the degree to which diversity in human culture can override or counter those propensities. About gender differences, if you were to consider Papua New Guinea, you would have an extraordinary range of different constraints and rules that govern gender differences and sexual behaviours, including some which seem remarkable in the extreme constraint they put on, for example, contact between men and women, and the age at which it starts. With regard to the sorts of things you are describing, for example, the fear of snakes, what interests the anthropologist is the degree of diversity.

Hinde: Concerning the issue of snakes, of course, snakes are sometimes used to symbolize other things, like royalty and health and medicine. I do not mean to imply that all symbols are based on basic propensities of the sort to which I referred. The thesis that I am trying to put across, namely, that these human universal propensities are important, is not necessarily contravened by one or two exceptions. I agree that cultural issues shape them, distort them, exaggerate them, or minimize them. The message I want to get through to anthropologists is that they will not understand these rapid changes unless they take into account the universal propensities and the dialectic relations between them and the culture.

13

Man's intelligence as seen through palaeolithic art
HENRI DELPORTE

Introduction

The standard hypothesis suggests that man underwent an important 'transformation' at approximately 35 000 BP. This transformation was manifested by various phenomena among which were the following three:

(1) the appearance of *Homo sapiens sapiens* following *Homo sapiens Neanderthalensis*

(2) the transition from the *Middle Palaeolithic* to the *Upper Palaeolithic*; and

(3) the appearance of *prehistoric art*.

The question is to determine whether there was a correlation between the above phenomena (and others) and to what extent they illustrate any significant advance in the evolution of human intelligence.

Intelligence

The concept of intelligence is obviously extremely complex. If we keep to the elementary dictionary definitions, *intelligence* is: 'the ability to know and understand' (the intellectual ability), i.e. 'all of the functions involved in knowledge (sensory perception, associations, memory, imagination, understanding, and reasoning)'. Similarly, *sensitivity* is defined 'as the set of affective phenomena', and *will* as 'pertaining to all the phenomena involved in movement and action'. The dictionary also adds that 'intelligence is essentially a synthetic ability, one of order and unity'. This, 'through the elaboration of general ideas and a multiplicity of judgements, which it imposes order upon by linking them one to the other through reasoning'. We oppose *instinct* to intelligence in a rather formal manner since it is 'a thought-out and voluntary adaptive ability'.

The ethnologist, A. P. Elkin (1939) studied the abilities of the Australian Aborigines, a population very similar to the ones in which we are interested. Among these abilities, he mentioned their knowledge of nature, which he broke down into two complementary aspects. The first was descriptive and allowed them to distinguish between and identify beings and facts. The second was correlational, recording the existing relationships between phenomena apparently different from each other. These two aspects reflect a mastery over nature and therefore a comprehension of the environment in which Aborigines live. In other words, a thought-out and voluntary form of intelligence as opposed to instinct.

This example gives us food for thought about the concept of intelligence. In absolute terms, it is an ability (like the concept of a 'Supreme Being'). However, its effective interpretation is highly subjective. Do the Aborigines, or did prehistoric men have an intelligence similar to our own? In any case, for a man, whoever he may be (can one speak of intelligence when referring to a human group?), so-called intelligence is a *relational function*. That is to say that intelligence can only be applied in relation to an environment, which is the sum total of the physical phenomena and living beings which give it and impose upon it its living structure, behaviour patterns and modes of thought. Whether the intelligence of the Aborigines or of the prehistoric men are or were identical to our own they can, or could, only have been fixed in space or time as responses to the demands of their environments. In the final analysis, is not intelligence the ability to know and understand the milieu and thus adapt to it whilst attempting to master it? We are thus authorized in applying Sahlins' (1972) formula to these manifestations, i.e. 'Want not, lack not'.

This is a problem, which is perhaps unanswerable for prehistoric men since we can only measure their intelligence through the few vestiges which have been conserved. These show the nature of his milieu as much as they demonstrate the nature of his intelligence.

Man

It was convenient to date and thus synchronize the three above phenomena at somewhere between 35 000 and 30 000 BP, i.e. the simultaneous appearance of *Homo sapiens sapiens* with art and to establish a significant correlation that reflects a decisive advance in intelligence: 'based on the hypothesis of a direct link between the appearance of art and that of *Homo sapiens sapiens* . . . art appears to be a manifestation of the fundamental biological progress of humanity' (Lorblanchet 1988).

It is clear, as Boule stated as early as the beginning of the twentieth century, that prehistoric *H. sapiens sapiens* had become anatomically, physiologically, and psychologically identical to modern man. However, the question remains as to whether there is any biological proof to suggest that *H. sapiens sapiens* was any more intelligent than Neanderthal man. Even considering that *H. sapiens sapiens* was different anatomically and physiologically, is there a single argument to suggest that the 'capacity for intelligence', i.e. for knowledge and comprehension, of Neanderthals was greater than that of the *H. sapiens sapiens*? Neither measuring the respective volumes of their brains nor the endocasts of their skulls shows any significant difference. Furthermore, Wynn (1979, 1981) maintained that Lower Palaeolithic men possessed intelligence because of their tool-producing skills.

The appearance of modern man seems to have been more or less simultaneous in many regions. There are, *a priori*, certain anomalies, the most significant being the presence of Neanderthal remains in the Castelperronian of Saint-Cézaire (Charente-Maritime), between 35 000 and 33 000 BP (Leveque and Vendermeersch 1980). However, bone deposits attributed to modern man have been dated at Niah (Indonesia) at *c.* 40 000 BP, at Tabon (Philippines) *c.* 24 000 BP, and at Bushman Rock Shelter (South Africa) *c.* 30 000 BP. Finally, in San Diego (California) the dates are spread out and uncertain, but average out at *c.* 30 000 BP (Genet-Varcin 1979). This relative harmony has been disrupted by the recent discoveries made by Bernard Vandermeersch (this volume). He indicated that in the Middle East, at Skuhl and Qafzeh (Vandermeersch 1981), men classified as being *H. sapiens sapiens*, but who manufactured a Mousterian-type industry, existed at *c.* 90 000 BP. During a period of 20 000 to 30 000 years, they must have been contemporary neons with Neanderthal man. Concerning South and East Africa Vandermeersch has shown that 'modern men (*H. sapiens sapiens*) were present approximately 100 000 years ago'. Furthermore, the results of molecular genetics, albeit still controversial, discussed by Rebecca Cann (this volume), place *H. sapiens sapiens* even further back in time.

Within the hypothesis of an 'African cradle', modern man would thus have emerged before 100 000 BP (Mellars 1989). Cann (1988), supported by Stringer and Andrews (1988), proposes that 'populations of modern man began to differ from their common maternal ancestors 140 to 290 000 years ago'.

In these circumstances where could an intellectual 'break' between Neanderthal man and modern man be situated, if, indeed it existed?

Material characteristics

Is the 'transformation' of tools, i.e. the transition from the Middle to the Upper Palaeolithic (especially the Mousterian), a sign of intelligence? The fact that this transformation or transition was not a revolution, but was doubtless a change, has been accepted for a number of years with respect to France as well as for other areas of Western Europe (Bordes 1958; Delporte, 1963, 1970; cf. Mellars 1973, 1989; White 1982 etc.). Furthermore, since the discovery at Saint-Cézaire the following, and perhaps hazardous hypothesis has been accepted as possible. This is that the craftsmen of the Castelperronian, and thus of the industry considered as the first expression of the Upper Palaeolithic, 'had been Neanderthal men' (Tillier 1990).

In fact, the break between the Middle and Upper Palaeolithic is perhaps only 'a question of habit'. It surely exists, but as we have seen, is not supported by the anthropological argument. Is it not then simply a question of its being accentuated and formalized or even dogmatized by examining the industrial entities of each period in the light of separate typological systems? In this case by using typological lists (Bordes 1950; Sonneville-Bordes and Perrot 1953).

The stone industries of the Upper Palaeolithic included tools which were more differentiated and better adapted than the Mousterian tools, e.g. the scraper, the chisel, and points, which already existed in the Middle Palaeolithic, although more rarely. Also typical of this period is the improved and 'profitable' chipping and producing of blades which had already appeared, more or less sporadically, before the Upper Palaeolithic. The only innovation for the stone industries (if indeed there was one) was perhaps that of the Levallois chipping technique. Here, there was a planning process for the product, which no doubt held within itself the germ of blade chipping. The following observation is fascinating: 'Contrary to all appearances, the evolution of chipping from MAT (Mousterian in the Acheulian tradition) of the Combe-Grenal type or Pech de l'Aze type to later Châtelperronian of the Roc-de-Combe type, hardly includes any new elements' (Pelegrin 1990). One tends to consider 'the chipping of blades of the Upper Perigordian and of the Magdalanian', described by Bordes (Boeda 1990), as blade chipping of the Upper Palaeolithic type.

In the final analysis it is clear that the stone industries experienced a progressive evolution since the beginning of the Palaeolithic period. This has not occurred with the appearance of successive 'monolithic civilizations'; but rather through the advent of new techniques and forms which were relatively spread out over time. These phenomena are signs of intelligence but it would be hazardous to consider any one of them as the indication of a decisive intellectual 'leap'.

The increasingly differentiated use of organic materials, especially bone, deer antlers, and ivory is typical of the industries of the Upper Palaeolithic. This is also the case for Castelperronian, e.g. in Arcy-sur-Cure (France) (Baffier and Julien 1990). Although stone tools appear in the European Mousterian, as is the case of Cueva Morin (Spain) (Freeman 1978), they have been questioned (Binford 1982) and include bones which have been reworked in the same way as stone tools (White 1982). They are certainly not part of any stereotyped or culturally differentiated series such as those characteristic of the Upper Palaeolithic. However, points and sculpted bones have been gathered in African sites dating from the 'Middle Stone Age', e.g. the Klasies River Mouth (Singer and Wymer 1982). It is impossible to imagine that bone-tool types of the European Upper Palaeolithic could have evolved uniquely as a result of the environment. Thus, it is highly probable that they are the result of imagination and therefore are a psychological form of what we could perhaps call intelligent action, in the framework of social behaviour and original techniques.

Facts illustrating behavioural differences between the Middle and Upper

Palaeolithic have been documented, e.g. in Arcy-sur-Cure: 'the latest inhabitants (the Castelperronians) built huts on solid foundations, cleared the ground on which they lived, used ochre, worked hides, had fur bedding, sculpted bone, and adorned themselves with pendants' (Farizy 1990). One must, however, be prudent and bear in mind that the use of ochre and fur bedding can also be found prior to the Upper Palaeolithic, e.g. in the Acheulian hut of Lazaret (Lumley *et al.* 1969). The proof of changing behaviour patterns between the Middle and Upper Palaeolithic periods has been questioned (Clark and Lindly 1988), but the arguments put forward in support of numerous aspects of this change seem to be convincing (cf. Mellars 1989). Perhaps the authors believe in the causal effects of the environment as much as they do in a possible intellectual transformation in explaining it.

Symbolism

Obviously, intellectual progress is more easily perceivable in the socio-psychological sphere of life but only, however, through its links with possibly changing behaviour patterns. Three major phenomena must be distinguished: symbolism proper (or religiosity), adornment, and art.

In the case of 'symbolism' proper there are two elements that are possible indicators of 'religiosity' or a metaphysical outlook and therefore a form of understanding and imagination. These are: (1) the use of dyes, especially ochre, and (2) the existence of tombs.

Ochre is to be found in dwellings which are several hundred thousand years old (Wreschner (1980; cf. Mellars 1989). This is an issue of some ambiguity as ochre might have been used in a symbolic fashion, a role which has frequently been advocated but not substantiated. However, it may have also played a technical role, e.g. in the preparation of hides (Audoin and Plisson 1982). The question of its utilization is also posed by its presence on large areas in dwellings. Its presence in tombs, on the rare occasions it has been noted, is hardly significant. It is possible that ochre acquired its symbolic role in the Upper Palaeolithic, but, as yet, no proof of this theory has been supplied.

Burial implies more convincingly an intention or a mode of thinking based on the existence of metaphysical concepts. This, of course, does not constitute an absolute affirmation of the belief in an 'after-life'. Burials follow strict ritual rules, e.g. positioning of the corpse, or the presence of material or offerings. These rituals are little known in France where the majority of the finds come from ancient excavations (Quechon 1976). These rituals have been described in detail, in other regions, e.g. Dolni Vestonice (Czechoslovakia) (Klima 1963, 1987).

It appears that burial was performed by Neanderthal man, although Gargett's (1989) objections are partially justified. Even if evidence, such as the presence of offerings in the 'tomb' at Chapelle-aux-Saints (Coreze), is open to question, it is likely that the burials of la Ferrassie (Dordogne) are not figments of Peyrony's imagination (Peyrony 1934). Also of significance are the flowers of a very specific nature, found in the tomb of Shanidar (Iraq) and identified by Leroi-Gourhan (1975). Burial then, does not appear to have been the indisputable invention of *H. sapiens sapiens*. Nor does the use of ochre bring any decisive proof in favour of the hypothesis of intellectual progress.

The case of 'adornment' is quite different, because even if there are a few indications that it existed amongst the Neanderthals (Gamble 1980; Combier 1988; cf. Mellars 1989), that is hardly convincing. Body ornaments tended to be widely used in the Châtelperronian, especially pierced teeth and seashells: this was demonstrated by the author in the Castelperronian of the Châtelperron site (Delporte 1957). Body ornaments have also been found in numerous excavation sites of this cultural group, especially in Arcy-sur-Cure. Y. Taborin (1990) wrote about this site that: 'adornment like artistic expression issues from two domains: that of reflected thought, from the realm of ideas and symbols, and, at least in the case of elaborated objects, that of applied techniques on a variety of materials'. On the technical level, piercing appears very rarely before the Castelperronian (Bordes 1969): it therefore possesses the characteristics of an invention. Concerning the symbolic aspect, adornment 'is created when the context associates social cohesion with the general tendency towards creativity' (Taborin 1990). Thus, we can state, on the basis of these two points of view, that adornment amongst the Castelperronians represents an authentic intellectual evolution. The Castelperronians therefore inaugurate in the classical sequence the Upper Palaeolithic. They were, however, as we have seen, still Neanderthals.

Art

Artistic expression, contrary to the symbolic phenomena such as the use of ochre and burial, appears, at least in the well-known archaeological context of European sites, as a pure innovation of the Upper Palaeolithic and more precisely of the Aurignacian. Some rare cases of older marked and engraved pebbles and bones have been noted (Bordes 1969), but they remain extremely questionable. It is possible that in a 'pre-figurative phase' man had mastered a certain mental image of animals which he translated through a spatial organization of vestiges (le Regourdou) rather than through drawings, no matter how rudimentary.

It has recently been suggested (Clottes 1989; Delporte 1992) that artistic creation necessitates the artist's acquisition of *an intermediary mental image*. This image is subsequently transcribed by the artist on to the walls of a cave or another object. This is usually without using a living model, especially since the models were mostly animals. This double phenomenon of acquiring and transcribing this mental image implies a *new memory capacity* as well as the training of this memory. Such a memory is undoubtedly more one of an individual instead of a group who made stone tools and tools from organic materials.

This intermediary image was largely 'modified' by Palaeolithic artists before or during its transcription. It was very often mutilated by them, e.g. illustrating only the head or forequarters of the animal, or on the contrary, depicting headless animals. They stressed details that they considered as being characteristics of the animals. But above all the artists organized these images in multiple forms of associations. This was done either on the same object or in sections of drawings on cave walls.

Initially, it had been proposed that art appeared in Europe between 30 000 and 35 000 BP, thus being well before the art created in Australia, Africa, or America, which has often been referred to as 'primitive'. Frequently, this has been in agreement with the theory of the 'miraculous invention', of art together with a suggestion of the white man's egocentricity.

However, we now know, primarily because of Lorblanchet's work (1988, 1989), that significant and ancient vestiges of artistic activity have been found on other continents. Several discoveries enable us to date the appearance of art at over 30 000 BP, e.g. Lake Mungo. According to Lorblanchet, 'Australia furnishes the oldest dates in the world for the appearance of art' (Lorblanchet 1989). We can say, therefore, that they are equivalent to the European dates. There are also paintings in the Apollo II cave in Namibia (26 000–28 000 BP), at Boqueirao do Sitio da Pedra in Brazil (more than 23 000 BP), etc. These observations led Lorblanchet to establish a correlation between the appearance of art and the arrival of modern man. However, as we have seen, the arrival of art is much more ancient.

The situation is clearer for Western Europe since the appearance of art is contemporary to the emergence of modern man (if we accept the fact that the Castelperronians were Neanderthals), and to the creation of the Aurignacian industries. We can distinguish two geographical groups when examining this newly arrived art, which had already used the media of painting, engraving, and sculpture:

(1) a Perigordian group (Delluc, 1978) dated at c. 27 000 BP had paintings on paving stones or slabs, which remind us of those mentioned by Lorblanchet. But, above all, they also had engravings on slabs depicting female vulvas (La Ferrassie) and sketchy depictions of a few animals (La Ferrassie, bouquetin de Belcayre). A carved phallus has been found in the Blanchard de Sergeac shelter (Delluc 1979); and

(2) a Central European group (Germany and Austria) no doubt a little older than the above (30 000–32 000 BP). Here we find statuettes, and very occasionally some three-dimensional sculptures featuring animals (Vogelherd) or humans (Geissenklösterle, Hohlenstein-Stadel). The most recent find is the 'Dancing Venus of Galgenberg' (Neugebauer-Maresch 1989), which is dated between 29 000 and 31 000 BP.

It is more obvious that artistic creation clearly marks a major intellectual evolution. Whatever the motivation underlying it, it is 'the expression of the intellectual appropriation of nature by man and therefore, at last, his mastery over it' (Vandermeersch 1990).

Even if, as we believed at the beginning of the twentieth century (Reinach 1903; Begouen 1924), Palaeolithic art was motivated by the 'magic' of hunting and fertility, a new intellectual process was none the less taking place. This took the form of an original belief, which assumes that the representation of an animal by one technique or another influences either its capture or its reproduction. From the point of view of association, similar to those under-lying imagination and reasoning, it is an undeniable proof of intelligence.

However, the studies of the last few decades, e.g. Raphael (1945), Leroi-Gourhan (1965), Laming-Emperaire (1962), and of their later disciples (Barriere 1982; Vialou 1986; Sauvet 1988, etc.), have shown that, at least in the case of mural art (and there is no reason to doubt that the same is not true *mutandis mutandi*, in the case of mobile or movable art), that art was not organized in a haphazard or fragmentary manner, but rather obeyed a code of organization. This set of norms was probably dualistic, the keys to and meaning of which we can only attempt to guess. However, all the above information obliges us to recognize the existence of this code.

If we examine the case of the Big Rotunda, also known as 'Salle des Taureaux', at Lascaux, we notice an extraordinary construction with a super-imposition of large bulls, average-sized horses, and small deer. This organiza-tion cannot be haphazard and is clearly the fruit of a system of thought and imagination and therefore of intelligence, and is certainly not an isolated case. L. R. Nougier and C. Barriere have brought to light the astonishing organization of the 'Grand Plafond' at Roffignac which they believe is struc-tured on the basis of the duality between Life and Death (Barriere 1982). Furthermore, mobile or movable art with its associations in lines, in con-frontations, in superimpositions, and in 'scenes' also probably illustrates not only intelligent thought but, beyond that, a translation of the mythological concepts of the artists, and consequently of the human groups of the Upper Palaeolithic.

This complex artistic expression with its 'sets' of equally as complex signs, not only corresponds to a system of myths but also reflects socio-cultural phenomena. C. Gamble (1980) maintains that 'the spectacular developments of material culture, such as the appearance of art, appear today to be more closely linked to the nature and the quantity of the information necessary to the Palaeolithic societies rather than their being conditioned by the evolution of the brain'. The birth of art corresponded, therefore, to an awareness of the new social structures and to a solution, based on reason, to the problems posed by them. Can these social changes be traced to modifications of the environment or an increase in the population (linked to the colonization of 'new territories')?

The need to maintain relationships can perhaps explain the appearance of new processes in language and communication, Gamble (1980) writes: 'Previously verbal communication might have been sufficient in order to exchange information, but now the possibility of differentiation of this language would make it more difficult to do so. This could more easily be overcome by calling upon a new form of communication based upon visual messages which could be stylistically coded'. The basic tenets of a polymorphic artistic expression are also demonstrated in Sauvet's (1990) semiological theory. It consists, on the one hand, of an important differentiation in the style and nature of animal depiction, and on the other, an often intensive falling back on signs. All of this constitutes a communicative activity, which, all things considered, represents one of the fundamental leaps forward of human intelligence.

References

Audoin, F. and Plisson, H. (1982). Les ocres et leurs témoins au Paléolithique en France: enquête et expérience. *Cah. C.R. Préhist. Paris* **8**, 33–80.

Baffier, D. and Julien, M. (1990). L'outillage en os des niveaux châtelperroniens d'Arcy-sur-Cure. In *Paléolithique moyen récent et Paléolithique supérieur ancien en Europe, Nemours, 1988* (ed. C. Farizy), *Mém. Mus. Préhist. d'Ile-de-France* **3**, 329–34.

Barriere, C. (1982). *L'art pariétal de Rouffignac*, 206 pp. Picard, Paris.

Begouen, H. (1924). La magie aux temps préhistoriques. *Mém. Acad. Sci. Inscr. Belles-Lettres de Toulouse* series 12, **11**, 417–32.

Binford, L.R. (1982). Comment on White: Rethinking the Middle/Upper Paleolithic transition. *Curr. Anthropol.* **23**, 177–81.

Boeda, E. (1990): De la surface au volume. Analyse des conceptions des débitages Levallois et laminaire. *Paléolithique moyen récent et Paléolithique supérieur ancien en Europe, Nemours, 1988* (ed. C. Farizy), *Mém. Mus. Préhist. d'Ile-de-France* **3**, 63–8.

Bordes, F. (1950). Principes d'une méthode d'étude des techniques de débitage et de la typologie du Paléolithique ancien et moyen. *L'Anthropologie* **54**, 19–34.

Bordes, F. (1958). Le passage du Paléolithique moyen au Paléolithique supérieur. In *Neanderthal centenary* (ed. G. von Koenigswald), pp. 175–81. Utrecht.

Bordes, F. (1969). Os percé moustérien et os gravé acheuléen du Pech de l'Azé II. *Quaternaria* **11**, 1–6.

Cann, R.L. (1988). DNA and human origins. *Ann. Rev. Anthropol.* **17**, 127–43.

Clark, G. and Lindly, J. (1989). The case for continuity: Observations on the biocultural transition in Europe and western Asia. *The human revolution: behavioural and biological perspectives on the origins of modern humans* (ed. P. Mellars and C. Stringer), Vol. 1. Edinburgh.

Clotttes, J. (1989). The identification of human and animal figures in European Palaeolithic art. *Animals into art* (ed. H. Morphy), *One Wld Arch.* **7**, 21–56.

Combier, J. (1988). Témoins moustériens d'activités non-utlitaires. In *De Neanderthal à Cro-Magnon* (ed. J. B. Roy and A. S. Leclerc), *Mém. Mus. Préhist. d'Ile-de-France* **2**, 69–72.

Delluc, B. and Delluc, G. (1978). Les manifestations graphiques aurignaciennes sur support rocheux des environs des Eyzies (Dordogne). *CNRS, Gallia-Préhistoire* **21**, 213–438.

Delluc, B. (1979). Le phallus de l'abri Blanchard (Sergeac, Dordogne). *Anti. Nat.* **11**, 23–8.

Delporte, H. (1957). La Grotte des Fées de Châtelperron. *Cong. Préhist. de France, Poitiers-Angoulême*, **15**, 452–77.

Delporte, H. (1963). Le passage du Moustérien au Paléolithique supérieur. *Bull. Soc. Mérid. Spél. Préhist.* **6–9**, 40–50.

Delporte, H. (1970). Le passage du Moustérien au Paléolithique supérieur. In *L'homme de Cro-Magnon* (ed. G. Camps and G. Olivier), pp. 129–39. Arts et Métiers Graphiques, Paris.

Delporte, H. (1992). La notion d'image dans l'art paléolithique. *Colloque l'Image et la science, Comité des Travaux Historiques et Scientifiques*, 115th Congress, Avignon, pp. 215–29. Comité des Travaux Historiques et Scientifiques, Paris.

Elkin, A.P. (1938). *The Australian Aborigines* [French translation, 1967, 451 pp.]. Gallimard, Bibliothèque des Sciences Humaines, Paris.

Farizy, C. (1990). Du Moustérien au Châtelperronien à Arcy-sur-Cure: un état de la question. In *Paléolithique moyen récent et Paléolithique supérieur ancien en Europe, Nemours, 1988* (ed. C. Farizy), *Mém. Mus. Préhist. d'Ile-de-France* **3**, 281–9.

Freeman, L.G. (1978). Mousterian worked bones from Cueva Morin (Santander, Spain): a preliminary description. In *Views of the past* (ed. L. Freeman), pp. 29–51. La Haye.

Gamble, C. (1980). Information exchange in the palaeolithic. *Nature*, **283**, 522–3.

Gargett, R.H. (1989). Grave shortcomings: the evidence of Neandertal burial. *Curr. Anthropol.* **30**, 157–90.

Genet-Varcin, E. (1979). *Les hommes fossiles*, 412 pp. Boubée, Paris.

Klima, B. (1963). *Dolni Vestonice. Vyzkum taboriste lovcu mamutu v letech 1947–1952* [D.V. recherches dans un campement de chasseurs de mammouths de 1947 à 1952], 427 pp. 108 plates. Ceskoslovakia Akademie Ved, Monumenta Archaeologica, Prague.

Klima, B. (1987). Une triple sépulture du Pavlovien à Dolni Vestonice, Tchécoslovaquie. *L'Anthropologie* **91**, 329–33.

Laming-Emperaire, A. (1962). *La signification de l'art rupestre paléolithique*, 424 pp. Picard, Paris.

Leroi-Gourhan, A. (1965). *Préhistoire de l'Art occidental*, 482 pp. Mazenod, Paris.

Leroi-Gourhan, A. (1975). The flowers found with Shanidar IV, a Neanderthal burial in Iraq. *Science* **190**, 562–4.

Leveque F. and Vandermeersch, B. (1980). Découverte de restes humains dans un niveau castelperronien à Saint-Cézaire (Charente-maritime). *C.R. Acad. Sci.* **291–D**, 187–9.

Lorblanchet, M. (1988). De l'art pariétal des chasseurs de rennes à l'art rupestre des chasseurs de kangourous. *L'Anthropologie*, **92**, 271–316.

Lorblanchet, M. (1989). Problèmes épistémologiques posés par l'établissement d'une chronologie d'art rupestre. *Préhist. Ariég.* **44**, 131–52.

Lumley, H. de, *et al.* (1969). Une cabane acheuléenne dans la grotte du Lazaret. *Mém. Soc. Préhist. Fr.* **7**, 234 pp.

Mellars, P. (1973). The character of the Middle–Upper Palaeolithic transition in south-west France. In *The explanation of culture change* (ed. C. Renfrew), pp. 255–76. Duckworth, London.

Mellars, P. (1989). Major issues in the emergence of modern humans. *Curr. Anthropol.* **30**, 349–85.

Neugebauer-Maresch, C. (1989). Zum Neufund einer weiblichen Statuette bei den

Rettungsbrabungen an der Aurignacien-Station Stratzing/Krems-Rehberg, Niederösterreich. *Germania* **67**, 551–60.

Pelegrin J. (1990). Observations technologiques sur quelques séries du Châtelperronien et du MTA B du Sud-ouest de la France. Une hypothèse d'évolution. In *Paléolithique moyen récent et Paléolithique supérieur ancien en Europe, Nemours, 1988* (ed. C. Farizy), *Mém. Mus. Préhist. d'Ile-de-France* **3**, 195–201.

Peyrony, D. (1934). La Ferrassie. *Préhistoire*, **3**, 92 pp.

Quechon, G. (1976). Les sépultures des hommes du Paléolithique supérieur. In *La préhistoire française* (ed. H. de Lumley). Vol. 1, pp. 728–33. CNRS, Paris.

Raphael, M. (1945). *Prehistoric cave paintings*. Pantheon Books (Bollingen series), Princeton University Press, New York.

Reinach, S. (1903). L'art et la magie. A propos des peintures et des gravures de l'âge du Renne. *L'Anthropologie* **14**, 257–66.

Sahlins, M. (1972). *Stone age economics*. Gallimard, Bibliothèque des Sciences Humaines, Paris.

Sauvet, G. (1988). La communication graphique paléolithique (de l'analyse quantitative d'un corpus de données à son interprétation sémiologique). *L'Anthropologie* **92**, 3–15.

Sauvet, G. (1990). Les signes dans l'art mobilier; *Colloque International d'Art mobilier, Foix-le Mas d'Azil, 1987*, pp. 83–99. Ministère de la Culture, Paris.

Singer, R. and Wymer, J. (1982). *The Middle Stone Age at Klasies River Mouth in South Africa*. University of Chicago Press, Chicago.

Sonneville-Bordes, D. and Perrot, J. (1953). Essai d'adaptation des méthodes statistiques au Paléolithique supérieur. Premiers résultats. *Bull. Soc. Préhist. française* **50**, 323–33.

Stoliar, A.D. (1985). *L'origine des arts plastiques*, 298 pp. Iskoustvo, Moscou. [Russian text, French résumé].

Stringer, C.B. and Andrews, P. (1988) Genetic and fossil evidence for the origin of modern humans. *Science*, **239**, 1263–8.

Taborin, Y. (1990). Les prémices de la parure. In *Paléolithique moyen récent et Paléolithique supérieur ancien en Europe, Nemours, 1988* (ed. C. Farizy), *Mém. Mus. Préhist. d'Ile-de-France* **3**, 335–44.

Tillier, A.M. (1990). Neandertaliens et origine de l'homme moderne en Europe: quelques réflexions sur la controverse. In *Paléolithique moyen récent et Paléolithique supérieur ancien en Europe, Nemours, 1988* (ed. C. Farizy), *Mem. Mus. Préhist. d'Ile-de-France* **3**, 21–4.

Vandermeersch, B. (1981). *Les hommes fossiles de Qafzeh (Israël)*, 319 pp. CNRS, Paris.

Vandermeersh, B. (1990). Réflexions d'un anthropologue à propos de la transition Moustérien/ Paléolithique supérieur. In *Paléolithique moyen récent et Paléolithique supérieur ancien en Europe, Nemours, 1988* (ed. C. Farizy), *Mém. Mus. Préhist. d'Ile-de-France*, **3**, 25–7.

Vialou, D. (1986). *L'Art des grottes en Ariège magdalénienne*, 432 pp. CNRS, Paris, Gallia-Préhistoire, supplement 22.

White, R. (1982). Rethinking the Middle/Upper Paleolithic Transition. *Curr. Anthropol.* **23**, 169–92.

Wreschner, E.E. (1980). Red ochre and human evolution: a case for discussion. *Curr. Anthropol.* **21**, 631–44.

Wynn, T. (1979). The intelligence of the later Acheulean hominids. *Man* **14**, 371–91.

Wynn, T. (1981). The intelligence of Olduwan hominids. *J. Hum. Evol.* **10**, 529–41.

DISCUSSION

Participants: J.P. Changeux, J. Chavaillon, C. Cohen, H. Delporte, G. Giacobini, J. Kozłowski, D. Premack

Premack: Your speculations or conjectures about the necessity of mental imagery is important. There is a body of evidence that children's representational drawings clearly show that mental image is required. If a child is asked to draw something in front of him, but other evidence indicates he lacks a mental representation for, despite the physical presence of the object, he cannot draw it. Some time ago, I was interested in discovering if representational drawing is among a number of faculties which are supposed to be unique to our species, such as language and social skills of various types. The question is to find out if it is cognitive in nature or motor. So, we reduced the motor demand of the task of representational drawing by simply giving a chimpanzee an outline of a face together with facial elements. The task was merely to re-assemble a face. One subject was really successful. It is not to say that the animal can produce representational drawing, it is a case of reconstruction as opposed to creation. If you give the animal the elements, it is capable of veridical accurate re-assembling. We made the same observations for analogies. The chimpanzee does not create analogies, but if you give it the elements, it can arrange them analogically. Furthermore, we also found pre-artistic skills in chimpanzees. The main point concerning the necessity for a mental image is that we already find in the chimpanzee the ability for re-creation if representational, but not creation. In addition, I can say that there is evidence for motivational factors in the chimpanzee. It is spontaneous manipulation because the chimpanzee is not reinforced or trained to do this. In conclusion, at least some elements of the mental image are already present in a rather distant precursor of *Homo sapiens sapiens*.

Changeux: You told us that the four lines on the lamp represent a mountain ship. This looks more like cuneiform signs than hieroglyphic-like representations of animals. So, they had the possibility to achieve symbolic representation of animals. Furthermore, we saw a series of incisions which resemble the ones we can see on a shepherd stick. I do not want to extrapolate too much, but would it be reasonable to recognize some process of writing or numbering?

Cohen: In his study, Leroi-Gourhan thought that the artistic representation was not arranged as a linear succession of signs, as in writing, but according to a pattern where organization radiates in all directions. This pattern was not representative of speech, but rather of supported speech.

Delporte: Many interpretations go too far. Nevertheless, these incisions may have a mnemonic function. We do not know yet if these representations should be called mythograms or pictograms. We do not understand their meaning, but we can understand their organization or pattern. We continue to make progress and I think that we are slowly getting closer to the meaning of these representations.

Chavaillon: All prehistoric human representations, but Brassempouy, are very caricature-like. We have to assume some kind of sense of humour or the fear of making realistic human representations, as is still frequent today in many populations, especially when you attempt to take a photograph. The Palaeolithic artists had all the technical capabilities to make realistic representations of humans as well as of bisons or horses.

Delporte: This is quite possible. I used to think that the Palaeolithic artist did not represent humans, i.e. man, woman, and child, because they did not want to expose their image to magic forces. I no longer believe this hypothesis. In prehistory, we are only able to suggest hypotheses and one should always bear this in mind.

Giacobini: You have presented many examples of extensive art of the Palaeolithic. But how do you explain, that by the end of this period, the art is still very poor.

Delporte: If artistic expression was conceived as a means of communication, we can explain this by its replacement by more symbolic means of representation. For instance, the Azilian pebbles show marks only. These changes are certainly related to new environmental conditions and new psychological processes. But we do not know the exact reasons.

Kozlowski: It is now clear that biological changes are not directly involved with the emergence of art. Both phenomena are well separated. I even think, especially after David Premack's comments that *Homo habilis* was able to construct images, but the conditions were not suitable. Conditions emerged because of the necessity to communicate when the frequency of inter-group encounters became important, together with seasonal gathering and the stability of the habitat. The Upper Palaeolithic societies were semi-nomadic. The emergence of artistic expression and communication took a long time, about 10 000 years during the Aurignacian. This necessity to communicate raises the question of the origin of writing.

14

The origins and evolution of writing
ANDRÉ ROCH LECOURS

Notes on the biology of reading and writing

As there are no indications that the human brain might have changed between the existence of ancient Babylon and modern Washington DC, beliefs
or facts concerning the neural substrate of written language do not really
apply to the subject of this book. However, given the presence of this chapter
within this book, one might infer that the biology of writing is to be considered as somehow relevant. There will thus be brain-focused comments,
occupying the place where one would normally expect an introduction.[1] Some
of these comments will bear on language disorders resulting from acquired
unilateral forebrain lesions during adult life and some will pertain to normal
and abnormal ontogenesis. In the latter case, particular attention will be paid
to myelinogenesis since: (a) it has been observed that gross retardation in
myelinogenesis is incompatible with normal cognitive development (Lecours
and Yakovlev 1966; Kemper *et al.* 1973), and (b) it has been suggested that
myelinogenetic maturity in a given axonal system can be taken as a marker
of functional maturity in that system (Yakovlev and Lecours 1967).

Storing sensory inputs ('stock' memories)

If an aptitude to store information of proprioceptive origin is an immediate
prerequisite to production abilities for both speech and writing, an aptitude
to store arbitrary auditory information is — in standard individuals of the
species — essential to decode speech[2]; and one to store arbitrary visual information is essential to decode writing. It might thus be 'significant', from the
point of view of one interested in the mutual relationships of brain and
language, that there exist major differences between the maturational cycle
of the auditory and that of the visual pathway. If myelogenesis from retina

---

[1] It would no doubt be more appropriate to introduce this chapter by discussing the social
pressures — urban and mercantile — that led to the invention of writing by a subgroup of the
human species, or else by disclosing the impact exerted on the evolution of seminal written codes
by their interactions with alien spoken codes, or again by revealing why Maya inscriptions should
or should not be considered as genuine 'writing' — what else might they be? — and, if yes, why
these inscriptions are likely or unlikely to represent some whimsical offspring of a Mesopotamian
or Egyptian invention. However, the author does not have the knowledge necessary to write
such an introduction.

[2] And writing in the case of sound-committed written codes (see below).

versus cochlea to colliculi and thalamus were to be taken as a functional marker, it could follow that the human auditory system has some capacity to subserve language at least 10 *weeks* before birth at term, and about 5 *months* ahead of the visual system; if, on the other hand, the functional marker were myelogenesis from retina versus cochlea to primary cortex, it could follow that the visual system is ready for service 2 to 3 *years* before the auditory one (Yakovlev and Lecours 1967; Lecours 1975). These are indeed intriguing facts but, as discussed at length elsewhere in this volume, it is known that the human organism evolved drawing and speech abilities at least 30 000 years (if not much more) before it 'invented' writing. All I might further suggest is then that, given an appropriate environment, the *Sapiens sapiens* (1500 cc) model of the *Homo* skull protects a brain that might perhaps learn to understand simple logograms — whatever the writing system — as early as it learns to understand simple spoken words, if not earlier.

This being said, it should be emphasized that, beyond these observations on standard maturation of the acoustic and optic pathways, the human brain retains the capacity to master abstract codes which are full homologues of spoken or written language even when congenitally deprived of acoustic or visual inputs (I doubt that it retains this capacity when deprived of proprioceptive inputs, if such a condition occurs in nature). One might also mention that these observations neither substantiate nor invalidate Critchley's (1964) suggestion that a lag or defect in brain maturation might be the cause of developmental dyslexia. In this respect, Galaburda's discovery of cytoarchitectonic anomalies suggestive of an early migrational deviance (around the 16th week of gestation) in the cortical mantle of individuals who had been diagnosed to have had developmental dyslexia is probably of greater significance (Galaburda and Kemper 1979; Galaburda 1989).

Organizing outputs ('action' memories)

If one defines 'telokinesis' as the motility of striated muscles when it is intended to exert influence upon the environment (Yakovlev 1968), one might suggest that the human aptitude to programme the telokinesis of speech production depends to a large extent on specialized proprioceptive (kinaesthetic) memories. It might thus be 'significant' that the myelinogenetic cycle of the somaesthetic pathways and that of the pyramidal tracts are cotemporaneous to a large extent, from the last trimester of gestation to the first months of the second year of postnatal life (Yakovlev and Lecours 1967). Likewise, given that the notions of hand preference and of writing have so far been intimately associated, and might remain so even after various types of bimanual keyboards have eventually replaced the various offsprings of the *calami* of old, it might be of interest that the ontogenesis of the motor pathway is such that, in about 80 per cent of human beings: (a) pyramidal axons originating from the left cerebral hemisphere cross the bulbar midline (decussate) above their homologues from the right, and (b) both the indirect and the direct pyramidal tracts are somewhat bulkier on the right than on the left side of the medulla oblongata at cervical (upper limb) level (Yakovlev and Rakic 1966). In addition, one might wish to consider the fact that, according to standard teaching, axial muscles, such as those of the phonoarticulatory apparatus, benefit from bilateral innervation to a far greater extent than appendicular muscles such as those of the writing hand.

The literate versus the illiterate brain

Turning now to acquired brain disease, one fact of potential interest is that, at least when they occur abruptly, focal lesions of the adult literate brain, usually left-sided but not infrequently right-sided, are more likely than otherwise to impair some aspect or another of written language; and also that certain of these lesions impair neuronal nets which are apparently dedicated to written-language behaviours in a more or less specific manner. For example, and beside a right visual field amputation, the clinical manifestations of certain lesions restricted to: (a) the left lingual and fusiform gyri, or (b) subjacent inferior longitudinal, geniculo-calcarine, and tapetal axons[3] can be limited — whatever the written code — to a reading disorder known as 'pure' or 'agnosic alexia' (Alajouanine 1960). This being said, I do not have any conception concerning what illiterates do, or not do, with the cerebral apparatus that others perhaps specifically devote to written-language behaviours. (However, I have met illiterate Amazonians with spatial orientation abilities so sophisticated that I cannot even imagine that I might have the capacity to devise a method for assessing these abilities on the basis of an explicit theory.)

Another fact of potential interest is that Sylvian lesions of the right half of the illiterate brain can cause lexical disorders which are not usually observed after similar lesions of the school-educated, more precisely of the alphabetized brain (Lecours et al. 1988). Whether or not the latter assertion should be extended to users of non-alphabetic codes, especially of the logographic type, remains to my knowledge an open question. [In this respect, it is of interest that Sasanuma (1975) suggested at one point that the right hemisphere of right-handed readers of Japanese might play some role in processing Kanji, although not Kana writing.] Similarly, I do not know whether greater functional lateralization of the human brain depends, in the form that I have just mentioned, on school-education as a whole rather than, specifically, on learning to read and write (Scribner and Cole 1981). In spite of this, one might wish to take this particularity of the unschooled illiterate brain as indicating that the genetic programme leading to left cerebral dominance for language is somehow less completely implemented in the brains of those who have not learned to read and write than in that of those who have (Cameron et al. 1971). From a diachronic point of view, this might in turn be taken as suggestive that, if the human brain has not changed in any significant manner

---

[3] According to Dejerine (1914–1926, p. 109), this triple axonal lesion in the lateral wall of the left occipital horn represents 'the localization of the lesion in pure verbal blindness'.

might wish to take this particularity of the unschooled illiterate brain as indicating that the genetic programme leading to left cerebral dominance for language is somehow less completely implemented in the brains of those who have not learned to read and write than in that of those who have (Cameron *et al.* 1971). From a diachronic point of view, this might in turn be taken as suggestive that, if the human brain has not changed in any significant manner since the advent of *H. sapiens sapiens*, important environmental changes have occurred — usually as the result of demographic growth and/or of technical advances — and have modified the manners in which human cerebrality accomplishes itself. And, in synchrony, it might be taken as suggestive that, if the anatomy of the human brain and its physiological and molecular properties as well as its cognitive potential are grossly comparable from one individual to another irrespective — in the absence of extreme environmental anomalies — of various cultural factors, the actualization of the cognitive potential of the human brain is in part determined, within and across collectivities, by cultural (social) factors.

Now, although I can attribute meaning to the term 'cognitive neurosciences', the distance between my knowledge of this field and my expectations remain gargantuan and I do not see how the above comments might coalesce into a gapless enlightening summary. Therefore, I will at this point change my course, i.e. I will forsake the title of this book and try to honour at best that of the present chapter. Following a preliminary on a few aspects of the pre-writing *Homo* achievements and some explanations as to the metaphorical summaries, i.e. 'models' (taking here the form of sets of boxes and arrows) accompanying my text, I will attempt to outline some of the main developments in the history of writing.

Boxes and arrows

Irrespective of the social constraints that engendered the need for city merchants to keep a record of their transactions, the invention of writing was obviously subserved by an already powerful brain–mind complex. In other words, it is quite clear that, when writing was invented, the human body had long been capable of storing within itself the representations of sensory inputs — including arbitrary sublexical and lexical ones (Fig. 14.1) — as well as of their conceptual counterparts, that sophisticated communication abilities had long been part of its inheritance, and that it had long learned to use its knowledge in order to master its environment and to ensure its survival.

Consider the type of 'formalizations' (or 'models' or 'diagrams') shown in this chapter. A first fact to underline is that these diagrams are obvious surface renovations of (most of) the 'classical models' of late nineteenth-century neurology. Another is that they should not be conceived as brain-based (Morton 1984), i.e. it should be understood that they would not change

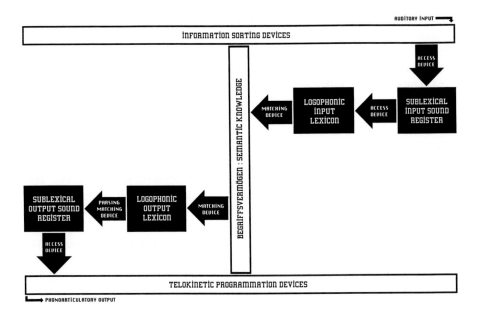

Fig. 14.1. *Homo*'s lexical achievement.

by one iota if writing were 'represented' in the spleen rather than in the brain. In spite of this, and beyond the subject matter of this chapter, a third fact is that the psychological realities behind their components—their boxes and arrows—have, to a greater extent in certain cases than in others, been recognized as the result of the observation of subjects with various types of spoken and written language disorders resulting from damage to their cerebral rather than splenic viscera. (For reviews, see Coltheart *et al*. 1980; Patterson *et al*. 1985; Lecours and Tainturier 1990; Morin *et al*. 1990.)

Still concerning boxes and arrows, let us take as an example the model summarizing the preSumerian word-processing *Sapiens* achievement (Fig. 14.1). All of the boxes in this model (and in the others in this chapter) are meant to represent *specialized memories*. These memories are of two types, those which specifically retain different categories of representations ('stock' or 'chunk memories'), and those which retain the procedures necessary to decode ('sorting memories') or encode ('action memories') incoming or outgoing information. 'Procedural memories' is a convenient term to collectively designate action and sorting memories. Arrows are also meant to represent particular procedural memories necessary to 'access' various stock components, or to 'parse' and/or 'match' to one another, or to 'convert' into one another, stock memory representations ('entries') related to a common reality.

The top and bottom boxes, those identified as 'INFORMATION SORTING' and 'TELOKINETIC PROGRAMMATION', correspond mainly to procedural memories. They are the depositories of the knowledge permitting, respectively, the initial phases of the decoding of afferent information (gnosis) and, at an abstract preparatory level, of the encoding of efferent ones (praxis). [Lordat (1843) qualified the latter as an 'instinctive memory', by which he meant that it functions automatically, without conscious control.] The five remaining boxes correspond to stock memories. Four of them, identified as 'LEXICONS' or 'REGISTERS', are stores of arbitrary representations of which the sensory origins are more immediate than in the case of the information in the fifth store, identified as the 'BEGRIFFSVERMÖGEN'.

One might wish to keep in mind that the INFORMATION SORTING and TELOKINETIC PROGRAMMATION boxes are shared by all animal species although the information to be sorted and the finality of striate muscle actions to be programmed vary from one to the other. The BEGRIFFSVERMÖGEN is also a shared component but it seems to be more sophisticated in the human species.

In my opinion, it is an essential characteristic of boxes of the LEXICON and REGISTER types that, once learning has been achieved and as long as these components are not debilitated or broken, the information that they keep in store, i.e. the formal representations of sublexical and lexical entities beyond the variability of their surface actualization, are stable and exhaustive; and it is no less an essential characteristic of the BEGRIFFSVERMÖGEN (the 'semantic box') that the information it holds is for the most part subject to perpetual revisions and modifications (even one's conception of a woman and one's conception of a chair are bound to change the day one first hears an expression such as 'Madam Chairman').

Also in my opinion, and since *nihil est in intellectu nisi primo fuerit in sensu* (although I reckon that the road is long and complex from inputs to the organs of the senses to a concept such as that of, say, 'excogitation'), the representations stored in boxes of the input-lexicon type are to be conceived as being of exteroceptive origin, whereas those stored in boxes of the output-lexicon type are to be conceived as being mainly — or at least for a fair half — of proprioceptive (kinaesthetic) origin. The main interest for emphasizing this opposition is that exteroceptive information (e.g. auditory, visual) is far more readily accessible to consciousness than is proprioceptive information (e.g. the information sent to the brain when the muscles within the pilars of the velum palati contract in a manner such that the posterior nostrils are occluded, thus permitting one to utter [d] rather than [n], for example).

The origins and evolution of writing

Under this heading, I will try to summarize the main steps that have, as far as documented history is concerned, marked the birth and evolution of writing systems. My summary will borrow more from Sampson's (1985) monograph than from others (Cohen 1958; Février 1959; Gelb 1962; Amiet *et al.* 1982). Following the single-case method, I will give gross indications about the Sumerian, the Akkadian, the Ugaritic, the Greek, and the Han'gŭl writing systems. In each case, I will attempt to identify the distinctive feature(s) of a presumed 'functional architecture'. Notes on a few offsprings of the Greek alphabet will also be included.

The Sumerian invention

It appears that the genesis of the very first writing system, naturally a pictographic one, initially began in Sumer about 5300 years ago, then the heart of a prosperous land located between the rivers Tigris and Euphrates. It is said that this development was independent of that in Egypt, which began two centuries later[4] (Amiet *et al.* 1982). Step by step, over a period that lasted from 900 to 1100 years, the Sumerian pictograms gave birth to the cuneiform writing system (Fig. 14.2). Let us stop history at an ideal hypothetical moment when pictograms had been replaced by cuneiform logograms; that is, by graphic signs of which the pictural origins had been (in order to give the system the capacity to represent concepts other than those linked to visible realities) deliberately removed, then forgotten. Each of these logograms — names excluded, there were about 500 to 600 of them (Cohen 1958; Jean 1987) — represented either a word and its meaning, or else, far more frequently if not always, a number of heterophonic words (up to 25 of them) and their context-dependent meanings [obtained by a slight modification of the logogram for 'head' or 'front', the one for 'mouth' could thus also represent 'tooth' and 'teeth', 'voice', 'speech', 'to talk', and 'to shout' (Cohen 1958; Sampson 1985) (Fig. 14.2).

One might add, for the sake of clarity, that the term 'logogram', as used in the present context, theoretically applies to the original set of Sumerian

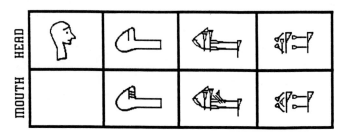

Fig. 14.2. From pictograms to cuneiforms.

---

[4] Beyond Sumer and Egypt, pictograms have been at the origin of other writing systems (Jean 1987), such as those of the Hittites, the Cretans, the Chinese, the inhabitants of the Indus Valley, the Mayas.

pictograms as well as to that of the non-figurative cuneiforms to which they gave birth, i.e. both were arbitrarily invested with the abstract power to represent all instances of given realities or sets of realities. The main difference between the two was, therefore, that eliminating the constraints inherent to figurative logograms permitted the system to become far more productive. In other words, no new invention took place between pictographic and cuneiform writing but the original invention was modified in a manner such that its efficiency was tremendously enhanced.

It is quite clear, as indicated above, that the Sumerians had long been enjoying (as did their ancestors before them) a healthy BEGRIFFSVERMÖGEN as well as two SUBLEXICAL REGISTERS and two LOGOPHONIC LEXICONS when the invention of writing was initiated. Of these registers and lexicons, two were of the input type and the two others of the output type, the former storing representations of exteroceptive (auditory) origin and the latter representations of proprioceptive (labioglossopalatolaryngokinaesthetic) origin (Fig. 14.1). What the Sumerian invention amounted to, therefore, was the addition—together with appropriate 'ACCESS', 'MATCHING' and 'PARSING DEVICES'—of four new stock memories, i.e. of new REGISTERS and LEXICONS (graphic ones this time), two of the input type and two of the output type, the former storing representations of exteroceptive (visual) and the latter representations of proprioceptive (cheirokinaesthetic) origins (Fig. 14.3). It is also quite clear, or at least very likely, that this invention turned out to be a formidable source of potential growth to all pre-existing cognitive

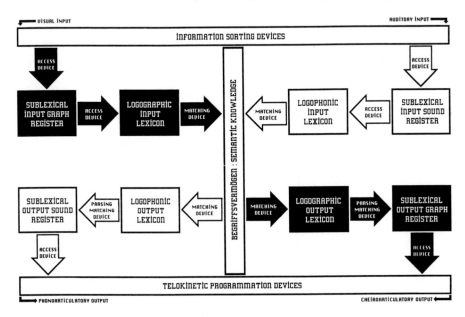

Fig. 14.3. The Sumerian invention.

Homo achievements, spoken language included (at least with regard to lexical expansion).

Thus, at the Sumerian point, one learned to read and write as one learned a second language; the original cuneiform symbols primarily represented meanings, i.e. they were not sound-specific. Just as written Chinese today, written Sumerian could and did subserve the visuographic transcription of different spoken languages, which was, no doubt, its main strength. And its main weaknesses were, no doubt, on the one hand, that expert readers/writers were very few (mastering the code indeed supposed a major learning investment) and, on the other, that Sumerian logograms were somewhat inauspicious to morphology (and adamantly inauspicious to the transcription of meaningless syllables or syllabic sequences). Even if there are no reasons to think that single-word written messages played a greater role 5000 years ago than now, one might wonder whether or not heavy context-dependency in decoding — great number of heterophonic homographs — should also be listed among handicaps. The second invention was to obliterate the latter phenomenon, temporarily for certain codes and definitely for others.

The Akkadian implement

Centuries before the Sumerian invention was completed, the logographic cuneiform writing system began to spread throughout the Middle East. About 4000 years ago, the Akkadians of Babylon implemented the second invention. One of the factors that led to this was probably that the language of Akkad had morphological constraints far more complex and imperious than those inherent to the language of Sumer (Sampson 1985). Be this as it may, and apparently in line with a process already initiated in Sumer, Akkad doubled the Sumerian invention by giving the system a potential option, that of primarily accessing semantic knowledge through output lexicons. In other words, the Akkadian invention consisted, essentially, in the addition of cuneiform-sound and sound-cuneiform 'CONVERTORS': a restricted subset of Sumerian cuneiforms was chosen by reference to the phonological structure of spoken Akkadian and came to represent sublexical syllabic sounds rather than meaning-bound words (Fig. 14.4).

In the metaphoric terms of my 'model', the Akkadian invention (Fig. 14.4) translates as being, on the one hand, a specification of the content of input and output graph registers, which became small-size stock memories of syllabic entries and, on the other hand, the addition of a set of pre- and postlexical procedures identified here as (a) CONVERTORS or else 'CONVERSION DEVICES', one phonographosyllabic and the other graphophonosyllabic, (b) ACCESS DEVICES to output lexicons, and (c) MATCHING DEVICES between the latter and the BEGRIFFSVERMÖGEN.

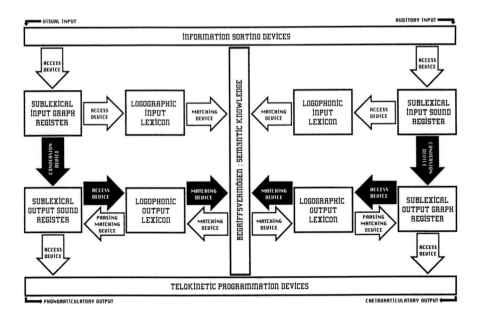

Fig. 14.4. The Akkadian implement.

Therefore, learning to read and write no longer means to learn a second language at the Akkadian point, but rather to learn a secondary code (see below), i.e. the new writing system is sound-specific, more precisely syllable-specific (without explicitation of the phonemic content of syllables, as in the Kana systems today.[5] The number of cuneiforms necessary to write has been drastically reduced as well as, consequently, the learning investment necessary to master it. Written language has become more accessible and, indeed, the number of readers/writers will progressively increase. Homography is now restricted to eventual homophones and the new code accommodates the transcription of the morphological components of the Akkadian language as well as that of its roots. The loss, of course, is that the new system, although it resorts to a subset of the same old cuneiforms, is spoken-language specific: one cannot read what the Akkadians write if one does not speak Akkadian, hence, no doubt, the 'confusion of languages' occurring at all levels of the Tower of Babel. And one who speaks Akkadian and has learned the new code can now read and write words which one does not understand (obviously a two-edged sword), as well as the names of distinguished foreign visitors (a definite commodity). Thus, Akkadian becomes the 'official language of diplomacy' about 38 centuries ago (Amiet *et al.* 1982).

---

[5] If one does not take into account the fact that a few Kana graphs represent phonemic (e.g. consonantal gemination) or even subphonemic (e.g. nasalization of a vowel) phenomena.

Modifying the Akkadian invention

Granted that, in the present context, an 'invention' is explicitly represented as a set of boxes and arrows in the 'models' (Figs 14.3 and 14.4), whereas a 'modification' is not (because I would not know how to formalize 'modifications'), what took place beyond the Akkadian point can be qualified as successive 'modifications' or refinements of the Babylonian convertor, which led to progressive specification of the phonemic content of syllables and, in one case (Sampson 1985), to specification of the featural content of phonemes.

The Ugaritic modification

About 34 centuries ago, in Palestine and Syria, Phoenician merchants began to use a 'consonantal' writing system — the Ugaritic alphabet — which initially comprised 22 graphemes (Lafforgue 1982), each of which corresponded to a simplified cuneiform and was meant to represent all of the syllables sharing the same consonantal phoneme.[6] Clearly, the Ugaritic modification has the advantage of further reducing the number of items to be learned and stored in the GRAPH REGISTERS and, therefore, of favouring further propagation of literacy. Other effects were a loss of the univocacy of the Babylonian CON-VERTORS, hence a massive resurgence of homography *à la sumérienne*, which might be taken as a 'weakness' of the system. For instance, the French words radis [ʀadi] (radish), rideau [ʀido] (curtain), radin [ʀadɛ̃] (mean), rude [ʀyd] (rude), ridé [ʀide] (wrinkled), roder [ʀɔde] (to grind), ardu [aʀdy] (difficult), etc., as well as the names Rodin [ʀɔdɛ̃], Redon [ʀœdɔ̃], etc, would all convert to ⟅symbols⟆ ⟅symbols⟆ if written using the original Ugaritic alphabet. But this is no doubt an abusive example without equivalents in languages now written using a consonantal system (the number of vowel sounds is three times greater in French than it is in Semitic languages, among others). Moreover, it is somewhat difficult for one without personal experience of such systems to appraise the extent in which such a high level of context-dependency in decoding indeed represents a weakness ('consonantal' *Hébreu carré* — as one uses it nowadays in Israel beyond the first phases of school education, at least for texts where some level of potential ambiguity is tolerable, e.g. newspapers — is apparently perceived as 'transparent').

The Greek modification

Although the Ugaritic alphabet eventually incorporated vowels (Healy 1990), it is often claimed that the first fully fledged alphabet — i.e. phonologically explicit (all consonants and all vowels represented) and 'regular' (one and only one specific letter for each phoneme) — appeared in Greece

---

[6] I do not know whether or not a few of these graphemes stood for syllabic vowels (an Israeli friend of mine identifies 'א' as a 'consonant').

about 28 centuries ago.[7] Beyond legitimacy, one of the reasons why priority is thus attributed to the original Greek writing system (at least in the West) is probably its lasting influence on virtually all the Occidental alphabets. This being said, the learning burden imposed upon the GRAPH REGISTERS at the Greek point is more or less comparable to that exacted by consonantal codes and the univocacy of the Babylonian CONVERTORS is being fully restored: context-dependency in decoding is restricted anew to eventual homophones. One may remain under the impression that perfection has been achieved and that no further modification to the CONVERTORS is worthwhile.

The Indian alternative

According to Coulmas (1992), the writing systems now practiced in India (about 200 altogether) stem from a common source which emerged some 29 centuries ago. Many of them, Dravidian ones in particular (Kannada for instance), can be considered as representing an alternative to the 'Greek modification' since, although their graphemes have a syllabic surface, con-sonantic and vocalic components are univocally specified within each written syllable. Such codes thus lend themselves to rule-governed perilexical reading and writing and they might, therefore, be rightfully qualified as 'pseudosyllabic'.

The Korean modification

In spite of the Greek and Indian achievements in this respect, an ultimate modification of the Babylonian convertor was yet to come, and an astonishing one at that. This took place 500 years ago, with the creation of the Han'gŭl writing system in Korea. Although this is challenged by V. Fromkin (personal communication, 1991), Han'gŭl almost qualifies as 'featural writing' according to Sampson (1985), i.e. not only consonants and vowels but also some of their sub-phonemic components are explicitly specified within syllable-like graph-emes (as the Dravidian ones, Han'gŭl thus belongs with the family of pseudo-syllabic writing systems). The features processed owing to the Han'gŭl modifi-cation are stop/continuant, lax/tense, etc., for consonants, and front/back, spread/rounded, etc., for vowels. If formal teaching of the Han'gŭl writing system is indeed founded on this set of features, the number of basic units to be stocked in the GRAPH REGISTERS comprises less than 10 featural opposi-tions; and the secondary units, the 28 letters (24 today), are two sets (con-sonants and vowels) of rather complex visual objects with a number of salient intricate formal similarities. Regarding (verbal) units eventually to be stocked within LEXICONS, the least that one can say is that they are visual objects of a greater order which are made of a string of rather complex visual objects of a lesser order: a standard Westerner can remain under the impression that lexicalizing a significant number of Han'gŭl logograms pertains to virtuosity.

---

[7] Past the Boustrophedonic menace to human brainedness.

Another characteristic of this system is that it permits adequate transcription of foreign names only in so far as they do not comprise phonemes that cannot be dissected by reference to Han'gŭl features (hence perhaps the Korean habit of transcribing European and American names using the Latin alphabet). Admirable as it is, the Han'gŭl modification somehow radiates (again to the eyes of a standard Westerner) an aura of excessive perfection.

Offsprings of the Greek writing system

Unlike modern Greek, a few of the Latin offsprings of the ancient Greek writing system have almost retained the 'regularity' of the mother code. Written Spanish is certainly an example in this respect and, according to A. Ardila (personal communication; Ardila and Ostrosky-Solis 1989), (a) such regularity not only favours but irrevocably prescribes graphophonemic CONVERSION as the sole reading/writing mode accessible to users of the Spanish writing system and, as a consequence, (b) the notion of the LOGOGRAPHIC LEXICON is not pertinent with regard to the Spanish writing system (thereafter 'Ardila's Law').

Other Latin offsprings of the ancient Greek writing system, such as French and English, have gone astray, i.e. have become 'irregular'. Not only has the one-to-one relationship between phoneme and letter disappeared in a significant proportion of the words of these languages but there are numerous instances in which graphophonemic conversion is not univocal (i.e. cannot be dealt with by a straightforward CONVERTOR such as the one invented in Greece 28 centuries ago). In this respect, the most salient characteristic of English is 'homographic heterophony'. This phenomenon is well illustrated by the following set of four English words: 'tough' [tʌf], 'though' [ððʊ], 'through' [θruː] and 'thought' [θɔːt]. In theory, a CONVERSION DEVICE of the Greek type might systematically manage with only one graphophonemic 'ough' correspondence (perhaps the most frequent in the language, say 'ough' = [ʌf] as in 'enough'), and homographic heterophony would therefore prescribe that 'though', 'through', and 'thought' can only be read and written globally, i.e. the Sumerian way, by accessing the LOGOGRAPHIC INPUT and OUTPUT LEXICONS (Fig. 14.3).

French also comprises instances of homographic heterophony, e.g. *mars* [maʀs] (March), but *jars* [ʒaʀ] (gander) and *gars* [ga] (lad). None the less, its most salient irregularity is rather 'heterographic homophony'. This phenomenon is well illustrated by the following set of four French words: *hareng* [aʀɑ̃] (herring), *champ* [ʃɑ̃] (field), *exempt* [ɛ̃gzɑ̃], and *dans* [dɑ̃] (in). Now, since 'eng', 'empt', 'amp', and 'ans' are always pronounced [ɑ̃] at the end of a French word, one might suppose that a robust Greek CONVERTOR could handle these four words [and at least a dozen other ways of writing a final [ɑ̃] in French (Lecours and Lhermitte 1979)] on the decoding side (reading). In writing, however, a Greek CONVERTOR might systematically manage with only one phonographemic [ɑ̃] correspondence (perhaps the most frequent one in the language, say [ã] = 'ans', as in 'dans'), and heterographic homophony would therefore prescribe that *hareng, champ*, and *exempt* can only be written globally, i.e. the Sumerian way, by going through the LOGO-GRAPHIC OUTPUT LEXICON (Fig. 14.3).

Another alphabet of Latin origin that certainly deserves particular attention in the Quoc-Ngu. This one was created around the middle of the seventeenth century by a French Jesuit, Alexandre de Rhodes (Robert 1982), and it has become the official writing system of Viêt-Nam at the turn of the twentieth. On the one hand, it is 'regular' and, on the other, it has the unique characteristic of comprising a set of five diacritics univocally specifying the pitches of a tonal language. The latter no doubt supposes the addition of a very special component to the Greek CONVERTOR.

More on boxes and arrows

Although what is universally God given (here, a metaphor for genetic programme) to bearers of a standard human brain is a potential to be fulfilled through interaction with one's environment, it is not uncommon, among aficionados of boxes and arrows, to act as if everybody within given school-educated subgroups of adult individuals sharing a given written code should have equivalent GRAPH REGISTERS (eventually) as well as equivalent LOGO-GRAPHIC LEXICONS. Provided, of course, that elementary schooling has been completed successfully, and also that one has ascertained that no individual within a given subgroup has a past history of developmental dyslexia/dysgraphia (see below), one might tend to agree with regard to REGISTERS. Not so (obviously, it seems to me) with regard to LEXICONS: here, the magnitude and stability of the stock should be linked to a number of factors, including characteristics inherent to writing systems, length of school education and teaching modalities, and also reading *and* writing habits past the period of formal learning. Let us investigate a little further on these various points.

The above assertion concerning the equivalence of GRAPH REGISTERS among those with standard school education is founded on the belief that these stock memories have never been quantitatively overloaded ever since the first one was switched on in Babylon. Although it is true that syllabaries have more units than alphabets, have not alphabets allowed for preregister proliferation of a sort by the introduction of capitals as opposed to minuscules, of script as opposed to longhand, etc.? Qualitatively, on the other hand, the units to be stocked in registers can be of a lesser or greater level of complexity as visual objects (e.g. the Greek versus the Ugaritic alphabet). Moreover, it seems to me that the modalities of formal teaching might (for codes other than the Sumerian–Chinese) have an impact on the organization of registers and functioning of convertors. For example, consider the Kannada writing system,[8]

---

[8] Conceivably, a remote scion of the Semitic achievements (Cohen 1958).

in which full phonemic specification is achieved within a pseudosyllabic set of graphemes (vowels are specified by a systematic indication appended to a consonantal matrix). Now, learning a system of this sort might, for instance, be done by first attending to the set of consonantic matrices and thereafter to vocalic indications and rules of association between the two, which would eventually lead to knowledge of a set of syllabic graphemes, or else it might be done by directly learning a set of graphemes standing for syllables. As a matter of fact, the former mode of learning (although not the latter) would suppose a duplication of the GRAPH REGISTERS (phonemic counterparts, on the one hand, and syllables, on the other), and this might have an eventual impact on reading/writing strategies, i.e. on the nature of subserving perilexical procedures. Other code-linked learning modalities might also have impacts of their own on reading/writing strategies (e.g. during the first years of school education, Japanese children learn only Kana writing and Israeli children learn to write using both consonants and vowels).

Concerning the magnitude of logographic lexicons [and bearing in mind that my impressions in this respect might to a large extent stem from ethnocentric considerations (Sampson 1985)], it seems to be that stocking large numbers of stable entries should require lesser or greater 'coding energy' and 'storing space' depending on the visual complexity of the incoming information (e.g. English versus Han'gŭl logograms). It is clear, on the other hand, that if input lexical storage is mandatory given a writing system such as the Chinese one, other systems (Spanish, for instance) can theoretically operate without an INPUT LOGOGRAPHIC LEXICON. Therefore, duration and modalities of school education, as well as reading and writing habits, might play a major role in the extent and stability of any individual's INPUT LEXICONS. Thus, in China, limited schooling should only yield mastery of very frequent logograms and, in Spain, being tutored in global as opposed to graphophonemic reading might be at the origin of substantial differences in lexical growth and organization.

When pathology reiterates history

As mentioned earlier, focal brain lesions occurring in fluent readers/writers can cause various types of reading/writing disorders. I will now illustrate this fact and attempt to oppose the notion of universal to that of potentially code-specific dyslexias/dysgraphias. It is the latter, code-specific disorders, that justify the heading attributed to the present subsection of this chapter.

Universal dyslexias/dysgraphias

Also as mentioned earlier, certain (occipital) lesions of the right-hander's left cerebral hemisphere can be the cause of 'agnosic alexia', a syndrome in which the subject: (a) can no longer understand written language (although he or she understands speech normally), (b) can no longer read aloud (although he or she can write), and (c) shows evidence of a disorder in copying written materials (Klippel 1908; Déjerine 1914/1926). By reference to Figs 14.3 and 14.4, this disorder witnesses to dysfunction of a particular subcomponent of the VISUAL INFORMATION SORTING DEVICE. There are no reasons to postulate that peripheral damage of this type might discriminate (qualitatively) between codes and, as a matter of fact, cases of pure alexia have been reported to occur among readers of logographic, syllabic, pseudosyllabic as well as of alphabetic codes. None the less, and this might be taken as providing support to the assumption attributing a certain weight to 'visual complexity' (see above), cases have been reported in which bilingual brain-damaged patients recuperated reading in a Latin alphabetic code (English) somewhat better than in Mandarin (Lyman *et al.* 1938) or Kannada (Karanth 1981), although their premorbid knowledge and use of the latter were at least equal if not greater than their knowledge and use of written English.

Code-specific dyslexias/dysgraphias

On the other hand, and given brain lesions of comparable locations as well as patients with comparable educational backgrounds, the clinical manifestations of other types of acquired disturbances of reading and/or writing might vary—at least by reference to the theoretical framework inherent to the formalizations presented here (Figs 14.3 and 14.4)—with the functional features specific to various writing systems. Let us consider, for example, 'surface dyslexia' (Marshall and Newcombe 1973; Patterson *et al.* 1985) and 'surface dysgraphia' (Beauvois and Derouesné 1981; Goodman and Caramazza 1986), which have been observed in a few readers/writers of irregular alphabetic codes. The salient manifestation of such disorders being one in reading and/or writing irregular words while regular ones are processed normally or nearly so, one might suggest that the 'functional lesion' responsible for such errors involves the Sumerian invention and leaves the Akkadian one intact (whatever the specifications of the CONVERTOR). Clearly, the impact of such a lesion should be dramatic in the case of a reader/writer of Chinese, whereas it might be minimal if not subclinical in that of a reader/writer of a regular alphabetic code such as Spanish, as well perhaps as in that of a reader/writer of a regular pseudosyllabic code such as Kannada or even Han'gŭl; and one might expect Kanji processing to be far more impaired than Kana processing in a user of the Japanese writing system.

Cases have also been reported in which brain-damaged subjects mastering an irregular alphabetic writing system could read words, be they regular or irregular, but could no longer read pronounceable (legitimate) non-words. This clinical picture is currently known as 'phonological dyslexia' (Beauvois and Derouesné 1979). The 'functional lesion' responsible for such a clinical picture might be characterized by saying that the Sumerian invention is not only intact but also fully operational, i.e. operational for all words, irrespective of the regularity/irregularity parameter, whereas the Akkadian invention (Greek modification) is broken. Clearly a lesion of this sort should have no impact on a reader of Chinese unless the biological space attributed to CONVERTORS in reader/writers of non-logographic codes is occupied in China by a phonological indices detector. Furthermore, a functional lesion of this type should leave Kanji processing intact while it might impair Kana processing, at least for infrequent words, in a user of the Japanese writing system. Finally, if Ardila's Law (see above) holds, the same lesion should result in absolute or nearly absolute alexia among readers of Spanish.[9]

Indeed, there are reasons to suggest that Ardila's Law might only be valid with regard to subjects with a limited school education and/or poor reading/writing habits. My argument in this respect is the following. According to Ardila's Law, 'deep dyslexia' [a total incapacity to read legitimate non-words associated to a production of semantic errors in reading real words (Colheart et al. 1980; Lecours et al. 1990)] is no more possible among readers/writers of Spanish than is 'phonological dyslexia'. Now, given that the patients investigated by Ardila and his associates had completed four years of school education on average, it seems to be of great relevance that Ruiz et al. (1994) have observed 'deep dyslexia' in highly educated brain-damaged readers/writers of Spanish. In other words, and I am now referring to a within-code rather than an across-code difference, a same brain lesion can provoke a nearly absolute alexia in a given speaker of a given code and a spectacular production of semantic errors in another reader of the same code.

Four arguments revisited

Let me now turn, at the end of this chapter, to four interrelated comments occupying the place where one would normally expect a discussion and conclusion. The first of these comments has to do with the notion of a 'primary' as opposed to that of a 'secondary' code, and each of the other three is linked to a classical assertion which, in one way or another, has to do with these two notions. One of these assertions maintains that certain writing systems are socially 'perfect' and others not, another, that writing systems are of no interest from the linguistic point of view, and the third that reading/writing abilities are not the expression of a species-specific genetic aptitude.

'Primary' and 'secondary' codes

One can define 'primary' codes as being those *that* are intransitively learned by reference to external and internal objects and events, and 'secondary'

---

[9] As to premorbid readers of Han'gŭl, they should be reduced to reading the names of famous foreigners and foreign brands (a condition which might deserve a label of its own, e.g. 'domestic alexia'?).

codes as being those learned by reference to their primary counterparts. One's mother tongue, whether it is spoken or belongs with the family of sign languages, is the prototype of the primary code. Second languages can be learned as secondary codes but, given my topic and since they come late in individual ontogenesis as they came late in human history, let us consider as prototypical the writing systems invented by humanity, in particular the sound-constrained ones.

In addition to the above, there are a number of differences between spoken and written codes. For example, spoken language comes first and writing second in the ontogenesis of the human species as well as (for the overwhelming majority) of human individuals. Or again, spoken language depends on ears and axial mouth, and written language more specifically on eyes and appendicular hand. But these differences have never, to my knowledge and beyond their historical or biological importance, instigated animated discussion. Others have: for example, great importance has been attributed to the fact that spoken codes are, until now and at least in early phases, nearly always learned by spontaneous imitation and within the relatively intimate social circle of family and kin, whereas written codes are, nowadays and at least in early phases, nearly always learned with a larger peer/social context and following the more formal and explicit fashions inherent to elementary school teaching. But it should be kept in mind that, beyond a certain period of early school education, further learning (mainly but not exclusively, lexical learning) can indeed occur without systematic subordination of one type of code to the other. Thus, a number of individuals (a large number in certain societies) will first learn through the eyes (rather than the ears) an important proportion of their words and manners of assembling words, and will thereafter resort to these words and manners in their spoken speech. What I am now suggesting is that it might be that the above opposition loses much of its cogency beyond early youth (and long before brain ontogenesis has reached its term). Great importance has also been attributed to the fact that written, but not spoken language, has the inherent property to permit one to externalize one's thoughts and memories, including post-mortem: 'a technique of representing speech by a durable trace', writes Hagège (1988).[10] But is not this the prototype of the argument that loses its momentum with the passage of time? As far as I am aware, autodafes were not invented yesterday: and durable recording techniques have provided many with the power to exert strong post-mortem spoken speech influence.

'Perfect' codes

In an essay on developmental dyslexia, Russell (1982) writes that, 'For a system of writing to be successful' (to be 'perfect' and to deserve greater endurance and dissemination) it should enable the greatest possible number

---

[10] *'Les paroles s'envolent mais les écrits restent.'*

of people to express their language in that system. Indeed, this is a generous and also perhaps a democratic idea. Russell later suggests that the existence of developmental dyslexia is the price to be paid by societies resorting to a 'perfect' writing system; that is, to an alphabetic system of Greek origin. However, I am not entirely convinced by the developmental dyslexia component of Russell's assertion. One may wonder, for example, why a 'perfect' code should generate a 'disorder', especially since a number of dyslexic children seem to fare much better with logographic ('global') than with graphophonemic ('analytic') reading (Lyon *et al.* 1982). Moreover, before incorporating Russell's opinion concerning developmental dyslexia to one's standard teachings, one might wish to scrutinize data on the incidence of developmental dyslexia among learners of different written codes, alphabetic or otherwise. For example, I would be surprised if it were to appear that the situation is the same for regular as opposed to irregular alphabetic codes, e.g. for Spanish as opposed to English. And I would be surprised if essential differences were to be observed in this respect between regular alphabetic and regular syllabic codes, e.g. between Japanese Kana and Spanish if the comparison were made during the first four years of school education. Further, I would also be surprised if groups of Chinese children of comparable 'intelligence' turned out to be homogeneous with regard to individual aptitudes to master a few thousand logograms.

There are other dimensions to the Russell assertion. For instance, could not one argue that, for a writing system to be perfect and/or successful (especially in periods of demographic turmoil), it should enable the greatest possible number of unilingual individuals speaking different languages to communicate through writing? Or else, might one not wish to ponder on the eventual advantages and disadvantages of mastering a code which permits one to read and write about what one does not understand?

On theoretical interest

Still considered ineluctable by a large number of linguists according to Sampson (1985), the second assertion was summarized by Householder in 1969 when he wrote that the first Bloomfieldian principle is that: 'Language is basically speech, and writing is of no theoretical interest'. Now, unless I have missed something very important, the basic reason why spoken language is of great theoretical interest is that it permits communication of thought by handling a set of arbitrarily selected abstract materials of progressively greater complexity following the rules of highly structured principled conventions. As far as I can understand, these conditions are met by written as well as by spoken language or, in more general terms, by secondary as well as by primary codes.

This being said, I readily recognize that I know very little about the arcanes of linguistics and that I might indeed have missed something very important concerning the justification of the first principle; that is, linguists may have their reasons to exclude written codes from their field of blooming. But, as a neurologist interested in language disorders, I would feel as comfortable as an illiterate working as a librarian if I had to take it for granted that the reading and writing disturbances of the aphasics are of no theoretical interest from the neurolinguistic point of view.

On innate abilities

As formulated by Hagège (1988), the third 'assertion' is that, unlike spoken speech, writing is not a 'defining property of the human brain' but rather a 'human invention'. According to this view, spoken speech abilities represent, given appropriate training, the immediate expression of an innate property of the human brain and body just as flying/diving abilities represent, given appropriate training, the expression of an innate property of the gannet's brain and body. In more abstract terms, the mapping between thought and sound is immediately brain-constrained, whereas, beyond biological properties, the mapping between phonology and written code is further specified by culture. In the absence of extreme environmental anomalies, one is biologically abnormal if one does not acquire a primary code during the first few years following one's birth, whereas, at this point in the evolution of humanity, it is clear that the acquisition of written language is not mandatory as is that of spoken language, i.e. one can be biologically normal and yet never acquire a written code. While this is correct, it is also potentially misleading, a fact of which I became aware in the context of a discussion about the many boys and fewer girls (see McGlone 1980) in a school for children with 'learning disabilities': if one fails to acquire a secondary code in a society where writing is mandatory and schooling immediately available, one is 'abnormal' and one's handicap could be brain-based. In other words, I am no longer as certain about defining properties of the human brain. I would at this point feel more secure with the notion of a defining property permitting the species to acquire knowledge about its changing environment by creating abstractions from the information it receives, and by organizing these abstractions in line with rule-governed arbitrary codes that it has the capacity to invent. Inventions of this sort would then represent actualizations of a defining property and occur in given circumstances as the result of an interaction between the individuals of given subsets of the species, i.e. as the result of 'social pressures'.

It might be that, about 35 000 (if not a few millions) years ago, environmental pressures exerted on the human brain led it to exploit its genetic potential and to invent spoken speech, and that, in due time, about 5300 years ago, other social pressures—of another nature, although generated to a large extent by the impact of the first invention—led the human brain to exploit its genetic potential anew, and to invent writing systems, then resorting to the same as well as to other of its neuroaxonal assemblies.

What has happened subsequently—from stylus and brush, to pen and pencil, to typewriters and word processors—might also be taken as witnessing to an aptitude of the human brain–mind complex to invent new techniques as the result of social changes and to adapt its modes of functioning accordingly.

Acknowledgement

The author's research on disorders of reading and writing is funded by the Human Frontier Science Program.

References

Alajouanine, Th. (ed) (1960). *Les grandes activités du lobe occipital*. Masson, Paris.

Amiet, P. *et al.* (1982). *Naissance de l'écriture*. Editions de la Réunion des musées nationaux, Paris.

Ardila, A. and Ostrosky-Solis, F. (1989). *Brain organization of language and cognitive processes*. Plenum, New York.

Beauvois, M. F. and Derouesné, J. (1979). Phonological alexia: Three dissociations. *J. Neurol., Neurosurg. and Psych.* **42,** 119–24.

Beauvois, M.E. and Derouesné, J. (1981). Lexical or orthographic agraphia. *Brain* **104,** 21–49.

Cameron, R.F., Currier, R.D., and Haerer, A.F. (1964). Aphasia and literacy. *Br. J. Disorders Commun.* **6,** 161–3.

Cohen, M. (1958). *La grande invention de l'écriture et son évolution*. Klincksieck, Paris.

Coltheart, M., Patterson, K., and Marshall, J.C. (1980). *Deep dyslexia*. Routledge & Kegan Paul, London.

Coulmas, F. (1992). *The writing systems of the world*. Blackwell, Oxford.

Critchley, M. (1964). *Developmental dyslexia*. Heinemann, London.

Déjerine, J.J. (1914–26). *Sémiologie des affections du système nerveux* (Revised edn 1926). Masson, Paris.

Février, J.G. (1959). *Histoire de l'écriture*. Payot, Paris.

Galaburda, A. (1989). *From reading to neurons*. MIT Press, Cambridge, MA.

Galaburda, A. and Kemper, T. (1979). Cytoarchitectonic abnormalities in development dyslexia; a case study. *Annals of Neurology*, **6,** 94–100.

Gelb, I.J. (1962). *A study of writing*. Chicago University Press, Chicago.

Goodman, R.A. and Caramazza, A. (1986). Aspects of the spelling process: Evidence from a case of acquired dysgraphia. *Language and Cognitive Processes*, **1,** 263–96.

Hagège, C. (1988). Writing: The invention and the dream. In *The alphabet and the brain*. (ed. D. de Kerckhove and Ch. J. Lumsden), pp. 72–83. Springer, New York.

Healy, J.F. (1990). *The early alphabet*. University of California Press, Berkeley, CA.

Householder, F.W. (1969). Review of Langacker: Language and its structures. *Language* **45,** 886–97 (quoted by Sampson 1985).

Jean, G. (1987). *L'écriture mémoire des hommes*. Gallimard, Paris.

Karanth, P. (1981). Pure alexia in a Kannada–English bilingual. *Cortex* **17** 187–98.

Kemper, T.L., Lecours, A.R., Gates, M.J., and Yakovlev, P.I. (1973). Retardation of the myelo- and cytoarchitectonic maturation of the brain in the congenital rubella syndrome. *Early Development* **51,** 23–62.

Klippel, M. (1908). Discussion sur l'aphasie. *Revue neurol.* **16,** 611–36.
Lafforgue, G. (1982). Alphabet. *Encyclopedia universalis*, Vol. I, pp. 961–3. Encyclopedia Universalis, Paris.
Lecours, A.R. (1975). Myelinogenetic correlates of the development of speech and language. In *Foundations of language development* (ed. E.H. Lenneberg and E. Lenneberg), Vol. 1, pp. 121–35. Academic Press, New York.
Lecours, A.R. and Lhermitte, F. (1979). *L'aphasie.* Flammarion, Paris.
Lecours, A.R. and Tainturier, M.J. (1990). Perturbations reconnues et perturbations pensables du lexique mental. In *Les agraphies* (ed. P. Morin, F. Viader, F. Eustache, and J. Lambert), pp. 243–75. Masson, Paris.
Lecours, A.R. and Yakovlev, P.I. (1966). Retarded myelination in the forebrain of a five-month old infant. In *Agenda of the meeting of the collaborative study on cerebral palsy, mental retardation and other neurological and sensory disorders of infancy and childhood*, pp. 19–20. McClelland, Washington, DC.
Lecours, A.R. *et al.* (1988). Illiteracy and brain damage. 3. A contribution to the study of speech and language disorders in illiterates with unilateral brain damage (initial testing). *Neuropsychologia* **26,** 575–89.
Lecours, A.R., Lupien, S., and Bub, D. (1990). Semic extraction behavior in deep dyslexia: Morphological errors. In *Morphology, phonology and aphasia* (ed. J.L. Nespoulous and P. Viliard), pp. 60–71. Springer, New York.
Lordat, J. (1843). Analyse de la parole pour servir à la théorie de divers cas d'alalie et de paralalie que les nosologistes ont mal connus. *Journal de la Société de médecine pratique de Montpellier* **7,** 333–53, 417–33; **8,** 1–17.
Lyman, R.S., Kwan, S.T., and Chao, W.H. (1938). Left occipito-parietal brain tumor with observations on alexia and agraphia in Chinese and in English. *Chinese Medical Journal*, **54,** 491–516.
Lyon, R., Steward, N., and Freedman D. (1982). Neuropsychological characteristics of empirically derived subgroups of learning disabled readers. *Journal of Clinical Neuropsychology*, **4,** 343–65.
Marshall, J.C. and Newcombe, F. (1973). Patterns of paralexia: A psycholinguistic approach. *Journal of Psycholinguistic Research*, **2,** 175–200.
McGlone, J. (1980). Sex differences in human brain asymmetry: A critical survey. *Behav. Brain Sci.* **3,** 215–63.
Morin, P., Viader, F., Eustache, F., and Lambert, J. (ed.) (1990). *Les agraphies.* Masson, Paris.
Morton, J. (1984). Brain-based and non-brain-based models of language. In *Biological perspectives on language* (ed. D. Caplan, A.R. Lecours, and A. Smith), pp. 40–64. MIT Press, Cambridge, MA.
Patterson, K., Marshall, J.C., and Coltheart, M. (1985). *Surface dyslexia: Neuropsychological and cognitive studies of phonological reading.* Erlbaum, Hillsdale, NJ.
Robert, P. (1982). *Le petit Robert 2.* Le Robert, Paris.
Ruiz, A., Ansaldo, A.I., and Lecours, A.R. (1994). Two cases of deep dyslexia in unilingual hispanophone aphasics. *Brain and Language* **46,** 245–56.
Russell, G.F.M. (1982). History of writing and its relevance for the linguistic disorder in dyslexia: Discussion paper. *Journal of the Royal Society of Medicine*, **75,** 631–40.
Sampson, G. (1985). *Writing systems.* Stanford University Press, Stanford, CN.
Sasanuma, S. (1975). Kana and Kanji processing in Japanese aphasics. *Brain Lang.* **2,** 369–83.
Scribner, S. and Cole, M. (1981). *The psychology of literacy.* Harvard University Press, Cambridge, MA.

Yakovlev, P.I. (1968). Telencephalon 'impar', 'semipar' and 'totopar'. *Intern. J. Neurol.* **6**, 245–65.

Yakovlev, P.I. and Lecours, A.R. (1967). The myelogenetic cycles of regional maturation of the brain. In *Regional development of the brain in early life* (ed. A. Minkowski), pp. 3–70. Blackwell, Oxford.

Yakovlev, P.I. and Rakič, P. (1966). Patterns of decussation of bulbar pyramids and distribution of pyramidal tracts on two sides of the spinal cord. *Trans. Am. Neurol. Assoc.* **91**, 366–7.

DISCUSSION

Participants: A. Caramazza, J.-P. Changeux, R.L. Holloway, A. Roch Lecours, L. Weiskrantz

Modularity and written language

Answering questions from **Changeux, Lecours** suggested: (i) that although reading and writing are subserved in part by specific brain structures, they cannot be considered as 'modular' in the Fodorian sense of this term, and (ii) that certain illiterates perhaps perform better on some visual tasks, using brain structures that, in literates, are dedicated to written language processing.

Written language in deaf or blind people

Weiskrantz and **Holloway** asked about the neuropsychology of written language in deaf or blind people.

Caramazza: To answer your question properly, one would first of all need to have some idea of the nature of the normal performance in these subjects. Deaf subjects are not normal in reading or in writing. For instance, deaf subjects are likely to make spelling errors, such as 'rhaic' for 'chair', which is an extremely unlikely event for a hearing child. The deaf population is ideal for studying the relationship between phonology and orthography. Regarding reading deficits in Braille, some cases have been reported, but they are mostly anecdotal description.

Part IV

Intelligence

15

The origins of consciousness
LAWRENCE WEISKRANTZ

Introduction

The term 'origins', in the topic that was assigned to me, is almost as ambiguous as 'consciousness'. I shall discuss the topic with reference to three different questions: Why?, How?, and Whether?. That is, what is it good for, how is it generated, and who has it. Implicitly, I shall have to say what I mean by 'it', consciousness. It has multifarious connotations, and I think little is to be gained by trying to impose a strict definition on a field before it is understood — like trying to define voltage before Ohm. But it will be seen that I will be referring to a relatively restricted domain, which derives from a neuropsychological context. The gist of the meaning is not only in being aware, i.e. being able to detect, but *knowing* that one is aware — acknowledged awareness.

Why?

If an attribute or capacity is worth having, why is this so? In an evolutionary sense, what advantage accrues to consciousness, and what kind of capacity is it that would be, or would have been, selected? There are various suggestions that stem from considerations of social complexity, of understanding and predicting on the basis of 'other minds', and the like. But I take my own starting point from the clinic, where in recent years there have been discoveries of a variety of syndromes following brain damage of patients who have lost the acknowledged awareness of capacities that they still retain, and of which we are all normally aware if not brain damaged. There are by now several examples, and they cover the spectrum from detection, to selective attention, to cognitive processing, to comprehension of language without awareness, and to memory — all without awareness. Many of these examples were well reviewed by Schacter *et al.* (1988) in a previous Fyssen Symposium, and I will review them only briefly here before turning to the question of what are the consequences of *not* having this species of awareness?

The most counter-intuitive is of detection without awareness, and the most studied example is of 'blindsight'. Lack of awareness of optical stimuli is caused by damage to the visual cortex in man. We know from animal work that, nevertheless, there can be intact visual discrimination after total removal of striate cortex; this is mediated by one or more of several non-striate pathways. But under certain conditions, discrimination is also found in patients to stimuli in their blind fields, even though the patients fail to acknowledge or admit that there were any stimuli. This is the phenomenon of blindsight. In particular, patients may be able to locate stimuli by reaching out for them, may be able to 'guess' whether or not a moving stimuli occurred, and so forth. There is now a considerable literature on this topic (cf. reviews by Weiskrantz 1986, 1988, 1990; Cowey and Stoerig 1991a). It is not yet clear just which discriminative properties are possible in which sort of patients—probably no patient ever has damage just to the striate cortex alone. But in some patients it has been possible to measure visual acuity, orientation discrimination, wavelength discrimination, and spectral sensitivity. There are also a range of non-verbal approaches to blindsight involving autonomic responses or interactions between the intact and blind fields. For present purposes, what we have is an example of intact functions in the absence of acknowledged awareness, caused by damage to the visual neocortex. In fact, it appears that these intact functions can be further fractionated so that, for example, movement might be preserved but not wavelength discrimination, and so forth.

An amusing and related example has recently been reported in *Nature* by Marshall and Halligan (1988). A patient with right parietal damage and consequential left-sided visuospatial neglect could not discriminate the difference between two pictures (placed one above the other) because the two drawings were identical in their right halves, even though they were markedly different on the left—one of the houses was on fire on its left side, the other was normal. The patient consistently judged the two pictures to be identical. But when forced to say which house she would prefer to live in—protesting that it was a pointless question because both drawings were the same—she reliably chose the one without the fire! So some part of this patient was *not* 'neglecting'.

A second example is in the field of prosopagnosia—the loss of conscious recognition of familiar faces following brain damage. It is a socially devastating and embarrassing disorder, fortunately rare. Given the electrophysiological evidence from primates, it seems reasonable to ascribe the brain damage to restricted regions of the neocortex of the temporal lobe. Despite this loss of conscious recognition of familiar faces, Bauer (1984), Tranel and Damasio (1985), and others have shown that in some cases the autonomic nervous system still continues to discriminate normally between old and new faces. More remarkably, DeHaan and colleagues have shown that various attributes of knowledge associated with particular faces are still discriminable when the patients are forced to guess, for example, whether the face is that of a politician or an actor, the basis of which must be past knowledge of the person in question, even though the patients are at chance in discriminating familiar from unfamiliar faces (DeHaan *et al.* 1987a,b). Of this phenomenon, too, there is an increasingly large literature.

A third example is from the field of aphasia, and is even more surprising given the density of the disorder as judged from clinical examination alone. A recent study by Lorraine Tyler (1988) in Cambridge is of a severely disabled Wernicke's aphasic patient. He could understand at best only some very simple questions and instructions, and could utter a few recognizable words, but no more. He was at chance in formal language tests. Nevertheless, Tyler showed that the patient was perfectly sensitive to the nuances of the English language, etc. One way she demonstrated this was to instruct the subject to respond as quickly as possible to a particular English target word. She then buried this word in appropriate or inappropriate sentences, the latter being degraded in very specific ways either semantically, syntactically, or pragmatically. The key comes from the fact that normal subjects spot the target word faster when it is in a normal linguistic structure than when it is in an anomalous sentence. And so does the aphasic patient.

A related example comes from spotting a word in a larger prose structure, where the meaning and syntax must be integrated over a much longer string. Here normal subjects show an increasing speed of response to the target word the later it falls in the prose section, and importantly they are also sensitive to the linguistic structure. And so is the patient. Overall, he is slower, but that is not in itself surprising. The important point is that he shows the normal pattern.

A final experiment by Tyler demonstrated that this is a genuine dissociation between covert comprehension (or what she calls 'online' comprehension) and overt comprehension ('offline'). She presented the same sentences as in the first experiment to the patient and simply asked him whether there was anything wrong with each of them by his saying either 'good' or 'bad'. Here R.H. was at chance—for the very same feature that had been appreciated implicitly. Overtly he did *not* comprehend, but covertly he did. This patient, R.H., was in no way untypical of a class of Wernicke's aphasics clinically, and I have no doubt that Tyler's results will not turn out to be unusual. Indeed, a study by Milberg and Blumstein in 1981 showed that aphasic patients were normally sensitive to priming by semantically related words in a lexical decision task.

One of the best and most thoroughly studied disconnections of awareness involves a system that appears to be organized critically as a whole subcortically, although, of course, it has cortical connections. This is in the field of memory loss, in the amnesic syndrome. The most striking feature of the syndrome is that persons report not remembering anything at all from minute to minute. You can be talking to such a patient, then leave the room and come back a minute later: the patient will not recognize you and admits of having no memory of having seen you. At the same time, the patient does not forget his well-established skills, and especially his verbal skills. Indeed, he or she may continue to tell you the same story or the same joke minute after minute, lacking any memory for having just told it. They cannot remember the day or time of day, unless they guess from looking at the amount of daylight or using some other cue. They are disconnected from any novel experience, and from the experiences that occurred even years before they became ill. Without going into the nuances of different putative types of amnesia, there is now agreement among all workers in the field that intact learning and retention can be demonstrated without any conscious recognition as such of the items being learned and retained. This is sometimes called implicit memory or procedural memory. Even though the patients fail to recognize any event as having occurred beyond a minute or so, they show good retention of material tested indirectly.

This first became clear from studies of learning of motor skills, such as the mastery of tracking tasks or mirror drawing (Corkin 1968). But even with verbal material they can be shown to have good powers of retention. For example, in a series of experiments by Professor Warrington and the author at the National Hospital in London initiated some 20 years ago (cf. Warrington and Weiskrantz 1968, 1973), we showed amnesic patients long lists of words, for example. If asked to recall or recognize the words from that list a few minutes later, the patients were at chance. But if we then showed them the first few letters of the word and simply asked them to *guess* what words they stood for, they were greatly helped by having seen the word in a list before. Needless to say, they were not helped if the word had not been in the list. The same technique can be used with pictures. With such a method one can study learning and forgetting in a systematic way. The phenomenon is called 'priming' in technical language, and is now much studied also with normal subjects in experimental psychology. There are now several other examples of intact learning and retention.

These examples are by no means exhaustive, and there is something of an epidemic of examples appearing not only in neuropsychology, but also from extensions into the psychology of normal subjects (e.g. Tulving 1983, pp. 100–20). We can say that in every major realm of cognitive achievement — perception, recognition memory and recall, language, problem-solving, meaning, acquisition of motor and other skills — residual capacities can be disconnected from awareness. The strength of these residual processes can be and often are equal to those shown by normal control subjects.

But what are the consequences for these patients? They are serious: the amnesic patient is severely impaired, and requires continuous custodial care, despite all of his or her primed retention. The blindsight patient continues to fail to identify objects and to bump into things in his blind field. The prosopagnosic patient fails to recognize his family, even though his autonomic nervous system does. The aphasic patient fails to comprehend speech even though implicitly he can process both semantics and syntax. In short, none of these patients can *think* or *imagine* in the terms of the capacity of which they are unaware. I am fond of a quote from Lloyd Morgan (1890, p. 375, quoting Mivart): 'If a being has the power of thinking "thing" or "something", it has the power of transcending space and time. Here is the point where intelligence ends and reason begins'. For present purposes, substitute 'conscious awareness' for 'reason'.

On this view, conscious awareness evolves because it is advantageous to be able to think and to imagine: to relate current events in a mental world that has both a past and a future. This includes, of course, a capacity for thinking of other minds and about social complexity. But I also think that there are costs: social consequences are not always positive. The fact that we have the capacity to think about what others might be thinking produces a condition that is probably species-specific to humans: namely, paranoia. And in the non-social domain, extreme obsessionality — failure to *stop* thinking — is perhaps more common in 'thinking' persons than in 'non-thinking' ones.

How?

In some of the examples I have reviewed, for example in blindsight and the amnesic syndrome, it is possible to say *which* lesions are critical. In amnesia, a system of limbic lobe structures are necessary for overt recognition. In blindsight, the primary visual cortex appears to be necessary for 'seeing with awareness', and knowledge is advancing about the properties and capacities of the extra-striate pathways that mediate residual visual functions (Cowey and Stoerig 1991*a*). But the primary visual cortex is not *sufficient*, removal of all *non*-visual cortex (but leaving motor cortex intact) is reported to render a monkey clinically blind (Nakamura and Mishkin 1980). Such an experiment, or such a state caused accidentally, cannot occur in humans, but on the assumption that the same result would occur, it would follow that non-visual cortex is also necessary, but not sufficient (because a striate cortex alone can cause 'blindness'). Therefore, to assign 'visual awareness' to a single place or structure, as in the style of 'the sensorium' popular with nineteenth-century physiologists, would appear to be unhelpful.

Neurones are neurones — full stop. What we have to ask is why in certain instances neurones are not sufficient for awareness in isolation (an extreme example being the unfeeling but *responding* isolated spinal cord of the paraplegic patient), whereas in other cases they are. The answer — if it is to be sought within a monistic realm of neural activity — must lie in their *organization*, either in its dynamic and/or structural connectivity.

There is a less specific view, that 'consciousness' *emerges* out of increasing neuronal complexity. But it is not clear why complexity, in itself, should lead to such a result. Does it do so inevitably and *de novo*? Is the blindsight subject 'visually unaware' because of a mere reduction in complexity? Or does the amnesic patient not recognize items he has stored because of a mere (quite small) reduction caused by a lesion in the mamillary bodies of the hypothalamus? Is a monkey, with a nervous system that is smaller than the human's, *proportionately* less aware, or is there a threshold of complexity? If there are certain critical organizational features that are necessary for awareness, they may be more likely to require and to occur in a complex nervous system than a simple one, but complexity *per se* does not reveal what they are.

There is another view that arises from a concern with issues of dualism and free will, and which appeals to phenomena at the level of quantum mechanics, with 'collapse of the wave function taking place through the presence of the observer' (Burns 1990), such a collapse allowing changes between two different domains or states. At a cosmological level all of us remain ignorant, but at a more mundane level one asks why quantal effects are so restricted to particular loci in the brain, or indeed why just to the brain? For example, quantum effects are well established as a limiting factor in the sensitivity of the retina, yet no one attributes consciousness to the retina *per se*, nor indeed to nerves in tissue culture. Indeed, why only to neurones? In terms of specific mechanisms, some theorists have apparently argued that special effects can occur through electrical (as opposed to chemical) synapses, and that 'electronic tunnelling, which provides a possible mechanism through which parts of the brain that are involved in active processing at any given time could be linked into a single quantum mechanical system' (Burns 1990). Perhaps such rare synapses happen to occur at just the critical points in the brain in which lesions cause the dissociations we have reviewed. This would be an extraordinary and *prima facie* unlikely coincidence, but at least open to empirical test. It might also imply that some simple organisms are super-conscious?

Quantum explanations fit most comfortably with dualistic (or even tri-istic, e.g. Popper and Eccles 1977) metaphysical assumptions. If we wish to remain within a monistic and mechanistic framework, dynamic or structural network arguments come to the fore. A dynamic hypothesis has been advanced by Francis Crick and Christof Koch (1990), who link the question of awareness to the 'binding problem' — the problem of how the multiplicity of specialized visual neurones allow a single coherent visual image to emerge. Attentional mechanisms selectively focus on particular details of the large variety of visual inputs to (primary visual cortex, V1), subsequently distributed widely to many separate specialized cortical areas (or modules). The relevant binding at any moment is associated with correlated firing of dispersed neurones in the 40–70 Hz range, as discovered in the cat cortex by Gray, Singer, and colleagues (Gray *et al.* 1989). That is, activity of neurones *some distance apart* display correlated oscillations when responding to single rather than multiple retinal stimuli. The hypothesis links the proposed solution to the binding problem to short-term attention, and hence to short-term working memory. While it is an intriguing hypothesis, the reported oscillations are still based entirely (as far as I am aware) on studies with *anaesthetized* cats. It has also been claimed that the relevant driving oscillators have a retinal, rather than a cortical origin. The hypothesis is thus at an early and perhaps precarious stage of development, but at least it does lead to testable predictions, e.g. that particular lesions would destroy the correlated 40–70 Hz oscillations, and hence would yield blindsight — visual processing without awareness.

Structural views appeal to parallel hierarchical processing (harking back, interestingly, to the kind of general view advanced influentially by Hughlings

Jackson more than a century ago, based on differential impairments of 'higher' functions with preservation of 'lower' in epileptic states.) Appeals to such arrangements have been advanced by theorists such as Johnson-Laird (1988). I have suggested (1986) that neuropsychological dissociations reviewed above may reflect disconnections of monitoring systems organized in parallel with lower-order processing systems. Monitoring systems would, in turn, also have communicative links, especially via language. This approach, interestingly, is also consonant with philosophical views of awareness involving 'higher-order thoughts', as advanced by David Rosenthal (1986). Just how such an approach translates into specific neural arrangements in, say, the visual system, is a challenge, but a realizable one.

Finally, some have argued for a specific locus in a box-and-arrow type of model for a 'conscious awareness module' (with a possible locus in the parietal lobe, given the evidence such as unilateral neglect) with specific connections to knowledge modules and an episodic memory system (among others). Schacter's (1989) DICE schema ('dissociable interactions and conscious experience') was deliberately designed to assimilate neuropsychological evidence of dissociations of awareness from residual function, with particular reference to multiple memory systems. Such a schema leaves unanswered the questions of how the awareness module achieves awareness, and why other modules do not (or is this a pseudo-problem?). Schacter (1991) also considers the alternative possibility that multiple subsystems will be sufficient without a special awareness module, e.g. that an episodic memory system is intrinsically sufficient to account for conscious recollection.

No one can maintain that an answer to the question of 'How?' is within sight. What is evident, however, is that the neuropsychological evidence has led to lines of thinking that simultaneously can draw on quantitative aspects of neurophysiological phenomena and neuroanatomical arrangements, on the one hand, and well-attested functional dissociations of awareness on the other. We at least have an appropriate tunnel at the end of which we might anticipate the light.

Whether?

Enquiries within that tunnel are bound to involve animal experiments. They, in turn, lead to the question of how one assesses *whether* animals have conscious awareness in particular situations, and of course that question has direct relevance to evolutionary issues in their own right.

In terms of the dissociations we have discussed above, the question can be made quite specific. For example, a monkey without striate cortex (V1) can be trained to make visual discriminations (using the 9 or so non-striate routes from the retina to the cerebrum) and these residual capacities are of the same order of magnitude quantitatively as seen in cases of blindsight (Weiskrantz 1986, 1990). The human blindsight subject discriminates but does not have awareness of 'seeing'. Does the monkey without striate cortex, similarly, discriminate without awareness?

A specific solution to the question can be proposed, which has been outlined elsewhere (Weiskrantz 1986, 1988). A similar, but independent solution has been proposed by Cowey and Stoerig (1991*b*). The basic manoeuvre is to put the discrimination and a commentary upon the discrimination under separate control. Thus, as an animal can discriminate between two stimuli with two keys, let us say, and independently indicate on two other keys whether it was 'guessing' or 'not-guessing'. It is clear how the human blindsight subject would respond. We still do not know how the animal would respond, but I have no doubt that an answer will be sought. The approach has an obvious family resemblance to the use of level-of-confidence judgements in psychophysics. A specific, response-bias-free signal detection procedure has been developed by Kunimoto *et al.* (C. Kunimoto, J. Miller, and H. Poshler, unpublished) to deal with the question of quantitative assessment of subliminal perceptual effects, a field in which the methodological issues are closely similar to blindsight. This is precisely equivalent to the use of 'commentary keys' in parallel with discrimination keys. Thus, specific quantitative procedures are available to put the human and animal enquiries on a common footing.

In fact, a behavioural homologue of the dissociation in the amnesic syndrome — between cued responding to stimuli in learned discrimination task but in the absence of recognition — was successfully employed by Gaffan (1974) with monkeys with fornix lesions (one of the structures involved in the circuits interrupted in human amnesia). The same set of visual stimuli that could be correctly discriminated, based on learned associations with food reward, could not be correctly discriminated in a recognition task from stimuli not seen before, thus closely paralleling a result with amnesic patients, who are normal with cues to previously primed stimuli but are at chance in forced-choice recognition with the same set of stimuli (Warrington and Weiskrantz 1974). The two procedures were carried out successively, rather than simultaneously, but the principle of independent examination of two modes of processing of a common set of stimuli — 'discriminating' and 'recognizing' — is the same as that outlined above.

Within the domain of normal animal behaviour, a closely similar approach was used some years ago by Beninger and colleagues (1974). They allowed behaving rats to perform freely, but four possible acts (face-washing, rearing up, walking, or remaining immobile) were selected for experimental enquiry. The question was, when the rat happened to perform any of these four acts, did it *know* what it was doing? The rats were trained to press a different one of four levers to indicate what they had just done. Rats could learn to do it. One would be surprised if monkeys could not, but equally surprised if a sea slug or a crab could. In principle, therefore, there is an approach to the question of whether an animal knows what it is doing, by freeing the methodology from a dependence on an act or a discrimination as such, and introducing an independent parallel 'commentary' response (either simultaneously, as suggested above, or sequentially, as in the Beninger *et al.* procedure) by the animal. Having said that, there are the attendant difficulties that arise from 'knowledge' becoming automatic, i.e. the animals with over-training might chain the two sets of responses together, much as we do when we ritualistically discuss the weather. But this, in its own right, is an interesting issue that has been addressed experimentally by Dickinson (1985, 1988).

The question of knowing 'whether?' an animal is aware, thus, allows both a link to be forged with human neuropsychological dissociations of awareness, the neurological arrangements of the systems that underly them, as well as enquiries into phytogenetic and evolutionary questions of animal awareness. If an animal has a capacity to know what it is doing, it has a capacity that can be selected. It is just as interesting, and just as open, to know when this capacity arises developmentally in children as it is to know when it arose phylogenetically, and to relate these, in turn, to the development of relevant brain circuitry. The 'Why?', the 'How?', and the 'Whether?', when brought together, can thus be seen to illuminate each other.

References

Bauer, R.M. (1984). Autonomic recognition of names and faces in proposagnosia: a neuropsychological application of the guilty knowledge test. *Neuropsychologia* **22**, 457–69.

Beninger, R.J., Kendall, S.B., and Vanderwolf, C.H. (1974). The ability of rate to discriminate their own behaviours. *Can. J. Psychol.* **28**, 79–91.

Burns, J.E. (1990). Contemporary models of consciousness: Part I. *J. Mind Behav.* **11**, 153–72.

Corkin, S. (1968). Acquisition of motor skill after bilateral medial temporal lobe excision. *Neuropsychologia* **6**, 255–65.

Cowey, A. and Stoerig, P. (1991*a*). The neurobiology of blindsight. *Trends Neurosci.* **14**, 140–5.

Cowey, A. and Stoerig, P. (1991*b*). Reflections on blindsight. In *The neuropsychology of consciousness* (ed. D. Milner and M. Rugg). Academic Press, London.

Crick, F. and Koch, C. (1990). Towards a neurobiological theory of consciousness. *Sem. Neurosc.* **2**, 263–76.

DeHaan, E.H.F., Young, A., and Newcombe, F. (1987*a*). Face recognition without awareness. *Cogn. Neuropsychol.* **4**, 385–415.

DeHaan, E.H.F., Young, A., and Newcombe, F. (1987*b*). Faces interfere with name classification in a prosopagnosic patient. *Cortex* **23**, 309–16.

Dickinson, A. (1985). Actions and habits: the development of behavioural autonomy. *Phil. Trans. Roy. Soc. (London)* **B308**, 67–78.

Dickinson, A. (1988). Intentionality in animal conditioning. In *Thought without language* (ed. L. Weiskrantz), pp. 305–25. Clarendon Press, Oxford.

Gaffan, D. (1974). Recognition impaired and association intact in the memory of monkeys after transection of the fornix. *J. Comp. Physiol. Psychol.* **86**, 1100–9.

Gray, C.M., Konig, P., Engel, A.K., and Singer, W. (1989). Oscillatory responses in cat visual cortex exhibit inter-columnar synchronization which reflects global stimulus properties. *Nature* **338**, 334–7.

Johnson-Laird, P.N. (1988). A computational analysis of consciousness. In *Consciousness in contemporary science* (ed. A.J. Marcel and E. Bisiach). Oxford University Press.

Kunimoto, C., Miller, J., and Pashler, H. (unpublished MS, Dept. Psychology Univ. California, San Diego.) Perception without awareness confirmed: a bias-free procedure for determining awareness thresholds.

Marshall, J. and Halligan, P. (1988). Blindsight and insight in visuo-spatial neglect. *Nature* **336**, 766–7.

Milberg, W. and Blumstein, S.E. (1981). Lexical decision and aphasia: evidence for semantic processing. *Brain Lang.* **14**, 371–85.

Morgan, C. Lloyd (1890). *Animal life and intelligence*. Edward Arnold, London.

Nakamura, R.K. and Mishkin, M. (1980). Blindness in monkeys following non-visual cortical lesions. *Brain Res.* **188**, 572–7.

Popper, K.R. and Eccles, J.C. (1977). *The self and its brain*. Springer, Berlin.

Rosenthal, D.M. (1986). Two concepts of consciousness. *Philosoph. Studies* **49**, 329–59.

Schacter, D.L. (1989). On the relation between memory and consciousness: dissociable interactions and conscious experience. In *Essays in honour of Endel Tulving* (ed. H.L. Roediger, III and F.I.M. Craik), pp. 355–89. Erlbaum, Hillsdale, NJ.

Schacter, D.L. (1991). Consciousness and awareness in memory and amnesia: critical issues. In *The Neuropsychology of consciousness* (ed. D. Milner and M. Rugg), pp. 179–200. Academic Press, London.

Schacter, D.L., McAndrews, M.P., and Moscovitch, M. (1988). Access to consciousness: dissociations between implicit and explicit knowledge in neuropsychological syndromes. In *Thought without language* (ed. L. Weiskrantz), pp. 242–78. Oxford University Press.

Tranel, D. and Damasio, A.R. (1985). Knowledge without awareness: an autonomic index of facial recognition by prosopagnosics. *Science* **228**, 1453–4.

Tulving, E. (1983). *Elements of episodic memory*. Oxford University Press.

Tyler, L.K. (1988). Spoken language comprehension in a fluent aphasic patient. *Cog. Neuropsychol.* **5**, 375–400.

Warrington, E.K. and Weiskrantz, L. (1968). New method of testing long-term retention with special reference to amnesic patients. *Nature* **228**, 628–30.

Warrington, E.K. and Weiskrantz, L. (1973). An analysis of short-term and longer-term memory defects in man. In *The physiological basis of memory* (ed. J.A. Deutsch), pp. 365–95. Academic Press, New York.

Warrington, E.K. and Weiskrantz, L. (1974). The effect of prior learning on subsequent retention in amnesic patients. *Neuropsychologia* **12**, 419–28.

Weiskrantz, L. (1986). *Blindsight. A case study and implications*. Oxford University Press.

Weiskrantz, L. (1988). Blindsight. In *Handbook of neuropsychology* Vol. 2, (ed. F. Boller and J. Grafman). Elsevier, New York.

Weiskrantz, L. (1990). Outlooks for blindsight: explicit methodologies for implicit processes. The Ferrier lecture. *Proc. Roy. Soc. (London)* **B239**, 247–78.

DISCUSSION

Participants: J.-P. Changeux, L. Cohen, M.P. Stryker, L. Weiskrantz

Stryker: Do studies of brain lesions give you insight into where the monitor is in the brain?

Weiskrantz: Actually, I am not even sure that it has to be cortical. Dan Schacter is very keen on the parietal cortex, mainly because of its attentional role. I think that because we tend to think in terms of fully formed prototypical stimuli, in terms of full objects, and not in terms of orientation columns, the monitor is going to be at the stage where processing has occurred. So I am very keen on inferior temporal cortex, which is where prototypical representations are constructed.

Cohen: In blindsight patients, you have one part of the brain which can talk and another part which performs the visual task but has no access to language. How is it possible that when a patient is verbally instructed for a task, the non-speaking part of the brain can understand what it is supposed to do?

Weiskrantz: You always must start these blindsight experiments by showing the subject stimuli in the good field, and give verbal instructions. I do not know what our results would be if we start completely in the blind field. The question of how you instruct the subjects is a very puzzling one.

A question from **Changeux** indicated the possible role of the reticular formation in consciousness.

Weiskrantz: This is a system which seems to control the level of vigilance, from coma to alertness. But I think blindsight patients can be extremely vigilant but still not know whether they are discriminating correctly or not. The information they need is cognitive.

16

The social mind
BERNARDO A. HUBERMAN

Distributed intelligence

Intelligence is not restricted to single brains; it also appears in groups, such as insect colonies, social and economic behaviour in human societies, and scientific and professional communities. In all these cases, large numbers of agents capable of local tasks that can be conceived of as computations, engage in collective behaviour which successfully deals with a number of problems that transcend the capacity of any individual to solve. In most instances they do so in the absence of global controls, while exchanging information that is at times inconsistent, often imperfect, and usually delayed. Equally important (and in contrast with the case of single brains) some of the mechanisms underlying such distributed computations are accessible to outside observers.

The economy is perhaps the most intuitively accessible form of a distributed computational system (Simon 1985). The calculations in the system are performed by countless individuals who operate with less than perfect information, with the supply and demand of particular goods determining their price. Based on their available information, agents in the economy choose strategies in order to gain access to resources, and their choices in turn influence the price and availability of those resources. In spite of its complexity, the system manages to stay in equilibrium, although some of its erratic swings are familiar to most of us. Most importantly, the agent's decisions are not only based on perceived pay-offs but also on guesses about what others will do when confronted with similar choices.

Another instance of a distributed intelligence is provided by the social insects, notably the honey-bee colony and certain ant societies (Wilson 1971; Seely 1989). In the case of the beehive, for example, individual bees with limited computational power co-ordinate their interactions through communication in the absence of any global controller. Once again, the information-processing ability of the colony as a whole is far superior to that of a single bee. Through continuous computations it gathers and updates diverse data about the quality of the nectar available, compares it with information about its own internal state, and chooses strategies that determine the amount of foraging that will take place. It is this organized form of distributed intelligence that makes the entomologist speak of the honey-bee colony as a 'superorganism'.

My last example concerns the scientific community. In any particular field, one can observe how individual researchers go about posing and solving problems, while benefiting from access to results obtained by their colleagues. Although scientific progress, when examined globally and for long periods appears purposeful and inexorable, it is not particularly coherent when observed at a local level and for short times. Faulty experiments, delays in the broadcasting of important new results, and the wrong assumptions about particular data, can hamper the progress of an endeavour regardless of the intelligence of the individual scientists. Nevertheless, at any given time a few scientists can make use of the diverse information produced by others, integrate it with their own, and generate novel results that represent true breakthroughs in our understanding of nature. As the field matures, specialization appears, a reflection of the fact that no single scientist can be aware of all the information being generated by the others.

Many of the features underlying distributed intelligence can also be found in the computing networks that are linking vast areas of this planet in ways resembling the spread of bacterial colonies. Within these systems processes are activated or 'born', migrate between machines, spawn other processes in remote computers of the network, die when finished, and often collaborate in the solution of problems while competing for resources contested by other processes. What is particularly striking is that once set in motion such networks operate and communicate in the absence of central controls, with information that is sometimes incomplete and often delayed. The processes in these networks become a community of agents which, in their interactions, strategies, and lack of perfect knowledge, behave like whole ecosystems (Huberman 1988).

The similarity between distributed social or biological intelligence and its computational counterpart suggests the possibility of using computer networks to study issues of collective problem-solving. To the extent that part of the language of collective intelligence is computational, one expects that some of the lessons learned in the study of distributed computing systems will provide clues for understanding the more complex issues of social organizations and their dynamics.

Distributed computation, like any other complex system, can be studied at various levels of abstraction; each level entails a different set of problems and its own descriptive language. Just as the study of human organizations can be performed at the level of the individual, the interactions between members of an organization, or the overall goals of a community, in distributed computation one can also ask questions about the nature of the processors, the architecture of the network, *or* the writing of programs so that given tasks can be accomplished in a co-operative fashion (Kornfeld 1982). One of the most promising approaches is the study of the global behaviour of large collections of agents from knowledge of their local procedures.

The slogan: *Think globally. Act locally* illustrates some of the difficulties encountered in moving from local procedures to global behaviour. The issue that this statement raises is not whether or not to do so is legitimate, but how it is that one goes about achieving collective goals. Since, in any large distributed organization, agents choose strategies on the basis of guesses that they make about what other agents will do, the global behaviour of the system can be very different from that anticipated by the agents. This endless regression about thinking what others think can lead to extremely complex behaviour which at times can have disastrous consequences.

Co-operative problem-solving

The aspect of distributed intelligence that I will address in this chapter has to do with the performance of a group of agents engaging in co-operative tasks. That such a group of agents does better than a single agent or process working in isolation on the same problem underlies the reasons behind the founding of a firm or the establishing of scientific journals, to give familiar examples. Otherwise, no one would bother with the complications and costs inherent in the operation of such institutions. If this is the case (and centuries of experience seem to confirm it), one can try to determine the overall improvement in performance that results from interactions among the agents of a given system.

Before stating the results it is important to remember that for a system of co-operating agents, their overall performance is determined by those making the most progress per unit time. A simple example will illustrate these ideas. Consider the problem of many agents searching for an item in a large database. The overall performance of the system, i.e. the time spent in finding the specific item in this case, is determined by the time spent by the agent that arrives at the answer first, thereby terminating all related processes. Notice that this does not imply that the existence of the other agents is superfluous, for their own searches and messages provide the hints that help the faster agent find the solution.

To determine the distribution of performance, one might expect it necessary to know the details of the co-operating processes. Fortunately, however, highly co-operative systems, when sufficiently large, can display universal individual performance characteristics, independent of the detailed nature of either the individual processes or the particular problem being tackled. This universal law, which we conjectured exists in distributed computation, predicts that rather than obtaining the familiar bell-shaped, Gaussian curve of performance, one observes an extended tail of high performance due to co-operative interactions. This universal distribution, which goes under the technical name of log-normal distribution, is expected to apply whenever the solution of a problem requires the successful completion of a number of nearly independent steps or subtasks, and describes systems as diverse as scientific productivity (Schockley 1957), species diversity in ecosystems (Krebs 1972), and income distributions in national economies (Aitchison and Brown 1957).

Such enhancements in performance become apparent when one considers the problem of many co-operating agents searching in tree-like problem spaces (Huberman 1990). Search, a central issue in problem-solving, is a prototypic problem in computer science, with well-established results for the case of single agent activities. In many heuristic search problems, combinatorial explosions, (i.e. the exponential growth of the search time with problem size) forces one to accept a satisfactory answer rather than an optimal one. In such cases, the search returns the best result found in a fixed amount of time rather than continuing until the optimal value is found. A well-known example is the travelling salesman problem, consisting of a collection of cities and distances between them and an attempt to find the shortest path which visits each of them. For this problem, a number of algorithms exist that provide satisfying solutions in spite of its computational complexity.

Since, for large parts of a problem, one must settle instead for paths that are reasonably short but not optimal, the effectiveness of a heuristic is determined by how well it can discriminate between states of high and low value. When faced with selecting among states with a range of values, a good heuristic will tend to pick those states with high value. Instead of focusing on the time required to find the answer, one can examine the distribution of values returned by the various agents in a given interval of time. In analysing this problem we allowed each agent to examine only one state, selected using the heuristic. The value returned by the agent then corresponded to this state.

The distributions in performance are compared in Fig. 16.1 for the case in

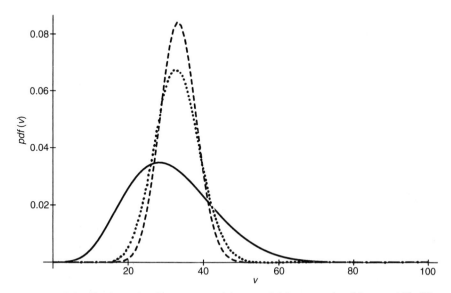

Fig. 16.1. Distribution of values returned in a satisfying search with $v = 100$. The dashed curve shows the distribution for the non-interacting case. The dotted curve corresponds to 10 collaborating agents, and the solid black curve is for 100 co-operating agents. The area under each curve is of value 1.

which the initial agents' heuristic has a bit better than random value discrimination and the hint fractions are normally distributed according to $N(1,0.05)$, giving a case in which the hints, on average, neither help nor hinder the search. In this case, the top 0.1 percentile level is at a value $v=52$ when the number of agents is 10 and $v=70$ when their number is increased to 100. This compares with the non-interacting case, for which this performance level is at a value $v=48$.

As can be seen, co-operation in the form of hints exchanged among agents leads to a nonlinear increase in the overall performance of the system. Thus, as a result of the enhanced performance tail, a collection of co-operating agents is far more likely to have a few high performers than the non-interacting case. This can be seen by examining the tail of the distributions, particularly the top percentiles of performance. In particular, for a system with n agents the expected top performer will be in the top $100/n$ percentile. This can be quantified by specifying the speed reached or exceeded by the top performers. With no hints, the top 0.1 percentile is located at a speed of 4.15. On the other hand, this percentile moves up to 4.89 and 7.58 when $n_{eff}=10$ and $n_{eff}=100$ respectively, where n_{eff} is the effective number of co-operating agents. Note that the top 0.1 percentile characterizes the best performance to be expected in a collection of 1000 co-operating agents. The enhancement of the top performers increases as higher percentiles of larger diversity are considered, and shows the highly non-linear multiplicative effect of co-operative interactions.

Combinatorial implosions

A more dramatic aspect of the value of co-operative problem-solving appears in the case when collaborating agents look for the optimal (as opposed to satisfactory) solution to a search problem. In this case, T. Hogg and the author discovered that there is a sharp phase transition from exponential to polynomial time to the solution as the number of co-operating agents is increased beyond a critical value. This combinatorial *implosion* is the combined result of both a drastic change in the topology of the problem as the heuristic improves, and the effect of the hints exchanged among the searching agents (Huberman and Hogg 1987).

To illustrate such an implosion suppose that the search process takes place in a tree with given branching ratio and depth, d. The search proceeds by starting at the root and recursively choosing which nodes to examine at successively deeper levels of the tree. At each node of the tree there is one correct choice, in which the search gets one step closer to the goal. All other choices lead away from the goal. The heuristic used by each agent can then be characterized by how many choices are made at a particular node before the correct one is reached. The perfect heuristic would choose correctly the first time, and would find the goal in d time steps, whereas the worst one would choose the correct choice last, and hence be worse than random selection. To characterize an agent's heuristic, we assumed that each incorrect choice has a probability, P, of being chosen by the heuristic before the correct one. Thus, for the case of a branching ratio of 2, the perfect heuristic corresponds to $P=0$, random to $P=0.5$, and worst to $P=1$. For simplicity, we assumed that the heuristic effectiveness, as measured by P, is uniform throughout the tree. Since the overall performance is related to the time needed, or the number of steps to reach a goal, one needs to evaluate the probability that any agent gets to the goal in a fixed amount of time as a function of both the number of co-operating agents and the number of diverse hints that they exchange. Such a probability is illustrated in Fig. 16.2 for the case of a binary tree of depth $d=20$.

As Fig. 16.2 shows, whereas for few exchanged hints the probability to reach the goal in finite time is essentially zero, there is a drastic speed-up in the search time as the number of hints increases beyond a value of about 70 (for this particular case). Conversely, for large numbers of exchanged hints, a moderate increase in the number of agents produces a vast improvement in the probability to reach the goal, as seen in the far right corner of the figure.

In summary, we have shown how the performance characteristics of interacting processes engaging in co-operative problem-solving can undergo a

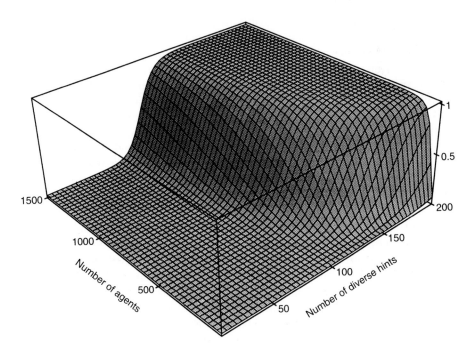

Fig. 16.2. A binary tree of depth $d = 20$.

highly non-linear and universal increase in performance due to the interactions between the agents. In some cases, this is further enhanced by sharp
phase transitions in the topological structure of the problem. These results
provide a quantitative indication of the power of collaboration among diverse
processes when engaging in distributed problem-solving.

Beyond symbols

Since its creation, the digital computer as a formal symbol-manipulating
system became a source of explanations for the workings of the individual
mind. But in spite of its crisp approach to a number of complex issues,
artificial intelligence systems are still far from duplicating human intelligence
in dealing with open-ended, real-world problems. Recently, a different
approach to the study of the brain, one that looks at the properties of large
assemblies of neurone-like elements instead of symbolic structures has become quite popular and promising among scientists. This so-called *connectionist* view of the mind, operates at a sub-symbolic level and can solve a
number of simple perceptual tasks that are hard to explain solely in terms of
symbol-processing. The problem facing this approach, however, is precisely
that of providing an account of how symbols appear out of the interaction of
many simple processing units operating at a sub-symbolic level.

The notion of a social mind discussed in this chapter shows that when large
collections of agents capable of symbolic-processing interact with each other,
new universal regularities in their overall behaviour appear. Furthermore,
these regularities are quantifiable and can be experimentally tested. Universal
features, such as the log-normal distribution of performing agents, or the
phase transitions that appear in heuristic searches, are devoid of symbolic
content *per se* and contain instead a few variables that act as descriptors of
the system. At this *supersymbolic* level, the myriad interactions among symbolic processes give rise to a much simpler description, based on average
properties of the system, as opposed to individual characteristics. Because of
its relative simplicity, this level could yield a satisfactory account of distributed intelligence without the philosophical problems that still plague the
study of the individual mind.

As these remarks indicate, progress in our understanding of a distributed
form of intelligence is being facilitated by the use of a new computational
paradigm with some characteristics of social and biological organization. This
is not an accident; social structures have been and are being used as inspira-
tion for the design of network systems. My point is that if our results are any
indication of future developments, distributed computation can also be used
to provide a flexible and controlled environment for the study of the complex
interactions that define natural social systems.

References

Aitchison, J. and Brown, J.A.C. (1957). *The log-normal distribution*. Cambridge University Press.

Huberman, B.A. (ed.) (1988). *The ecology of computation*. North-Holland, Amsterdam.

Huberman, B.A. (1990). The performance of cooperative processes. *Physica D* **42**, 38–47.

Huberman, B.A. and Hogg, T. (1987). Phase transitions in artificial intelligence systems. *Artific. Intell. J.* **33**, 155–71.

Kornfeld, W.A. (1982). Combinatorially implosive algorithms. *Commun. ACM* **25**, 734–8.

Krebs, C.J. (1972). *Ecology*. Harper & Row, New York.

Schockley, W. (1957). On the statistics of individual variations of productivity in research laboratories. *Proc. IRE* **45**, 279–90.

Seely, T.D. (1989). The honey bee colony as a superorganism. *Am. Sci.* **77**, 546–53.

Simon, H.A. (1985). Cohabiting the planet with computers'. In *Cohabiting with computers* (ed. J.F. Traub), pp. 153–71. Kaufman, Los Altos.

Wilson, E.O. (1971). *The insect societies*. Harvard University Press.

DISCUSSION

Participants: R. Boyd, A. Caramazza, J.-P. Changeux, S. Dehaene, M. Hardy, R.L. Hinde, B.A. Huberman, P. Picq, P. Rakic, M. Stryker

Hinde: I understand why you get a different distribution if you get co-operation. But I do not understand the reverse, that if you get that distribution, you can assume co-operation. There are surely all sorts of mechanisms.

Huberman: My conjecture is that the signature of a social mind is this distribution. Of course, that does not mean that we cannot find examples of log-normal distributions where there is no such co-operation.

Dehaene: It seems obvious that the log-normal distribution could arise for a range of reasons, the most obvious one probably being the choice of the measurement unit. If you measure decibels instead of using the linear measurement of the energy of sound, then you find a different distribution of the sound. So, the real evidence for your model is not the repeated finding of the log-normal distribution in the world, finding a transition from normal to log-normal in the same system, with the same measurement unit, when co-operation increases.

Huberman: Correct. In co-operative search experiments carried out at the University of Pittsburgh, and at Xerox P.A.R.C. they started with a Gaussian distribution and then, after switching on the interaction, they saw a log-normal distribution appearing.

Dehaene: I suppose the problem is what we consider as being data. You discuss a type of computer experiment, but I would like to know whether the kind of things that you predict had actually happened in a real co-operative system, for example, agents in an economic society.

Huberman: I agree that one should be very careful when one extrapolates from computer models. Actually, it would be extremely hard to invent an economy in which it would not be allowed to ever interact and to behave in such a way, that you would get a Gaussian distribution, of incomes, and then allow the interactions to take place. This is why I am advocating this approach. I have proposed a hypothesis that a log-normal distribution, a rather abnormal distribution, is extremely indicative of a co-operative system.

Dehaene: Therefore, outside the computer, you cannot measure experimentally the effect of the co-operation on the system in the way your model predicts? Your model cannot be tested.

Changeux: Concerning the validity of the theoretical model, could you, for instance in economics, have rules for co-operation which are imposed?

Caramazza: I thought that economists always emphasize competition, as opposed to co-operation, to explain the superiority of the capitalist system. The basic argument is that, if you impose forced co-operation, you get disasters.

Huberman: Co-operation does allow for free riders, and it is not always easy to impose, but when it works, we see the effects. Let me give an example. In the United States, as opposed to Italy, you pay very high penalties if you do not pay your taxes. Now, let us assume that co-operation means that you pay your taxes because it is the way everybody benefits. Now, the Italian economy, in spite of relative lack of co-operation compared to the US apparent anarchy, also generates incomes that are log-normally distributed. But the mean and variance of the distribution is different from those of the US.

Boyd: The result of your model is interesting, but I would like to get a better understanding of what you mean by 'hint' and how the 'hint' affects the search of the receiver of the hint. In what sense is the hint likely to help in the search?

Huberman: The search process consists in the following. Suppose you start searching for an object in a big tree. If you are alone, at any given time you have one correct choice and all the others are incorrect. If you make an incorrect choice you have to retreat and start again. Now, suppose you introduce a group of processes that are also involved in the search, and these processes provide information to each other about their own partial results. For example, they tell you that the data you are looking for is not found on their branch. What the information does essentially is to pull out branches of the tree. Each time you get a hint, you chop off a whole branch. So, the hints enhance the probability that at any given node, you will choose correctly, and therefore they increase the likelihood of your finding the answer sooner.

Rakic: If a person is co-operative, he can provide information. But if there is competition, he would lie. So, basically, there is no information available. It works on computers because you decide whether it is co-operation or competition, but in reality, you do not know.

Huberman: Let us go back to the example of the tree. Say that out of a collection of N agents, half of them will send you information that is going to increase your search time, on average. Now, let us put some intentionality into the system, meaning others are cheating, so that you do not find the answer. When I say 'on average' it means that from the information that I am receiving, some of it is taking me backwards instead of forwards. But I maintain that out of, say 50 items of information, on average 25 will send you down the wrong path. You can choose any of them: let us say 75 per cent will send you on the wrong path. But interestingly enough, even in that case, there are some agents that, in spite of the fact they are getting the wrong information, somehow randomly, they manage to find the right answer faster than if they were not getting any information at all.

Stryker: How is it that information, which is equally likely to be true or false, can carry anything useful at all, and is this any different from not getting any information?

Rakic: Do you want to say that more knowledge is better than less knowledge?

Huberman: Exactly. Diverse knowledge in particular. Information in general is likely to help. For instance, you or I can read a scientific journal. But the information acquired might be helpful to me but not to you, because you are in the wrong branch of the search. But on average, you will find in another issue of the journal another piece of information that will be helpful.

Hardy: Regarding the punctuated equilibrium example that you have shown — the constraints that have to be put into the system in order to recruit agents, which produce the quantitative jump out of a qualitative recruitment, have to be imposed suddenly, or are the constraints placed before in that experiment? Is it possible that the equilibrium, as studied here, only shows movement forward towards higher levels? What happens if it gets to lower levels?

Huberman: The jump appears long after the constraints appeared. If you wait long enough, the system will reach an equilibrium. The question is 'suddenly', I mean that at any given time, the constraints change faster than any characteristic time scale in the system, for example, a currency conversion, a new law concerning taxes, or whatever. Then you ask: how sensitive is the system to that? If it goes to the lower levels, the optimal result is decreased, and therefore it does not, eventually, benefit the individual agents to do that. There is a case where a few agents actually crossed over to the new strategy, but it was too expensive to stay there, and they go back to the old strategy.

Stryker: About your model, the issue is not the spreading of the distribution which is log-normal but the flattening of the distribution that is spreading. It seems to me that what you are attempting to do is to increase diversity.

Huberman: Correct. The whole point is that you get a few individuals who now can do it and they could not do it before. For example, you are trying to discover the laws of the universe, and here is Albert Einstein. He did not invent calculus, but he was a lucky man because he lived at a time when most of the mathematical techniques he needed were developed by other people. He was a smart man and at that particular time he could integrate all this knowledge and come up with the answers. So, the basic question is why do I need all these people? And yes, a very strong notion of diversity underlies it all.

Picq: Let us go back to human evolution. Perhaps we can use something from your model. I am thinking about hunting. Monkeys are good hunters, but there are limitations for them regarding the size of the available prey. There is some kind of low level of co-operation for stalking the prey, but the monkeys still meet technical problems in order to catch and kill large prey. Now, about human evolution, one used to attribute hunting success to an increased level of co-operation. Then, perhaps with *Homo habilis*, humans were more efficient through co-operation to capture larger prey or a larger variety of prey. Would your model account for that?

Huberman: It would be nice. When Professor Coppens talked about technology lagging behind biology, and then suddenly switching; I wondered if there was evidence that social co-operative behaviour suddenly provided the possibility that all the elements would start moving faster? Your question raises that as an intriguing possibility, but I do not know whether the switch was determined by increased levels of co-operation or not.

17

Facts about human language relevant to its evolution
STEVEN PINKER

Introduction

Ever since it was formally banned from discussion by the Société de Linguistique de Paris in 1866, the evolution of language has been considered a disreputable scientific topic. Words do not fossilize, it is said, so all attempts to reconstruct its phylogeny are doomed to be exercises in story-telling. Of course, now and again bolder scientists have tried to eke conclusions from tangential sources: from chimps' ability to learn artificial sign systems, in the hope that they may reflect the abilities of our common ancestor; from children's language development, in the hope that ontogeny recapitulates phylogeny; from fossil cranial endocasts, in the hope that homologues of the language centres of the modern brain are visible; from fossil basocrania, in the hope that vocal tract anatomy can be reconstructed and the neural control circuits inferred (see Harnad *et al.* 1976; Parker and Gibson 1979; Lieberman 1989; Bates *et al.* 1990, for reviews.) Many of these investigations are ingenious; some of them are informative; all of them are controversial. But in this chapter I will argue that an important source of evidence, perhaps the most important source, has been neglected in the study of language evolution. That evidence is language itself.

Many people equate evolutionary evidence with fossils, or more recently, molecules. But although such evidence is important in reconstructing phylogenetic sequences, it is not always helpful in understanding the evolutionary forces that shaped a biological system. Neither fossils nor genes distinguish the adaptations of a system, shaped by natural selection, from the byproducts or spandrels that were not directly selected for (Williams 1966; Gould and Lewontin 1979). Either can fossilize or fail to, and above the level of specific gene products, neither is directly identifiable in molecules. In Darwinian theory, the most important criterion for identifying adaptation and the selective forces producing it is the presence of *design* — complexity in the phenotype organized to achieve some otherwise improbable state of affairs conducive to reproduction. As George Williams put it in his classic book *Adaptation and natural selection* (1966): 'For the same reason that it was once effective in the theological "argument from design", the structure of the vertebrate eye can be used as a dramatic illustration of biological adaptation and the necessity for believing that natural selection for effective vision must have operated throughout the history of the group'.

During the past 35 years there has been an explosive increase in our knowledge of the structure, development, use, and breakdown of human language. Although it would be foolhardy to claim that any of these topics is understood, much has been discovered. Oddly, little of this knowledge has been brought to bear on the study of human language evolution. Frequently, evidence such as stone tools or primate communication have been discussed in the context of ideas about human language abilities that are clearly wrong: out of date, at best, or without ever having enjoyed a shred of support at worst.

In this chapter I will review some basic facts about the structure, psychology, neurology, and genetics of the human language faculty relevant to its evolution. These facts do not support anything like a full theory of language evolution, much less a phylogenetic sequence, but they are enough to refute many popular assumptions, have strong implications about the relevant evolutionary forces, and offer some suggestions about where to look further.

Some terminological preliminaries

Before reviewing these facts there are some terminological distinctions that are crucial to explain, or else confusions can easily arise. In the formal study of grammar, a *language* is defined as a set of sentences, possibly infinite, where each sentence is a string of symbols or words. One can think of each sentence as having several representations linked together: one for its sound pattern, one for its meaning, one for the string of words constituting it, possibly others for other data structures such as the 'surface structure' and 'deep structure' that are held to mediate the mapping between sound and meaning. Because no finite system can store an infinite number of sentences, and because humans in particular are clearly not pull-string dolls that emit sentences from a finite stored list, one must explain human language abilities by imputing to them a *grammar*, which in the technical sense is a finite rule system, or programme, or circuit design, capable of generating and recognizing the sentences of a particular language. This 'mental grammar' or 'psychogrammar' is the neural system that allows us to speak and understand the possible word sequences of our native tongue. A grammar for a specific language is obviously acquired by a human during childhood, but there must be neural circuitry that actually carries out the acquisition process in the child, and this circuitry may be called the *language faculty* or *language acquisition device*. An important part of the language faculty is *universal grammar*, an implementation of a set of principles or constraints that govern the possible form of any human grammar.

In the rest of this paper I will use the terms 'language' and 'grammar' more or less interchangeably to refer to the human language faculty. I take pains to point this out because there are two other possible referents of these terms

I will not be referring to. One is *particular* languages, living, dead, or reconstructed (such as English or proto-Indo-European). Currently there is a burst of new research activity attempting to trace the ancestry of existing languages and to reconstruct the ancestors over much greater historical spans than had previously been thought possible (e.g. Greenberg 1987; Renfrew 1987; Cavalli-Sforza *et al*. 1988; Kaiser and Shevoroshkin 1988; see also Allman 1990). However, these time spans are still so short on evolutionary scales that it is highly likely that the investigations will be relevant only to the cultural 'evolution' (really, history) of particular languages in humans with a constant biology, not biological evolution of the human language faculty itself. Secondly, when I use the term 'grammar' I will not be referring to 'prescriptive' or 'correct' grammar such as that taught in school or expounded by language guardians; these are largely a set of arbitrary conventions of a standardized written dialect. Rather, I refer to grammars in the generative linguists' sense of mental structures or programmes underlying the ability to string words together in patterns recognized by other native speakers as belonging to the language (see Pinker 1994).

1. The capacity for language is part of human biology, not human culture

Language is often discussed as a human 'invention' or an aspect of human 'culture'. But when we carefully compare language to systems that we know to be cultural inventions, and whose history is well documented, we see that human language is different. Though any obvious invention such as agriculture or the wheel can be used as an illustrative contrast, Liberman *et al*. (1967) point to one that is particularly striking: alphabetic writing (where each symbol corresponds roughly to a unit of sound such as a consonant or vowel). Although in literate societies we often think of alphabetic writing as being intimately tied to language itself, the differences between the two systems in terms of human evolution could not be more dramatic. Let me list some things that are true about language but false about writing and most other clear examples of cultural inventions.

First, language is found in all human societies. No anthropologist has ever come across a 'primitive' tribe lacking language. But the vast majority of cultures did not have writing of any sort until recently.

Second, language has been found in all societies that have ever been known to exist. There is no record within historical or reconstructable times that any society or region served as the 'cradle' of language from which it spread to previously language-less groups. Alphabetic writing, in contrast, appears to have been invented in one place over a relatively circumscribed period of time (see Liberman *et al*. 1967; Chapter 14) and spread to other groups in a clear sequence by diffusion or conquest.

Third, there is no correlation between grammatical complexity and cultural complexity. So-called primitive or illiterate cultures do not have primitive grammars by any standard; the ubiquitous impression that, say, working class people speak a simpler or coarser language is an untutored impression based on ignorance or on superficial dialect differences. Intuitively 'simple' sentences easily mastered by any speaker, like *Where did he go?* or *The guy I met killed himself*, require underlying rule systems that are more complex than any existing computer language system can handle. European languages, such as English, are in many areas quite primitive. For example, Cherokee has many more personal pronouns than English, allowing one to distinguish, for example, among 'you and I', 'another person and I', 'several other people and I', and 'you, one or more other persons, and I', which English speakers crudely collapse into the single pronoun *we* (Holmes and Smith 1977). American Black English systematically differentiates a currently ongoing activity from a current state in the distinctions between *He be working* versus *He working*, a grammatical device that is unavailable to speakers of Standard English (Labov 1970). See Pinker (1994) for other examples.

Fourth, there seem to be many *nonfunctional, nontransmitted universals* of language. The existence of universal properties of grammar is uncontroversial but is also uninformative about the biological basis of language: the fact that no human language contains a word a million syllables long, or that all human languages have words for 'hand' and 'food' merely shows that people everywhere use language for similar purposes and under similar real time constraints. It is no more informative than the fact that people everywhere eat with their hands.

But crucially, there are many properties that are universal across languages, or at least found widely in many historically unrelated languages, whose form is not predictable from considerations of communicative power or efficiency. For example, forms related to the English double-object construction *give me a book* are widely seen and everywhere have a meaning related to giving or benefiting someone (see Pinker 1989*a*). Questions and other sentence manipulations related to its truth value are formed in many unrelated languages by inverting a restricted class of words with a narrow range of meanings, called auxiliaries, from a customary second position to the first position in the sentence. In these cases and many like them, the grammatical system is useful but countless equally useful alternative systems could be envisioned (e.g. use the double-object form to express the motion of objects toward locations; signal questions by switching the first and last words of a sentence.) The fact that languages cluster around the same partially *arbitrary* solutions to communicative problems suggests that the source of these solutions is in the common structure of the brains of the speakers, not in a universal law of efficiency or in a unique arbitrary historical decision conveyed in an unbroken chain of cultural transmission. In contrast, it is inconceivable that we would discover that the same arbitary spelling irregularities (e.g. the pronunciations of *tough* and *though*) were found in historically unrelated orthographies.

Moreover there are universal properties of language that *limit* people's expressive power. No language systematically uses analogue representation — quantity in language (e.g. number of words, length of words, loudness) conveying quantity in the world. Not even the sign languages of the deaf, with their seeming natural links to the gesture system (Newport 1982; Petitto 1987). Indeed, languages are not terribly good at conveying analogue information such as precise magnitudes, directions, and forces by any means (Talmy 1983; Pinker 1989a; Jackendoff 1990), although the historical contingencies that would make such communication useful are easy to imagine. Rather, grammar only provides speakers with the means to convey crude topological and qualitative spatial relations such as contact, containment, proximity, alignment, and attachment (Talmy 1983).[1]

In general, language universals need not even be thought of as a list of common properties and constraints. Rather, it has become possible to lay out the universal computational structure of all existing or, by hypothesis, possible human languages in precise formal theories of universal grammar. While there is no consensus on the correct variant of these theories, a number have been proposed and defended in the face of cross-linguistic data (e.g. Perlmutter 1980; Chomsky 1981; Bresnan 1982; Gazdar *et al.* 1985), and there are enough commonalities among these theories to instil confidence that there is a unique computational architecture underlying human languages.

A fifth point to keep in mind is that there is, as far as we know, complete equipotentiality among racial and ethnic groups in their ability to learn human languages. Although there are common folk theories that French gender distinctions can only be mastered by those with French blood (Hebb *et al.* 1972) or that assimilated Jewish students innately outperform their Gentile classmates in college Hebrew courses, they can be dismissed.

A sixth fact is that there appears to be a fairly poor correlation between language typology and historical relatedness of languages, except over very short time spans (a few millennia or less). That is, variable properties of language do not cluster within a single branch or clade of the tree of languages. Nor do they generally fall into clusters that correspond to cultural type, such as industrial, peasant, nomadic, or hunter–gatherer. For example, languages generally fall into three types in terms of how they package components of verb meaning (e.g. direction of motion, manner of motion, destination of the motion, kind of entity undergoing a motion) into different verbs. One pattern is found in Romance, Semitic, Polynesian, and Nez Perce; a second is found in Chinese, English, and Caddo; a third is found in Navajo and American Sign Language. Conversely, cultural and spatiotemporal continuity does not imply linguistic continuity: in a few hundred years English changed

---

[1] This fact may have implications for hypotheses about the exact selection pressures shaping language; in particular, what kinds of messages language evolved to communicate efficiently. If grammars are poor at conveying precise analogue information about forces, directions, shapes, sizes, directions, and material properties, it seems unlikely that they evolved under selection pressure for the transmission of the tool-making skills, or at least the sensorimotor components of such skill. I return to the relation of language to tool-making on p. 274 and in the conclusion.

typologically from a relatively free-word order, case-marked language, to a rigid-order language with an impoverished morphological case system. It appears that many features of languages change according to a random walk within a circumscribed set of possibilities; they do not necessarily diverge according to their family tree (Kiparsky 1976; Slobin 1977).

In contrast, the fact that shared and non-shared aspects of writing and other obvious cultural systems can be satisfactorily explained in terms of their historical patterns of transmission is almost too obvious to mention.

A seventh fact is that language is found universally across individuals within a society. One does not need to have a high IQ or high socio-economic status to be a competent language user. Intuitive impressions that some individuals speak a computationally simpler language (professional athletes and juvenile delinquents, according to two stereotypes) are without foundation, for similar reasons to those discussed in connection with the absence of 'primitive' cultures.

Eighth, no formal instructions, lessons, or tutelage is necessary for language acquisition to take place successfully. While some parents may lavish attention on their children's language, such exertions are certainly not necessary. Indeed, in some cultures parents do not regularly converse with pre-linguistic children at all (Heath 1983), and are incredulous that we do — why waste your breath on someone who cannot understand a word you say? they ask. The children first learn language from overhearing streams of adult speech.

Of course, reading and writing are highly dependent on education or tutoring, and even then are vulnerable skills, with dyslexia occurring in a high proportion of people with normal language functioning (Geschwind and Galaburda 1987).

A ninth fact: language acquisition in children is not propelled by attempts to improve communicative power. Children quickly come to respect the fine points of grammar (apart from the previously mentioned irrelevant differences between spoken dialects and standard written English), and spontaneously unlearn the few errors they do make, even when correct grammar affords no improvement in communicative power (Maratsos 1983; Pinker 1984). Children enjoy no increment in expressive power when they stop saying *Don't giggle me*, *Where he can go?* and *I breaked it* in favour of the grammatical alternatives, but they stop anyway. Indeed, in some cases when children improve their grammar they decrease the semantic distinctions they can convey. Verbs like *hit, cut, put*, and *cost* have the same form in the present and past tense in English. A child who allows himself to say *hitted* or *putted* to refer to past tense events is availing himself of a distinction unavailable to adults.

Indeed, most grammatical development seems to occur in a burst (plausibly related to the maturation of the brain) between the ages of 2 and 3 years (Pinker 1984, 1989b, 1994; Bickerton 1992). At the end of this period the child respects, at least probabilistically, most of the grammatical distinctions in the spoken version of the language to be acquired. For example the English-speaking 3-year-old supplies the seemingly frivolous third person singular verb agreement marker -s (as in *he walks*) in more than 90 per cent of the sentences that require it (Brown 1973).

A tenth fact is that children obey many or all language universals by this age, and avoid many tempting errors, ones that seem to be unavoidable generalizations of patterns exhibited elsewhere in the language. There is by now a fairly large experimental and naturalistic literature on children's conformity to universal grammatical constraints (see e.g. Pinker 1984, 1989*a,b*, 1994; Crain 1991; Marcus *et al.* 1992), and I will present a single example to illustrate the claim.

In English, and many other languages, an embedded clause can have a missing or 'understood' subject, whose position is often symbolized with the mnemonic 'PRO' (as in 'pro'nouns). For example, in the sentence *John tried PRO to leave* the position of the subject of the verb *to leave* is not filled by an overt subject, but by the silent placeholder PRO. We understand the sentence as meaning that John was attempting for himself to leave, not for someone else to leave. The subject of the main clause is said to 'control' PRO, the subject of the embedded clause. In contrast, when the first 'matrix' verb is transitive, as in *John told Mary PRO to leave*, we understand the subject of *leave* as being Mary, not John: Mary is the one who should be doing the leaving. Children hear sentences of both kinds from their parents, presumably in contexts where they can independently infer who is doing the leaving in each case. The question is, what do they learn?

A tempting rule would be: 'the noun phrase closest to PRO in the sentence controls it'. Proximity between *John* and the missing subject position in the first sentence, and *Mary* and the missing subject in the second, is easily computed and surely noticed. In fact it is incorrect — in English and other languages. The correct rule, in the jargon of linguistic theory, is that the Noun phrase that 'c-commands' PRO controls it, where 'c-command' means roughly, higher in the phrase structure tree. The correct rule is harder to formulate and the difference between it and its simpler counterpart does not matter for the two simpler sentences used as examples. Only for more complex and rarer sentences does it make a difference; but that's just how human language (at least in adults) works.

Maratsos (1974) did the crucial experiment to see whether children take away the simple rule referring to proximity in the string of words, or the complex rule referring to the geometry of the phrase structure tree (the 'c-command' rule). The diagnostic sentence is a passive such as *John was told by Mary to leave*. Adults immediately see that it's John, not Mary, who was to leave. This interpretation is mandated only by the c-command rule, as the following diagram shows: while *Mary* is closer to PRO than John is (in fact, they're adjacent), only *John* is at a higher level of branching in the tree and thus c-commands it.

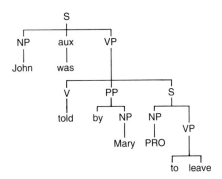

Maratsos (1974) ran an experiment with 4 to 5-year-old children who understood simple passive sentences (ones without embedded clauses). He read sentences to them, and asked them to act it out with toys. He found that they interpreted the missing subject of the embedded clause in the sentence exactly as adults would, with the distant but commanding matrix subject, not the adjacent but buried prepositional object, as being the one interpreted as leaving. This suggests that children obey the complex c-command rule, not the simple proximity rule (many other experiments bear this out; see Pinker 1984; Crain 1992; Sherman and Lust 1993). They do so even though sentences with passivized *tell* are extremely rare in everyday speech and the more common sentences they do hear are equally compatible with the simple and the complex rules.

This suggests that models of cultural transmission that treat bits of culture as simple traits passed on by direct exposure from parent to child are not appropriate to understanding the transmission of language from generation to generation. The complexity of grammar comes from the mind of the child, and does not reside in the stimuli transmitted from parent to child.

The most dramatic manifestation of children's language-creation abilities come from rare cases where the normal parent-to-child transmission channel is disrupted, and this brings us to an eleventh fact about language. If current claims are correct, children can create a language, themselves, in a single generation. Two unusual situations inspire this conclusion. In many tropical areas plantation owners deliberately mixed slaves or indentured servants from different language backgrounds. The workers, like many people in contact with speakers of other languages, developed a 'pidgin', a limited communicative system borrowing words from the superstrate language but without much in the way of grammar. Their children, growing up in groups and separated from their parents most of the time, were exposed to the pidgin, which served as the lingua franca of the community. But that is not what they ended up speaking. Rather, they developed a new language, a 'creole', displaying much of the organization found in long-existing languages. Indeed, geographically and historically separated creoles display commonalities that Bickerton (1981, 1984) claims reflect the natural language-forming capacity of the child, the 'language bioprogramme', uncontaminated by the overlay of historical accretions found in long-existing languages. The claim is controversial; see the exchange in Bickerton (1984) for details.

A similar process may be observable in real time. J.L. Singleton and E.L. Newport (unpublished manuscript) and Singleton (1989) studied a deaf child who acquired a sign language from an impoverished and distorted source: his parents, who had themselves acquired sign language late in life and, for reasons we will come to later, imperfectly. Amazingly, his own language abilities were similar to those of other native signers and more complex than those of his parents. The parents, for example, used some grammatically complex signs and sequences as fixed memorized wholes, but the child correctly analysed them into their constituent grammatical components.

Needless to say we would suspect telepathy or trickery if children spontaneously developed mastery of English orthography or Western agricultural practices without having been exposed to them.

Some implications of the innateness of the language faculty for language evolution

I have tried to summarize evidence suggesting that the foundations of human language ability are part of the biology of humans, not the product of modelling by parents and pick-up by children accumulated over the generations. Are there implications of this fact for current proposals about the evolution of language? I believe there may be. Consider the following chain of inference:

1. All modern biological humans we know of have human language, regardless of cultural transmission (according to the evidence reviewed in this section).

2. The first biologically modern humans were biologically human (tautology).

3. Therefore: the first biologically modern humans had modern human language.

This is not a valid syllogism: the first biologically modern humans need not have been biologically modern in every aspect. But massive parallel evolution is unlikely in the case of language, because, as mentioned, the universal language code, like any code, has arbitrary aspects that are unlikely to have been settled in identical ways by independent groups, even if the selection pressures for communication were identical. An analogy would be different societies inventing their own versions of Morse code. Basic properties such as a dot–dash organization or shorter sequences for more common letters might be expected to develop in each system, but if there is no way (except for uncanny coincidence) that they would develop identical codes letter-for-letter. I suspect that similar logic can be applied to the common computational structure, including arbitrary aspects, underlying human languages, and to the total equipotentiality across races and ethnic groups. The human language faculty and universal grammar seem to have been inherited from an ancestor common to all living humans.

If so, several additional conclusions follow. First, human language ante-dated the divergence of modern races, so we have a temporal lower bound on the appearance of language. Current estimates would place this on the order of 100 000 years ago and certainly no later than 50 000 years ago.

That would mean that language did not first appear in the Upper Palaeolithic beginning about 30 000 years ago, contrary to claims frequently seen in archaeological (e.g. Marshack 1976; White 1989) and popular science treatments (e.g. Begley 1986; Diamond 1989). Indeed, the reasoning behind this estimate will be undermined on other grounds in the next section.

Finally, the first human language was not the oldest reconstructable common ancestor of modern human languages, if that appeared about 20 000 years ago (see Cavalli-Sforza et al. 1988; Keiser and Shevoroshkin 1988; Allman 1990). There is nothing paradoxical about this: if every word of an individual language has some probability of disappearing per unit time, then given enough time, none of the original words of it will remain. This fate could have befallen 'Evish' long before 'Proto-World', the grail of language reconstructors, showed up.

2. Language is a module (a 'mental organ')

Language has been treated within the empiricist tradition as one manifestation of a single capacity for 'learning' or 'intelligence' or 'information-processing' that also underlies other cognitive and social skills, such as tool use, social interactions, physical problem-solving, and non-verbal communication. Work on human language has challenged that tradition. Several lines of evidence suggest that language is a 'module' or 'mental organ' that obeys many of its own principles and that is neurobiologically distinguishable from the neural systems underlying other cognitive and social systems. This should not be surprising: the parcelling of algorithms into modules each dedicated to solving one problem well is a ubiquitous principle of design (Simon 1969), both in computer science (e.g. Kernighan and Plauger 1978) and in the evolution of neural systems (Tooby and Cosmides 1990; Jacobs et al. 1991).

First, there are 'eccentric' principles underlying successful acquisition and use of language, principles that are not motivated by general inductive power or computational efficiency. That is, the learning algorithms for language are at cross-purposes with learning algorithms for other cognitive domains. Here is a simple example. Imagine a general-purpose learning programme faced with extracting the patterns underlying the following set of strings, in order to generate and recognize new strings obeying the pattern.

Dax bok blick.
Dax bok coob.
Dax wug blick.

It is hard to imagine such an algorithm designed to make generalizations that did not pick up on the apparent interchangeability of *bok* and *wug* in second position, and *blick* and *coob* in third position, and thereby generalize to the reasonable *Dax wug coob*. Indeed a number of artificial intelligence programs designed to acquire rules of language using general pattern-matching mechanisms, summarized in Pinker (1979), made just such leaps. Unfortunately, when the example is converted to the identical problem in human language acquisition, the generalization turns out to be ill-advised:

John likes fish.
John likes apples.
John might fish.

---

*John might apples.

Lest the example seem misleading because of the lack of a reasonable meaning for the fourth sentence, here is another one:

Dax bock blick po bof.
Bof Dax bok blick po?
Dax bok blick sa bof.

Bof Dax bok blick sa?

John eats toast with bananas.
What does John eat toast with?

John eats toast and bananas.
*What does John eat toast and?

The fourth sentence here is perfectly reasonable on semantic grounds but is clearly not natural English. In both these examples our simple inductive pattern-matcher ran afoul of the inherent idiosyncratic logic of human grammar. Grammar couches its rules and principles not by directly specifying the positions and ordering of words, but of abstract grammatical categories and structures, such as noun, verb, auxiliary, and phrase structure geometries. Only a mechanism designed to look for examples of these abstract symbols and for certain regularities among them is likely to succeed at generalizing properly from a sample of parental speech to the correct infinite language (Chomsky 1972, 1975, 1980; Pinker 1979, 1984; Wexler and Culicover 1980).

A second body of evidence that language is a module separate from other aspects of cognition is the fact that the two can dissociate. Given certain neurological and genetic conditions, a person can have normal intelligence and severely impaired language, or — strikingly — the reverse.

Intelligence without language. An impairment of language in the presence of normal intelligence can occur in certain cases of stroke-induced damage structures within the language areas (left hemisphere perisylvian regions) of the brain. This form of aphasia, called Broca's Aphasia, can include lack of fluency, omission or misuse of inflections and grammatical words such as articles and prepositions, and difficulty in using syntactic cues in understanding complex sentences (see Caplan *et al.* 1984). However Broca's aphasics clearly need not be retarded, demented, or otherwise cognitively impaired.

'Specific language impairment' (SLI), sometimes called 'developmental dysphasia' (Gopnik 1990*a*; Gopnik and Crago 1991), is a syndrome of impaired language not attributable to cognitive problems like retardation, social problems like autism, or perceptual problems such as hearing impairments. The most noticeable symptoms include delayed language onset, articulatory difficulties in childhood, and persistent difficulties in controlling inflections productively (when asked for the plural of 'a zag', for example, they shrug). Many SLI children are free of cognitive impairments. One boy, described in Gopnik (1990*b*), was top of his mathematics class and was a proficient computer user (at least for a 9-year-old). Interestingly, some forms of SLI may be inherited: the syndrome runs in families and is concordant in monozygotic twins to an extent that suggests a strong genetic influence, perhaps in some cases a single defective gene.

Finally, we are all, in one sense, specifically language impaired. Recent evidence has corroborated Lenneberg's (1967) hypothesis that there may be a critical period for the acquisition of language. The most sensitive period for language learning typically lasts between the ages of about 18 months to 5 or 6 years, and a gradual decline then ensures, bottoming out by puberty (Curtiss 1989; Johnson and Newport 1989; Newport 1990). This is most striking in the area of phonology (in particular, accent) but also can be measured in syntax and morphology. The evidence comes in several forms. The proficiency of immigrants with English declines with age of arrival into the United States between the ages of 5 years and puberty, but is unrelated to age on arrival before or after (being unimpaired for younger immigrants, and randomly distributed around a low value for older ones), and is unrelated to the sheer number of years spent in the country when this is statistically separated from age of arrival (Johnson and Newport 1989). Deaf people who first acquire a sign language as adults attain a much cruder mastery of the grammar than those who acquired it as children (Newport 1990). And in one case study (Curtiss 1989), a hearing-impaired languageless adult in her thirties was discovered and fitted with a hearing aid that brought her hearing to within normal levels for the first time. Despite intensive language therapy, and the acquisition of many words, her ability to order those words grammatically was pathologically poor. Since we adults like to think of ourselves as cognitively competent compared to children, our relatively poor language acquisition skills must be attributed to something else, presumably related to changes in the adult brain (Pinker 1994).

Language without intelligence. The dissociation that is most devastating to the notion that language is an exercise of general cognitive processes comes from case studies of individuals who are severely cognitively impaired but none the less maintain sophisticated language skills. Syndromes with this general kind of profile have sometimes been noted in certain neurological and psychiatric populations. For example, the speech of schizophrenics, Alzheimer's disease patients, some autistic children, and some aphasics (such as 'transcortical sensory aphasia') can be deficient or impaired in content for various reasons but grammatically well-formed (Curtiss 1989; Cromer 1991). Recently two quite dramatic dissociations of this kind have been documented. One is Williams syndrome, a form of retardation thought to be caused by a defective gene also involved in calcium metabolism (Bellugi *et al.* 1990). A second occurs in children whose retardation is caused by a number of conditions, most often hydrocephalus caused by spina bifida (Curtiss 1989; Cromer 1991). Many of these children are described as 'hyperlinguistic' or as exhibiting 'cocktail party conversation': their speech is fluent, grammatically well-formed, and socially appropriate, and they score well in formal psycholinguistic tests where they have to act out complex sentences. However, the content of the speech is childlike and often confabulated. These syndromes show that language does not depend on full complement of intelligent faculties; indeed, that it may be a partially autonomous input–output channel that expresses the products of intelligence but is not itself one of those products.

Some implications of the modularity of language for the evolution of language

The evidence that language and intelligence are doubly dissociable, hence largely independent mental faculties, shows that there is no such thing as a general 'symbolic' or 'semiotic' capacity underlying language and other forms of symbolic expression, such as art. Nor is language sufficiently explained as the product of some general capacity for 'planning' or 'co-ordinated action' or 'hierarchical organization' that also underlies the creation and use of tools. Therefore, the appearance of art or tools in the archaeological record, often interpreted as evidence for analogous abilities to manipulate symbols or plan complex action sequences in language (e.g. Hewes 1973; Isaac 1976; Marshack 1976; Parker and Gibson 1979; Kitahara-Frisch 1980; Wynn 1985*a*), may in fact say nothing about language (see also Atran 1982; Wynn 1985*b*; Lewin 1986). Based on what we know about the possible fractionation of mental abilities in modern humans, talking hominids could in principle have had no art or complex tools, and tool-using art-making hominids could have had no language.

This is not to deny that there are connections between language and other mental abilities. Surely language would have been most likely to evolve among organisms that had something to say, and complex technologies were most likely to have developed among groups that could efficiently pass on knowledge from member to member and generation to generation (although see note 1, p. 266). I am only calling into question the practice of making direct inferences about the presence or form of grammatical abilities that are based on an assumption that the same information-processing mechanisms underlie language and the cognitive activities that leave material remnants.

3.   Language is an 'organ of extreme perfection and complication'

In *The origin of species* Darwin paid special attention to 'organs of extreme perfection and complication', such as the eye and wing. These organs are extremely improbable as naturally occurring arrangements of matter, as they are composed of many parts in precise arrangements serving to accomplish difficult engineering tasks. The clear presence of design in their organization requires a special explanation. In Darwin's time this was considered a major challenge to the theory of natural selection since the hand of a divine designer was the most obvious explanation. Nowadays, complex design is interpreted as the best evidence for natural selection (Williams 1966; Dawkins 1986) because selection is the only known physical process that can lead to design. That is because selection is the only natural process in which the design of a system — measured in its effects on reproduction — can play a causal role in its evolution.

The search for evidence of design can play the same role in understanding the evolutionary forces shaping mental systems as it does for physical systems (Cosmides 1989; Tooby and Cosmides 1990). Let me review some evidence that the human language faculty shows signs of complex design. See Pinker and Bloom (1990) where the evidence and its implications are laid out in more detail.

(a)   Organization of grammar into submodules

Although a grammar maps sounds and meanings, it does not do so in a single step. Grammars are composed of several submodules, each complex, and each dedicated to computing some data structure intermediate between sound and meaning. For example:

Phrase structure rules and principles map the data structures representing 'meanings', which often have a complex hierarchical structure, on to the necessarily serial strings of words produced or perceived. They do so by dictating possible choices and orderings of words within phrases and of phrases within clauses and sentences. For example, they dictate the different meanings underlying *Large trees grow dark berries* as opposed to *Dark trees grow large berries*.

Tense systems demarcate the relative time of the event directly referred to, the moment the sentence is spoken, and often some third reference event, as in *John has arrived, John had arrived (when Mary was speaking)*, and *John will have arrived (by the time Mary speaks)*.

Grammatical relations and case encode the relations between predicates and arguments—information about who did what to whom. For example, they are used to differentiate *Man bites dog* and *Dog bites man*.

Morphology defines operations on word forms that express tense, case, and certain other features. For example, in English a morphological rule adds *-d* to the end of a verb to signal that the event or state described by the verb took place before the moment of speaking, and another one adds *-z* to the end of a noun to signal that more than one example of the referent of the noun is being talked about.

Phonology. Effective use of the vocal-auditory channel requires that the meaningful sound sequences of a language be rapidly articulable but reliably discriminable—desiderata that are often at cross-purposes. Phonological rules and principles define a consistent set of possible sequences in a language, as if they enforced an agreement among speakers and hearers as to which sequences they would 'agree' to use. For example, the 'add *-d*' rule alluded to in the preceding paragraph is fully general—it can apply in principle to any verb—but it can yield difficult-to-pronounce sequences for some verbs: *pat* + *-d* and *walk* + *-d* are virtually unpronounceable without modification. Two rules of English phonology that we are totally unconscious of modify such sequences of sounds: 'Epenthesis' inserts a neutral vowel *e* between the *t* of *pat* and the added *d*, eliminating the need to tap the tongue twice without an intervening pause; Voicing Assimilation changes the *d* to a *t* following *walk* and other verbs ending in an unvoiced consonant, eliminating the need to turn the vocal cords off and then on in close succession.

(b)   Complexity of neural circuitry and underlying language

Language is produced by complex software, implemented in the wiring of the neural circuitry in left perisylvian regions of the brain. It is not a direct product of the mass action of a large brain or of gross neuroanatomical properties such as possessing a particular macroscopically visible gyrus or fibre tract.

First, absolute brain size is not directly related to language. It is well known that there is considerable variation among normal individuals in brain size (Gould 1981). Even beyond this range language can be subserved by a small brain, as long as it is a human brain. Lenneberg (1964, 1967) noted that nanencephalic ('bird-headed') dwarfs, with brain sizes not much larger than those of chimpanzees, had complex language abilities.

Second, language can be computed by brains with unusual gross geometry.

John Lorber (see Lewin 1980) surveyed cases of hydrocephalic children who had greatly expanded ventricles and only a few millimetres of cerebral tissue lining the skull. Many of them were cognitively and socially indistinguishable from normal individuals; some were superior college students.

Third, there are surprising individual differences in the precise localization of language abilities in the brain. For example Lecours *et al.* (1984) review literature showing that about 10 per cent of aphasics show highly unexpected language deficits given the location of their brain lesions, for example, Wernicke-like symptoms following damage near Broca's area or vice versa. These phenomena are poorly understood and difficult to interpret for a number of reasons: actual measurement of neuroanatomical pathology is difficult in living patients; testing is usually done following recovery and hence often assesses what can be re-learned with therapy and effort, not what was lost; linguistic tests administered by neurologists are often crude and uninformative; and children's brains are somewhat plastic so undetectable early lesions could have triggered a relocation of language circuitry in some people. But it seems clear that language abilities do not invariably map on to brain regions on the scale at which we can now measure them in humans. Most likely, it is the microcircuitry that is crucial and the developing brain may have some freedom to move these circuits around within certain regions of the brain.

(c)   Genetic complexity of language

Specific language impairment is 80 per cent concordant in monozygotic twins, 40 per cent concordant in dizygotic twins and non-twin siblings, and shows high family aggregation rates (Tallal *et al.* 1989; Tomblin 1989). A recent case study (Gopnik 1990a; Gopnik and Crago 1991) described a syndrome of specific language impairment (SLI) in about half the members of three generations of a large family that patterned among the family members as if it was the result of a single dominant autosomal gene.

As mentioned, SLI is not a global loss of all language functions but seems to affect phonology and morphology in childhood with the morphological deficits lasting throughout life. Other cases of SLI have slightly different patterns. For example, while the members of the family Gopnik studied made few errors with verb argument structures (i.e. they used transitive verbs correctly with objects, and intransitive verbs without them), and they made few or no over-regularization errors like *breaked* and *foots*, such errors do occur in other SLI children (M. Rice, personal communication). Moreover, other inherited individual differences in language abilities exist. For example, Bever *et al.* (1989) showed that right-handers with left-handed relatives show a variety of subtle yet measurable differences in how they process words and sentences. All of this evidence suggests that language is under the control of several genes.

Implications of language being a complex organ for the evolution of language

The computational, neurological, and genetic complexity of language renders a number of widely expressed proposals about the evolution of language suspect.

First, language evolution is unlikely to have been an automatic byproduct or 'spandrel' of the evolution of a bigger brain, of general intelligence, of hierarchical or serial information-processing, of the way neurones are packed in the skull, or of strategically placed surplus neural tissue, for example, between the centres for vision, audition, and vocal tract control; see Chomsky 1972; Jerison 1973; Gould and Lewontin 1979; Wilkins and Dumford 1990, for examples of such claims. This is not to deny that such developments may have been prerequisites for the evolution of language, only that they were not sufficient. Some degree of overall reorganization, as suggested by Holloway (Chapter 3), and specific changes in circuitry seem to be necessary for language.

Second, current evidence implies that language is unlikely to have evolved exclusively by genetic drift, hitch-hiking, or other non-selectional processes. Such processes can result in a single gene or a functionally random set of genes becoming fixed without specific selection pressures, and if the entirety of language were the product of massive pleiotropy of a single gene, they would be difficult to rule out (see Lewontin 1990). But if language is the complex product of a large set of genes acting in concert, it seems less likely that just that set, or an equivalent set with just the right functional outcome (complex grammar) was coincidentally fixed in the human genome by processes that are insensitive to the functional outcomes of the genes.

Finally, fossil cranial endocasts must be interpreted with caution. Given the complex, poorly understood relationship between sulci and gyri and language functions in modern humans, inferences from the presence of such features in the brains of other human species must for the moment remain highly tentative.

Summary of language facts and their implications

I have reviewed evidence from contemporary linguistics, psycholinguistics, and neurolinguistics for a number of conclusions. First, language is part of human biology, not human culture. Secondly, language is best thought of as a neural system, a computational module, and a mental organ. Thirdly, language is complex computationally, neurologically, and genetically. The implications are that language is as old as modern humans, that it did not necessarily appear at the same time as tools, art, or other cultural products, and that it did not fully evolve as a byproduct of something else.

Although some of these implications may seem more negative than constructive, I think a positive conclusion may be forthcoming from the facts. Paul Bloom and I (Pinker and Bloom 1990) argued for the following proposition: language is a system that evolved via conventional neo-Darwinian mechanisms (see also Lieberman 1984; Brandon and Hornstein 1986; Hurford 1989; Bickerton 1990; Newmeyer 1991; Pinker 1994). Specifically, modern language abilities are in large part the products of a history of selection for neural circuitry enabling efficient communication of an unlimited set of messages of a certain kind (basically, hierarchical propositions involving human actions, beliefs, desires, and obligations; objects and their rough relative locations, motions, and forces; and the durations and relative times of events and states; see Pinker 1989a; Jackendoff 1990). Although it may seem uncontestable, even banal, to say that the language faculty was selected for its ability to communicate thought, the claim is controversial; see Pinker and Bloom (1990) and the accompanying commentaries, which discuss the issues in detail.

Acknowledgements

The preparation of this chapter was supported by NIH grant HD 18381. I thank Paul Bloom, Leda Cosmides, Misia Landau, and John Tooby for helpful comments and discussions.

References

Allman, W.F. (1990). The mother tongue. *U.S. News & World Report* 5 November, 60–70.

Atran, S. (1982). Constraints on a theory of hominid tool-making behavior. *L'Homme* **XXII**, 35–68.

Bates, E., Thal, D., and Marchman, V. (1989). Symbols and syntax: A Darwinian approach to language development. In *The biological foundations of language development* (ed. N. Krasnegor, D. Rumbaugh, M. Studdert-Kennedy, and R. Schiefelbusch). Oxford University Press.

Begley, S. (1986). The way we were [Cover story]. *Newsweek* 10 November.

Bellugi, U., Bihrle, A., Jernigan, T., Trauner, D., and Doherty, S. (1990). Neuropsychological, neurological, and neuroanatomical profile of Williams syndrome. *Am. J. Med. Gen. Suppl. 6*, 115–25.

Bever, T.G., Carrithers, C., Cowart, W., and Townsend, D.J. (1989). Language processing and familial handedness. In *From reading to neurons* (ed. A. Galaburda). MIT Press, Cambridge MA.

Bickerton, D. (1981). *The roots of language*. Karoma, Ann Arbor.

Bickerton, D. (1984). The language bioprogram hypothesis. *Behav. Brain Sci.* **7**, 173–212.

Bickerton, D. (1990). *Language and species*. Chicago University Press.

Bickerton, D. (1992). The pace of syntactic acquisition. In *Proceedings of the 17th Annual Berkeley Linguistics Society Meeting*. (ed. L. Sutton, E. Johnson, and R. Shields). Berkeley Linguistics Society, Berkeley, CA.

Brandon, R.N. and Hornstein, N. (1986). From icons to symbols: Some speculations on the origin of language. *Biol. Phil.* **1**, 169–89.

Bresnan, J. (ed.) (1982). *The mental representation of grammatical relations*. MIT Press, Cambridge, MA.

Brown, R. (1973). *A first language: the early stages*. Harvard University Press, Cambridge, MA.

Caplan, D. *et al.* (ed.) (1984). *Biological perspectives on language*. MIT Press, Cambridge, MA.

Cavalli-Sforza, L.L., Piazza, A., Menozzi, P., and Mountain, J. (1988). Reconstruction of human evolution: Bringing together genetic, archeological, and linguistic data. *Proc. of Natl Acad. Sci. USA* **85**, 6002–6.

Chomsky, N. (1972). *Language and mind*. Harcourt, Brace, and World, New York. [Extended edition.]

Chomsky, N. (1975). *Reflections on language*. Pantheon, New York.

Chomsky, N. (1980). *Rules and representations*. Columbia University Press, New York.

Chomsky, N. (1981). *Lectures on government and binding*. Foris, Dordrecht, The Netherlands.

Cosmides, L. (1989). The logic of social exchange: Has natural selection shaped how humans reason? Studies with the Wason selection task. *Cognition* **31**, 187–276.

Crain, S. (1991). Language acquisition in the absence of experience. *Behav. Brain Sci.* **14**, 597–650.

Cromer, R. (1991). The cognition hypothesis of language acquisition. In *Language and thought in normal and handicapped children* (ed. R. Cromer). Blackwell, Cambridge, MA.

Curtiss, S. (1989). The independence and task-specificity of language. In *Interaction in human development* (ed. A. Bornstein and J. Bruner). Erlbaum, Hillsdale, NJ.

Dawkins, R. (1986). *The blind watchmaker: Why the evidence of evolution reveals a universe without design*. Norton, New York.

Diamond, J. (1989). The great leap forward (Cover story). *Discover* May.

Gazdar, G., Klein, E., Pullum, G.K., and Sag, I.A. (1985). *Generalized phrase structure grammar*. Harvard University Press, Cambridge, MA.

Geschwind, N. and Galaburda, A.M. (1987). *Cerebral lateralization: Biological mechanisms, associations, and pathology*. MIT Press, Cambridge, MA.

Gopnik, M. (1990a). Feature-blind grammar and dysphasia. *Nature* **344**, 715.

Gopnik, M. (1990b). Feature blindness: A case study. *Lang. Acquisit.* **1**, 139–64.

Gopnik, M. and Crago, M. (1991). Familial aggregation of a developmental language disorder. *Cognition*, **39**, 1–50.

Gould, S.J. (1979). Panselectionist pitfalls in Parker & Gibson's model of the evolution of intelligence. *Behav. Brain Sci.* **2**, 385–6.

Gould, S.J. and Lewontin, R.C. (1979). The spandrels of San Marco and the Panglossian program: A critique of the adaptationist programme. *Proc. Roy. Soc. London* **205**, 281–8.

Gould, S.J. (1981). *The mismeasure of man*. Norton, New York.

Greenberg, J. (1987). *Language in the Americas*. Stanford University Press, Stanford, MA.

Harnad, S.R., Steklis, H.S., and Lancaster, J. (eds.) (1976). Origin and evolution of language and speech. *Ann. N.Y. Acad. Sci.* **280**.

Heath, S.B. (1983). *Ways with words: Language, life, and work in communities and classrooms*. Cambridge University Press, New York.

Hebb, D.O., Lambert, W.E., and Tucker, G.R. (1972). Language, thought, and experience. *Mod. Lang. J.* **55**, 212–22.

Hewes, G.W. (1973). An explicit formulation of the relationship between tool-using, tool-making, and the emergence of language. *Vis. Lang.* **7,** 101–27.

Holmes, R.B. and Smith, B.S. (1977). *Beginning Cherokee* (2nd edn). University of Oklahoma Press, Norman, OK.

Hurford, J.R. (1989). Biological evolution of the Saussurean sign as a component of the language acquisition device. *Lingua* **77,** 187–222.

Isaac, G.L. (1976). Stages of cultural elaboration in the Pleistocene: Possible archeological indicators of the development of language capabilities. In *Origin and evolution of language and speech* (ed. S. R. Harnad, H. S. Steklis, and J. Lancaster). *Ann. N.Y. Acad. Sci.* **280,** 275–88.

Jackendoff, R. (1990). *Semantic structures.* MIT Press, Cambridge, MA.

Jacobs, R.A., Jordan, M.I., and Barto, A.G. (1991). Task decomposition in a modular connectionist architecture: The what and where vision tasks. *Cog. Sci.* **15,** 219–50.

Jerison, H.J. (1973). *Evolution of the brain and intelligence.* Academic Press, New York.

Johnson, J. and Newport, E. (1989). Critical period effects in second language learning: The influence of maturational state on the acquisition of English as a second language. *Cog. Psychol.* **21,** 60–99.

Keiser, M. and Shevoroshkin, V. (1988). Nostratic. *Ann. Rev. Anthropol.* **17,** 309–29.

Kernighan, B.W. and Plauger, P.J. (1978). *The elements of programming style* (2nd edn). McGraw Hill, New York.

Kiparsky, P. (1976). Historical linguistics and the origin of language. In *Origin and evolution of language and speech* (ed. S.R. Harnad, H.S. Steklis, and J. Lancaster). *Ann. N.Y. Acad. Sci.* **280,** 97–103.

Kitahara-Frisch, J. (1980). Symbolizing technology as a key to human evolution. In *Symbol as sense: New approaches to the analysis of meaning* (ed. M. Foster and S. Brandes). Academic Press, New York.

Labov, W. (1970). *The study of nonstandard English.* National Council of Teachers of English, Urbana, IL.

Lecours, A.R., Basso, A., Moraschini, S., and Nespoulous, J.-L. (1984). Where is the speech area, and who has seen it? In *Biological perspectives on language* (ed. D. Caplan, A.R. Lecours, and A. Smith). MIT Press, Cambridge, MA.

Lenneberg, E.H. (1964). A biological perspective on language. In *New directions in study of language* (ed. E.H. Lenneberg). MIT Press, Cambridge, MA.

Lenneberg, E.H. (1967). *Biological foundations of language.* Wiley, New York.

Lewin, R. (1980). Is your brain really necessary? *Science* **210,** 1232–4.

Lewin, R. (1986). Anthropologist argues that language cannot be read in stones. *Science* **233,** 23–4.

Lewontin, R. (1990). Evolution of cognition. In *Thinking: an invitation to cognitive science,* Vol. 3 (ed. D.N. Osherson and E.E. Smith). MIT Press, Cambridge, MA.

Liberman, A.M., Cooper, F.S., Shankweiler, D.P., and Studdert-Kennedy, M. (1967). Perception of the speech code. *Psychol. Rev.* **74,** 431–61.

Lieberman, P. (1984). *The biology and evolution of language.* Harvard University Press, Cambridge, MA.

Maratsos, M.P. (1974). How preschool children understand missing complement subjects. *Child Dev.* **45,** 700–6.

Maratsos, M.P. (1983). Some current issues in the study of the acquisition of grammar. In *Carmichael's manual of child psychology* (ed. P. Mussen) (4th edn). Wiley, New York.

Marcus, G., Pinker, S., Ullman, M., Hollander, M., Rosen, T.J., and Xu, F. (1992). Overregularization in language acquisition. *Monogr. Soc. Res. Child. Devel.,* **57.**

Marshack, A. (1976). Some implications of the paleolithic symbolic evidence for the origin of language. In *Origin and evolution of language and speech* (ed. S.R. Harnad, H.S. Steklis, and J. Lancaster) *Ann. N.Y. Acad. Sci.* **280**, 289–311.

Newmeyer, F. (1991). Functional explanation in linguistics and the origin of language. *Lang. Comm.* **11**, 3–96.

Newport, E. (1990). Maturational constraints on language learning. *Cog. Sci.* **14**, 11–28.

Newport, E. and Supalla, T. (in press). A critical period effect in the acquisition of a primary language. *Cog. Psychol.*

Newport, E.L. (1982). Task specificity in language learning? Evidence from speech perception and American Sign Language. In *Language acquisition: The state of the art* (ed. E. Wanner and L. Gleitman). Cambridge University Press, New York.

Parker, S.T. and Gibson, K.R. (1979). A developmental model for the evolution of language and intelligence in early hominids. *Behav. Brain Sci.* **2**, 367–408.

Perlmutter, D. (1980). Relational grammar. In *Syntax and semantics* (ed. E. Moravcsik and J. Wirth) Vol. 13: *Current approaches to syntax.* Academic Press, New York.

Petitto, L. (1987). On the autonomy of language and gesture: Evidence from the acquisition of personal pronouns in American Sign Language. *Cognition* **27**, 1–52.

Pinker, S. (1979). Formal models of language learning. *Cognition* **7**, 217–83.

Pinker, S. (1984). *Language learnability and language development.* Harvard University Press, Cambridge, MA.

Pinker, S. (1989*a*). *Learnability and cognition: The acquisition of argument structure.* MIT Press, Cambridge, MA.

Pinker, S. (1989*b*). Language acquisition. In *Foundations of cognitive science* (ed. M.I. Posner). MIT Press, Cambridge, MA.

Pinker, S. (1994). *The language instinct.* Morrow, New York.

Pinker, S. and Bloom, P. (1990). Natural language and natural selection. *Behav. Brain Sci.* **13**, 70–84.

Renfrew, C. (1987). *Archeology and language: the puzzle of Indo-European origins.* Cambridge University Press, New York.

Sherman, J.C. and Lust, B. (1993). Children are in control. *Cognition*, **46**, 1–51.

Simon, H. (1969). The architecture of complexity. In *The sciences of the artificial* (ed. H. Simon). MIT Press, Cambridge, MA.

Singleton, J.L. (1989). Restructuring of language from impoverished input: evidence for linguistic compensation. PhD dissertation, University of Illinois, Urbana.

Slobin, D. (1977). Language change in childhood and in history. In *Language learning and thought* (ed. J. Macnamara). Academic Press, New York.

Tallal, P., Ross, R., and Curtiss, S. (1989). Familial aggregation in specific language impairment. *J. Speech Hear. Dis.* **54**, 167–73.

Talmy, L. (1983). How language structures space. In *Spatial orientation: Theory, research, and application* (ed. H. Pick and L. Acredolo). Plenum, New York.

Tomblin, J.B. (1989). Familial concentration of developmental language impairment *J. Speech Hear. Dis.* **54**, 287–95.

Tooby, J. and Cosmides, L. (1990). On the universality of human nature and the uniqueness of the individual: The role of genetics and adaptation. *J. Person.* **58**, 17–67.

Wexler, K. and Culicover, P. (1980). *Formal principles of language acquisition.* MIT Press, Cambridge, MA.

White, R. (1989). Visual thinking in the Ice Age. *Sci. Am.* **261**, 92–9.

Wilkins, W. and Dumford, J. (1990). In defense of exaptation. *Behav. Brain Sci.* **13**, 763–4.

Williams, G.C. (1966). *Adaptation and natural selection: A critique of some current evolutionary thought*. Princeton University Press, Princeton, NJ.

Wynn, T. (1985a). The intelligence of later Acheulian hominids. *Man*, **14**, 371–91.

Wynn, T. (1985b). Piaget, stone tools and the evolution of human intelligence. *World Archeology*, **17**, 32–43.

DISCUSSION

Participants: R. Boyd, J.-P. Changeux, A. Roch Lecours, S. Pinker, M.P. Stryker

Selection and language evolution

Stryker: Can you give us some idea of how something like language, but more primitive, might be a path via which language might evolve?

Pinker: There are two questions there. First, can you imagine a language less complex than a modern human language? My abilities in French would be a good example of an intermediate communication system. I think that a whole range of communication systems are conceivable, although I would not want to speculate as to what they might be. Secondly, could language have evolved? The human language faculty is not the only kind of system that is capable of processing language. We can also use general intelligence, for instance, as adults do when acquiring a second language. This process is however much slower, more taxing, and less reliable than our native language competence. Acquisition, through genetic change, of automatic dedicated machinery, could therefore be a way in which language might follow a gradual course of evolution in a conventional Darwinian way.

Answering questions from **Changeux**, **Pinker** suggested that language in young children goes through a succession of consistent communication systems. It cannot be excluded that these stages may correspond to phylogenetically early stages of language evolution.

Answering a question from **Boyd**, **Pinker** analysed what could be the evolutionary advantage of the fact that there are many different languages, and not just one: 'We have only two guesses. One of them is that you have to synchronize your output with the listener's input. Since you cannot guarantee that every listener is genetically identical to you, there may be a necessity for a tuning process, to make sure that we agree on words, word order, and things like that. The other possibility is that our ability to learn the differences between languages is related to more general abilities. They were there first and we evolved only as much specific language machinery as we needed to do the rest'.

Explicit and implicit language knowledge

Lecours: What about the relationship between implicit learning of language rules by children, and explicit learning of grammar at school? Could the latter include more complex rules?

Pinker: I think that what the 3½-year-old knows is more complex than what your grammar teachers teach you. For instance, the past tense of the verb 'to highstick' is 'highsticked', not 'highstuck'. Basically, it is because any verb that comes from a noun in its derivation is automatically regular. We find that 4-year-olds obey this principle, as well as people without college education. The only people who do not understand this principle are the people who write grammar books.

18

Cause/induced motion: intention/ spontaneous motion
DAVID PREMACK

Introduction

The human infant, according to a model (Premack 1990) I shall follow here, uses an analysis of motion to divide the world into two kinds of objects, those that are intentional and those that are not. The objects the infant interprets as intentional start and stop their own motion, whereas those not perceived as intentional, move only when acted upon by another object. 'Start and stop their own motion' is shorthand for a psychophysical account of the stimuli we need. Fortunately, this qualitative description suggests the distinction we require; it gives an account of the intuitive distinction that could be illustrated with appropriate videotapes. One could be shown abstract objects, such as balls that, in one case, never move, and in the other, not only move (starting and stopping recurrently), but move in goal directed ways.

Objects that do not themselves move spontaneously can none the less be moved when acted upon by another object. The infant interprets motion of this kind — motion induced in a non-intentional object — as cause. The infant's interpretations are automatic, the interpretation of spontaneous motion as 'intentional' no less than its interpretation of (appropriate) induced motion as 'causal'. Intention can be seen as the imputation of an internal cause; an explanation of motion that lacks an external cause.

On this view, cause and intention are parallel concepts and are therefore profitably discussed together. Let us begin by defining object and event, since both are presupposed by the concepts of cause and intention. An object is an item that is extended in time and space, and whose parts move together. An event is an object that instantiates a change, either in state, location, or both. Thus, an apple is an example of an object, while an apple that turns red is an example of an event (change in state), as is one that falls from a tree (change in location), as is one that is eaten (change in location and state). An event, therefore, is simply a non-stable object, one 'photographed' while undergoing a change. There may be objections to this proposal on grounds that not all the changes that constitute events are instantiated by objects, for example, the transition from silence to sound. Even in such a case, however, there is an implicit object — one that is the locus of the sound and is responsible for the auditory change.

Cause: natural and arbitrary

Let us start with a discussion of cause, before moving to intention. We may distinguish two kinds of cause, *natural* and *arbitrary*. Regrettably, the English language does not make this distinction but conflates the two cases, and in so doing has given rise to or exacerbated a longstanding confusion. For example, a dispute between those who hold cause to be a learned relation and those who hold it as innate. The protagonists are talking about different causal relations — for example, Michotte (1963) (a proponent of causality as an innate relation), is talking about *natural* causality; Hume (1748) (a proponent of causality as a learned relation), about *arbitrary* causality.

Natural causality concerns the relation between a special, highly restricted set of events. Let us examine one example of what is meant by natural causality: the collision of a moving object with a stationary one, followed by the movement of the stationary object. The relation between these two events — a moving object setting a stationary one into motion — is a prototypic example of natural causality. Such relations have the following special properties:

1. A natural relation is one for which there is an underlying theory that explains the relation, as for example, Newtonian physics explains the launching of one object by the impact of another. (This particular case is of additional interest because humans interpret it as causal on the basis of a single perception.) Unfortunately, this condition may not apply to all natural causal relations; some may not be directly perceived. For instance, human infants may not interpret thermal transformations — colour changes produced by heat; viscosity changes produced by cold — as causal on the basis of a single perception. There is a general need to study natural causal relations in domains removed from Newtonian physics in terms of perception based on single versus multiple exposures.

2. Although one-trial 'perceivability' cannot be made the defining condition of a natural causal relation, we can still exploit this condition (it applies to some natural causal relations, never to arbitrary ones). In some cases, the perception of a *single* relation of this kind is sufficient to produce the interpretation of causality; the interpretation does not depend on *repeated* associations of the events in question.

3. In the one-trial case, the perception leads inexorably to the interpretation, i.e. the interpretation is automatic or hard-wired.

4. Humans (and perhaps other species) believe natural causal relations to be permanent. If a moving object strikes a stationary object, causing it to move, the observer expects such a cause to have the same effect on all subsequent occasions. If by some laboratory contrivance we arrange that it does not, the observer will not construe the outcome as evidence against natural causality. No experience — nothing short of chemical or surgical

intervention — can divest the human (and perhaps other species) of a belief in natural causality.

5. The perception of causality is accompanied by a distinctive subjective experience; there may even be a preference for the perception of natural causal relations.

In marked contrast, *arbitrary causality* concerns the relation between events that are in no way special. Indeed, one could put all the events in the world into a bowl, draw pairs of the events at random, and arrange any such pair to provide the basis of an arbitrary causal relation. While the pairs of events making up natural causal relations must be chosen with utmost care, those forming the basis of arbitrary causal relations may be chosen blind.

For example, one could use the movement of an object as one event, a gun shot as another; or use the gun shot as cause and the movement of an object as effect; or use a lever press as one event, the appearance of a pellet of food as the other; or the appearance of a pellet of food as one event, a gun shot as the other; etc. There is no restriction on either the kind of events that can be paired, or the role that they can play; within the same causal pair, an event A could occur either as the cause of event B, or as caused by B. Such relations have the following special properties:

1. An arbitrary causal relation can be constructed by pairing any two events in either time or space. It is possible that a proportionality in intensity between two events will have the same effect as temporal or spatial contiguity, but the former is more difficult to arrange (Watson 1966). In order to pair two events in time or space, both events can be constant, whereas to demonstrate covariation in their magnitude, both events must vary in magnitude and vary sufficiently to make their covariation evident. Proportionality of intensity has been little studied; almost all arbitrary causal relations are based on temporal contiguity. There is, however, some work showing that spatial contiguity can be substituted for temporal (Watson 1966, 1979).

2. One instance of temporal or spatial contiguity between two events does not lead to an interpretation of causality; such events must be paired repeatedly.

3. Humans (and perhaps other species) do not believe that arbitrary causal relations are permanent. Although a human would be surprised if shown the breakdown of such a relation, his surprise would readily habituate. The breakdown of arbitrary causal relations is accepted, while that of natural causal relations is not.

Infants interpretation of natural causality

How do infants react to natural causality? Do they interpret it as does the adult? Leslie and Keeble (1987) sought to answer this question by using habituation/dishabituation with 6-month-old infants. They exposed one group to the case in which a moving object struck a stationary one, propelling it forward — 'direct launching'; and a second group to a similar case except that the moving object stopped short, and the stationary object did not move except after a brief delay — 'indirect launching'. Thus, in 'direct launching' the critical events were both temporally and spatially contiguous, whereas in 'indirect launching' they were neither. After habituating each group to its respective case, they tested both groups on a reversal of the action.

Rather than have object A strike or approach object B (sending it forward either immediately or after a brief delay), object B now struck or approached object A (and A moved forward either immediately or after a brief delay). Reversal had a significantly greater effect on 'direct' than on 'indirect' launching. That is, infants in the direct launching group looked for a longer period when shown the action reversal.

This outcome can be used to defend of the claim that a 6-month-old infant interprets the perception of direct launching as does an adult. What do we mean by 'interpret' or 'interpretation'? Consider two examples.

Contrast the effect of reversal on 'Mary *pushes* John down the slide' with that of 'Mary *precedes* John down the slide'. After habituating one group of infants to 'Mary pushes . . .', and another to 'Mary precedes . . .', we test both on reversal of the action. John *pushes* Mary down the slide in one case, and John *precedes* Mary down the slide in the other. We can be reasonably confident that the reversal of 'push' will produce greater dishabituation than the reversal of 'precede'. Why?

Because the perception of 'push' will lead to *interpretation* whereas the perception of 'precede' will not. Precede is no more than a temporal relation, one of indeterminately many physical distinctions that many species can discriminate. In the case of push, by contrast, an infant not only perceives the action but interprets it; he will interpret the action as *intentional*.

We find the same distinction in an experiment by Roberta Golinkoff (1975) in which she contrasted donorship with spatial location. Infants were first habituated to a scene in which Mary gave Jane an apple. Then, half of them were shown a reversal of donorship (Jane now gave Mary the apple), whereas the other half were shown a reversal of spatial location (the left–right position of Mary and Jane was interchanged but donorship was not). The results corresponded with those predicted for push versus precede: Reversal in donorship produced greater dishabituation than reversal in spatial position.

Donorship, who gives to whom, belongs to the same domain as does push; it implicates the same concepts, intention, and other mental states that humans attribute to social agents in accounting for their action. Left–right position, on the other hand, is like temporal order in that, unless otherwise specified, it is a theoretically neutral physical distinction, one of indeterminately many that every species makes. It is a perceived, not interpreted, distinction.

The Leslie–Keeble (1987) outcome can be understood by the same principle. Causality is to physical objects what intention is to social objects. It is the major interpretation (the conceptual primitive) that applies to the domain of physical objects. One perceives a causal relation between colliding objects only when there is spatial and/or temporal contiguity between the impact of one and the movement of the other. If there is both a temporal and spatial gap between the events, as there was in the case of 'indirect launching', one perceives only temporal order (one object moved before the other), not a causal relation. Temporal order is only a physical distinction; hence, its reversal produces less dishabituation than does that of a causal relation.

We can formulate two general rules that will help identify interpreted relations. *First*, the reversal of an interpreted relation will produce greater dishabituation than the reversal of an uninterpreted one (the case already described). *Secondly*, the effect of reversing an (old) relation will be greater than that of presenting a *new* instance of the relation only in the case of *interpreted* relations; in the case of *uninterpreted* relations, a new case and the reversal of an old one will produce equal amounts of dishabituation.

For example, the dishabituation produced by changing from A pushed B to C pushed D (new case) will be less than that produced by changing to B pushed A (reversal of the old case); whereas changing from A preceded B to C preceded D (new case) will not be less than that produced by B preceded A (reversal of the old case). Only in the case of interpreted relations does reversal maximize dishabituation.

In a second test, Leslie and Keeble (1987) again habituated two groups of infants, one to direct launching and the other to indirect, and then tested both, not on reversal but on disruption of the action. That is, although object A struck or approached object B, B did not move forward, either immediately or after delay. The effect of disruption was like that of reversal, greater in the case of direct launching than in the case of indirect.

This outcome supports the same claim. Natural causal relations (of which direct launching is an example) are thought to be permanent; whereas mere correlations between events (of which indirect launching is an example) come and go. The disruption of a relation regarded as permanent should produce greater dishabituation than the disruption of one regarded as temporary. In brief, the effects of both reversal and disruption indicate that 6½-month-old infants not only perceive direct launching distinctively but also interpret the perception, and in a manner compatible with the adult interpretation.

Two senses of causality or only one?

Is the sense of causality that applies to the natural case different from that which applies to the arbitrary case? Or is there one sense of causality only? The fact that a single word is applied to both cases suggests that the interpretation is the same in both. Granted the difference in aetiology of the two —

genes in one case, experience in the other — once the arbitrary case has been formed, it may not differ from the natural one. The neural basis of an arbitrary association (once formed) and of a natural causal relation may be indistinguishable.

One way to find out is to repeat the Leslie–Keeble (1987) tests on infants, while at the same time introducing a third condition, an arbitrary causal relation. Leslie–Keeble (1987) provide the comparison of a natural causal relation with and without temporal and spatial contiguity; we require such a comparison be made with an arbitrary causal relation. Infants shown repeated pairings between two arbitrary events, should form a causal relation between the events. The subsequent disruption or reversal of this relation should occasion dishabituation. The magnitude and pattern of dishabituation should be the same as in the natural case — if the interpretation of causality is the same in both cases. Major differences in dishabituation would indicate the contrary, differences in the interpretation of causality in the natural and arbitrary case.

Are learning and arbitrary causality equivalent?

What we have called 'arbitrary causal relation' and 'learning' are different names for the same thing. The pairing in time of arbitrary events (the basis of arbitrary causality) is the basis of learning — it is found in classical and instrumental conditioning, the fundamental procedures of learning. In classical conditioning, an individual is given an auditory or visual stimulus, followed by an arbitrary event (e.g. electric shock, puff of air to the eye, meat powder or weak acid introduced into the mouth). In instrumental conditioning, an individual presses, say, a lever resulting in an arbitrary event (e.g. a pellet of food, electric shock, etc.), the response and ensuing event being paired exclusively in time. An individual experiencing events paired in this fashion, forms an association between the events. But what is the nature of the association and does the individual put an interpretation on it?

Interestingly, when humans are asked to make judgements of causality in arbitrary causal cases, their judgements are sensitive to the same parameters that affect learning in non-human species (Shanks and Dickinson 1987). As Dickinson and his associates have shown, parameters that 'lead' rats to press a lever that is followed by food also 'lead' humans to say that pressing the lever caused food to appear. Should we propose that rats — and all of the many others species capable of conditioning — also make causality judgements, and if they could talk, they would tell us as much? Does the study suggest that the interpretation of causality is the basis of learning?

The ability to make causality judgements, however, does not depend on language. If rats are capable of such judgements, they can 'tell' us as much even if they cannot talk. If an individual has a concept and can use it instrumentally, he can respond differentially to positive and negative exemplars of the concept. Thus, in the present case, we simply require the rat to respond one way to a positive instance of a causal relation, a different way to a negative instance.

The ability to use a concept instrumentally makes demands that exceed those of the habituation/dishabituation test. Individuals who can be judged 'to have a concept' on the basis of habituation/dishabituation data often fail to use the concept instrumentally. For instance, 18-month-old chimpanzees can detect the sameness/difference not only of objects but of relations. When habituated to pairs of like or unlike objects, e.g. AA or CD, and then tested on either BB or EF, they respond more to BB when it is preceded by CD than AA; and more to EF when it is preceded by AA than CD (Premack 1988a). Thus, they respond more to the heterogeneous cases (same followed by different, different followed by same) than to the homogeneous ones (same followed by same, different followed by different); indicating an ability to recognize the sameness/difference of relations. Nevertheless, they can not match either like or unlike pairs of objects. When given match-to-sample tests with the same stimuli used in the habituation/dishabituation tests (e.g. either AA or CD as sample, BB and EF as alternatives), they cannot match AA to BB or CD to EF. The passive recognition of a distinction, registered by habituation/dishabituation, is not equivalent to the ability to make instrumental use of the distinction as in judgement or choice. Thus, rats may pass an habituation/dishabituation test for causality, but fail one requiring differential responding to positive and negative exemplars of causality. Even if rats could talk, we should not expect them to make causality judgements.

Genesis of causality

One is tempted to view the natural case as the genesis of our idea of causality, and to treat the arbitrary causal case as a degeneration of the natural. Reasonably enough, for the natural case involves (among other things) a relation of time, space, and intensity; whereas only time remains in the typical arbitrary case, space and intensity having been discarded. However, it would be a serious mistake to construe the relation between natural and arbitrary cases in this light. Almost certainly, the evolutionary course was quite the opposite; the arbitrary case of 'cause' preceded the natural one.

Although all species are sensitive to the arbitrary causal relation (i.e. to learning), not all are sensitive to the natural causal relation, nor can they distinguish it from the arbitrary one. Indeed, for most species, I suspect, the category 'natural causal relations' is a null class; all associations are arbitrary and learned. We should be able to detect such species on grounds that they have no special 'event pairs', events that are 'associated' in one trial. For instance, were we to show a bee, fish, bird, rat, etc., a moving object strike a stationary one, propelling it forward, which of these species would immediately recognize the causal relation between the two events? Which will require repeated trials to form an association between these events (quite as in the arbitrary case)?

Our failure to find special pairs for certain species might simply reflect ignorance, of course, an inability to identify what is 'natural' for the species. This is not an irrelevant consideration, for even in the human case the concept of 'natural' is problematical. For instance, when a moving object strikes a stationary one, launching is not the only 'natural' outcome. The stationary object could shatter to bits, as it would if the object were made of glass. Is this natural? could an infant recognize it as such?

One object will not, in fact, launch another, unless a number of conditions are met. The stationary object must be of a certain weight relative to the moving one, not be attached to the surface, not be fragile, etc. All natural cases presuppose default values; which is to say, the infant's default values for the Leslie–Keeble (1987) example must be appropriate, or it would not react as it does.

The inability to find natural cases for many species is not likely to be explained simply by ignorance. The sea snail and the human share at least some aspects of classical conditioning, both forming an association — whose exact nature need not be the same — between arbitrary events that are repeatedly paired in time. Despite this similarity, we do not expect to find sensitivity to natural causal relations in aplysia, or in most of the other species that also share with the human a capacity for classical conditioning or simple learning. Indeed, we must contemplate species that lack not only the concept of natural causal relation, but all concepts. These species have no interpretative level of any kind. Like all species, they have the capacity to form associations between events that are related in time; learning does not depend on a conceptual or interpretative level. They have, however, no higher-order level, no level in which the associations formed can serve as inputs to units that produce interpretations as outputs. The whole interpretative level is lacking; it has not yet evolved in the case of certain species.

The advantage of the interpretative level: causal reasoning

What is the advantage of having concepts and an interpretative level? — learning does not depend on either. The answer may lie in the kind of problem-solving that such a level makes possible. Causal reasoning, which is the basis of explanation and thus a major hallmark of human mentation, depends on this level. Causal reasoning is difficult to find in non-human species. To demonstrate an approximate of causal reasoning, one must, even with the chimpanzee, extract the causal sequence from its natural setting, and present the isolated elements in a linear format. For example, physical causality can be depicted by a three-item sequence: an object in a base condition, an instrument that can transform the object, and the transformed object, e.g. an intact apple, a knife, and a cut apple. When given incomplete sequences of this kind chimpanzees can choose the missing item from a set of alternatives (Premack 1976). But when the causal sequence is not extracted from its natural setting, and the animal must, itself, both identify the critical elements and recognize the relations among them, it fails.

For example, four juvenile chimpanzees were tested in a simple two-step reasoning problem. First, they were trained to run down a path to a consistently positive, hidden goal-object. When their running speed stabilized, we interpolated negative trials, replacing the food with a rubber snake. The unpredictability of the negative trials profoundly changed the character of their run. They no longer dashed full speed to the goal, but slowed midway, approaching the goal hesitantly.

In the second step of the experiment, we offered the animals the opportunity to reason — to escape the uncertainty of the negative trial using the affective state of a conspecific to infer the condition of the goal box. Before starting a run, each animal was put into contact with an informant, a conspecific that had just finished a run. The informant was in either a positive or negative state (depending on whether it had encountered food or a snake at the end of the run). The informants' state was, we have reason to believe, successfully communicated to the recipient. Blind or uninformed human judges shown videotapes could discriminate not only the positive and negative states of the informants, but also the state of the recipients following their contact with informants in either a positive or negative state.

Although humans would profit from a similar experience — readily infer the goal-box condition from an informant's state, using this information to decide whether to accept or reject the opportunity to run — the chimpanzees did not. They accepted all opportunities to run, doing so in the same hesitant way whether: (i) the informant was in a positive state, (ii) a negative state, and (iii) on control trials, when they had no contact with an informant at all (Premack and Premack 1994).

Before testing the chimpanzees, we assumed the problem was a simple one, and could be used as a base condition on which to impose variations that would permit answering fundamental questions about reasoning. For example, it had been our plan to expose recipients to informants that did and did not have the experience of running on the path. How would the recipient respond to an 'informant' that it knew never ran on the path? Would it react in the same way to any 'happy' conspecific, or would it take the 'origins' of an emotional state into account, and be affected only by individuals it had grounds for relating to the goal condition of the path? What would constitute *grounds* for a chimpanzee? We were not able to address this question, or to study the character of chimpanzee reasoning because we could not find a minimal condition in which to obtain reasoning in the chimpanzee (see Premack (1983), and Premack and Premack (1994) on chimpanzee reasoning).

Why is reasoning so difficult? What does it require that learning does not? To answer these questions, consider a case of observational learning in monkeys that is essentially analogous to the above reasoning problem. Two monkeys are seated across from one another, one the model, the other the observer. Both have two buttons, one leading to electric shock, the other to food. The observer benefits from watching the model, quickly learning to avoid the button that leads the model to shriek and grimace.

In the monkey experiment, the observer can see not only the model push one button or the other but also its subsequent reaction to having done so; thus the model's response and reaction are observable and contiguous. Neither of these basic conditions obtains in the reasoning experiment. The chimpanzee could not see the model's response — its run down the path — nor was there temporal contiguity between this response and its outcome. Instead, reasoning requires that the chimpanzee, never actually seeing the informant's run, imagine, or infer the run. In addition, the chimpanzee did not contact the informant in the immediate vicinity of the path but in a small room near the path.

Hence, whereas learning requires only that two temporally contiguous events be associated, reasoning requires that an imaginary event be associated with an experienced event. The difference between the conditions of learning and reasoning is far greater than meets the eye!

Consider a three-step progression from simple learning to reasoning. First, in simple learning, two events must be temporally contiguous; essentially all creatures are sensitive to this condition. Secondly, the individual itself need not experience the temporal pairing of the two events, it needs only to observe them as experienced by another. This must not require much beyond simple learning for, although observational learning has traditionally been associated with monkeys, it has recently been reported in the octopus.

Thirdly, reasoning extends beyond observational learning, relaxing all the constraints of learning. While, as in the case of observational learning, an individual need not itself experience the temporal pairing of the two events; more important, the individual need not even observe both the events. One of the two events — the response in the present example — can have an imaginary or inferred status. And the other event, the 'reaction' to the response, can occur at both temporal and spatial remove.

In the course of reasoning, an individual must picture one event while experiencing another, e.g. picture an informant running down the path while experiencing (or being exposed to) its emotional state. But what are the temporal constraints on associating an *imagined* event with an experienced one? They are unlikely to be as severe as those for associating actual events.

Reasoning escapes the narrow constraints of learning, enabling imagined events to be associated with others that are temporally and/or spatially removed. In all likelihood, reasoning depends on a capacity for mental models, for example, the ability to picture a physical circumstance — the path and its goal box — and another's reaction to the circumstance. But it demands more than the ability to form representations of unperceived events.

Reasoning depends on a disposition to seek the causal origins of 'unexplained' events. An individual who encounters a conspecific in a deflected state, highly positive or highly negative — will reason — or has the possibility of reasoning, only in so far as he is disposed to ask why is he deflected? What caused him to be in this state? Above all else, reasoning depends on recognizing the 'unexplained' event, and being disposed to seek its causal origin. Which of the several requirements for reasoning is weak or lacking in the chimpanzee is not easily decided at this time.

We can find further evidence of the difference between species with and without an interpretive level if we attempt to teach the animal a name for the concept of cause. In principle, one can give a name to causality in the same ostensive way one gives a name to an apple, Jane, red or, anything else that can be exemplified. We present examples of causal action, associate a would-be word with them, and then make the appropriate transfer tests.

If the individual has been taught 'cause' with such exemplars as a person painting a wall, someone cutting an orange, etc., he should be able to apply the word to comparable cases — those in which an individual changes the *state* of an object with the use of an *instrument*. He may be able to apply it more broadly, to cases of comparable change where no instrument is used, to cases in which it is not the state but the *position* of an object that is changed, even to cases in which for the first time the recipient is animate, e.g. to Mary washing her child's face. On the other hand, his transfer may be constrained by entirely trivial details, e.g. he may fail to transfer if the agent is female (all the agents in the teaching exemplars having been male). The transfer data will reveal all — the nature of the underlying conceptual structure, its strengths and weaknesses.

The difference between species with and without conceptual levels may also appear in the capacity for consciousness. Consciousness can be treated as a recursion on knowing — an individual uses information, knows that he uses information, knows [that he knows (he uses information)]. We can equate consciousness with the third state — knows that he knows — and grant it, albeit in lesser degree, to the second state as well.

Of course, every species uses information, but which know that they do? Do chimpanzees? To answer this question we used a two-step procedure (Premack 1986), first teaching four young chimpanzees to use plastic 'words' in a simple way, and then interfering with the information on which their use of the words depended. Each animal was shown a plastic word displayed on a writing board and trained to select one of the three objects that was 'named' by the word. When the animals attained criterion on the task, we introduced a critical step: the writing board was rotated 180 degrees, concealing the word and depriving the animals of information. How did they react to this loss?

One animal acted as though nothing had been changed, choosing among the objects as promptly as ever, performing, of course, at chance level. A second animal performed in the same way, but with great reluctance, whining and 'complaining'. It appeared to 'know' that something was wrong, but not what it was or how to correct it. The third and fourth animals redeemed the species: they proceeded directly to the rotated board, turned it around, and only then made their choice. Shyness or a reluctance to handle the board could not explain the behaviour of the two animals that did nothing. On alternate trials when the board was rotated only 90 degrees — so that the animals could see the word simply by adjusting their posture — they again did nothing. Two of four animals gave clear evidence that they not only *used* information but also *knew* that they used it. When critical information was taken from them, they acted to restore it. Individuals who do this can be said to be aware of the conditions on which their successful behaviour depends.

The several abilities shown by the chimpanzee, but lacking in the monkey, may require a mind that can make copies of its own circuits. In carrying out a causal sequence, such as cutting an apple with a knife, an individual may form a neural circuit enabling him to carry out the act efficiently. But suppose he is not required to actually cut an apple, but is instead shown a representation of cutting, an incomplete depiction of the cutting sequence such as was given the chimpanzees, could he use the neural circuit to respond appropriately, i.e. to complete the representation by choosing the missing element? Probably not, for the responses associated with the original circuit are those of actual cutting; they would not apply to repairing an incomplete representation of cutting. Moreover, the representation of a sequence can be distorted in a number of ways, not only by removing elements as in the chimpanzee test, but also by duplicating elements, misordering them, adding improper elements, or combinations of the above. To restore distorted sequences to their canonical form requires an ability to respond flexibly, e.g. to remove elements, add others, restore order, and the like. Flexible novel responding of this kind is not likely to be associated with the original circuitry (that concerned with actual cutting), but more likely with a copy of the circuit. Copies of circuits are not tied, as are the original circuits, to a fixed set of responses, and they may therefore allow for greater novelty and flexibility. For this reason, it is essential for the mind to be able to make copies of its own circuits.

In brief, the interpretive level may be associated with several hallmarks of human mentation: causal explanation, concepts that can be named and used instrumentally, and consciousness or the recursive use of knowing.

Intention

As suggested earlier, an infant divides the world into two kinds of objects, those that are — or appear to be — self-propelled, and those that move only when acted upon by another object. The brain may have separate circuits for dealing with the two kinds of events. If the infant is shown objects that move only when acted upon by another object, the circuits for intentional objects will be inhibited; all computations will be made by circuits for *non*-intentional objects. Moreover, these will concern only parameters of a certain kind, e.g. speed, distance, force, and the like.

By contrast, when the infant is shown objects that start and stop their own motion — and move in goal-directed ways — the circuits for non-intentional objects will be inhibited. Computations will be made by circuits for intentional objects, and will concern parameters different from those above. For example, the circuits will compute charge or valence, i.e. whether the action of one object on another is positive or negative. The simplest criterion used in assigning valence is intensity. Gentle motions, one object rubbing another, are coded positive; hard motions, one object striking another, negative. More complex formulae, functionally equivalent to 'helping/hurting', are also used in the assignment of charge; and further computations are made on the basis of this assignment. For example, the infant expects that acts of intentional objects will be reciprocated, and that reciprocation will preserve charge, e.g. that if object A acted positively on B, B will act positively on A. By contrast, when a non-intentional object is struck and set into motion, the motion is not coded positive or negative. Circuits for non-intentional objects do not compute either valence or reciprocation.

Comparison of human and chimpanzee theory of mind

In 1978, Premack and Woodruff suggested that the chimpanzee, like the human, has a 'theory of mind', by which they meant '. . . imputes mental states to himself and others . . . states . . . not directly observable . . . that can be used to make predictions . . . about the behavior of others . . . ' (p. 515). What are the implications of having a theory of mind? How does the behaviour of an individual who has such a theory differ from that of one who has none?

Being a social species does not guarantee a theory of mind. Members of social species interact extensively, of course, but social interaction hardly depends on a theory of mind. One individual can act on another for quite different reasons; in the simplest case, to affect what the other one *does*. This is apparently the goal of most species, and as such it makes no requirements on a theory of mind.

On the other hand, one can act not only to affect what another *does* but also to affect what he *believes* (*sees, knows, thinks*, etc.). Acting to affect the other one's state of mind, which is apparently the goal of exceedingly few species, makes strong requirements of a theory of mind.

One cannot distinguish species that do and do not have theories of mind simply by observing them in the field. Acts undertaken to affect what another *does* look no different from acts undertaken to affect what he *believes*. Special tests are required to distinguish the two.

Consider the original test on which the claim for a theory of mind in the chimpanzee was based. An adult chimpanzee was shown brief videotapes of a human actor confronting a problem (e.g. jumping up and down below bananas out of reach overhead), then given photographs of which one represented a solution to the problem (e.g. in this case, the human actor stepping up on to a chair). On nine of ten such problems, the animal chose the correct solution on the first trial.

The ability to choose 'solutions' presupposes the ability to perceive 'problems'. But what is a problem? A videotape is not a problem but simply a sequence of physical events. For the videotape to become a problem, someone must read or interpret it in a certain way. For example, the chimpanzee must 'see' the actor not merely as jumping up and down below the bananas, but as *wanting* bananas and as *trying* to get them. When these mental states are attributed to the actor, the videotape changes from a sequence of physical events into the representation of a problem. At this point, the chimpanzee can examine the alternatives offered it and choose the one that constitutes a solution to the problem. In brief, an individual who systematically chooses 'solutions' when given these tests can be said to have a theory of mind. For the consistent choice of solutions presupposes the identification of 'problems', and this in turn the attribution of states of mind to the actor.

Not everyone shown the videotapes chooses 'solutions'. Over 50 per cent of 3-year-old children chose, not solutions, but photographs of items physically resembling salient items on the videotape. For example, they chose a yellow bird, presumably because it was the same colour as the bananas in the videotape. While choosing 'solutions' requires that the videotape be interpreted, choosing 'physically similar items' does not. Children of this age either do not interpret the videotape, or more likely, the interpretations they make are overriden by the disposition to put like things together. Sorting, or putting similar things together is a strong disposition in young children, far more so than in the chimpanzee (Premack 1988a).

A popular objection to this argument is that the animal is not attributing states of mind at all but is simply choosing the event that comes next (e.g. Bennett 1989). The animal is familiar with the problem, understands how it should be solved, and therefore chooses the photograph showing the next step in the solution. Although this appears to be a seemingly plausible argument, it can easily be faulted, because the concept of 'next', on which the argument turns, is not as simple as it may seem.

For example, suppose we show the individual a videotape in which the actor (again) repeatedly jumps to obtain bananas, intermittently hitching up his pants while doing so. We thoroughly familiarize the subject with the videotape and then test him as follows. We stop the videotape at exactly the point where the next event depicts the actor hitching up his pants, and then offer him three alternatives: (1) the actor hitching up his pants (next), (2) stepping up on to a chair (relevant), (3) reaching out with a stick (irrelevant).

Three- to five-year-old children, given six different videotapes of this general kind, chose the relevant alternative overwhelmingly (Premack and Dasser 1991). This outcome was the more impressive because the children were, in fact, told to choose 'What comes next'. Nevertheless, fewer than 10 per cent of the children followed the instruction. Most children, rather than choosing the 'next' act, chose the relevant one, and did so on all problems.

Relevance is defined, of course, by the state of mind attributed to the actor. Thus, stepping up on to a chair is a relevant act provided one believes the actor *wants* the bananas; if one thinks he is jumping up and down below the bananas as a form of exercise, then stepping up on to a chair is not a relevant act.

In a more recent test, given to both children and chimpanzees, the subject was faced with a choice between two containers into one of which food was placed. However, the subject could not see which because a barrier blocked their view. What the subject could see instead were two onlookers standing near the containers. One of them was in exactly the same predicament as the subject; his view was also blocked by a barrier so that he could not see where food was placed. By contrast, the other onlooker had an unobstructed view and could easily see where the food was placed.

Before the subject chose between the containers, she or he was given the opportunity to seek the advice of the onlookers. The children chose by simply pointing to one individual or the other, the chimpanzees (who do not point) by pulling a string attached to each onlooker. In both cases, the chosen onlooker pointed to a container, the correct one if he was the 'clear view' onlooker, the incorrect one if he was the 'blocked view' onlooker.

Three of four young chimpanzees chose the correct individual from essentially the first trial (Premack 1988*b*); whereas to our considerable surprise, children below 4 years of age performed at chance (V. Dasser and D. Premack, in preparation). Another group tested in a similar fashion also performed at chance; these were rhesus monkeys (Povinelli *et al.* 1991). Their failure was not a surprise.

The monkeys' failure on this problem is in keeping with the rest of the data for this species. There is no experimental evidence that the monkey, unlike the chimpanzee, attributes mental states to others. Although social attribution has been claimed for monkeys on the basis of field observation, no such claim has survived actual test. A recent failure is of special interest. Years ago Kummer reported an observation he made in the field, of a young female baboon observed peering over a rock, 'hiding' her body behind the rock, while copulating with an alien male. This anecdote proved to be nearly irresistible, many readers not only accepting the 'fact' of hiding, but adding the interpretation that the young female acted deceptively so that her mate would not *know* what she was doing. Recently, Kummer and his associates have attempted to produce 'hiding' and perspective-taking in the monkey in the laboratory — all their attempts have failed (H. Kummer personal communication).

Hiding, in the human sense, is not a simple act. It presupposes the ability to take into account the other's perspective, to understand the conditions on which seeing depends, and at the very least, to relate seeing to action. It is doubtful that monkeys 'hide'.

Although the monkey has neuronal sensitivity to a variety of social features

—gaze direction, face, facial expression, etc. (e.g. Perrett *et al.* 1982; Perrett *et al.* 1989; Brothers and Ring 1992)—what use it makes of these cues is not well known. Evidently, these cues are not used as a basis for attributing states of mind to the other. Detecting social cues is one thing, *using* social cues as a basis for complex computations is another.

How do we account for the surprising disparity between the chimpanzees and the 3½-year-old children? Anomalies of this kind are meant to teach us something, of course, and this one can offer either of two lessons. First, apes solve the problem because, as neoteny teaches us, they mature earlier than children. That is to say, apes and children use the same mechanisms in solving the problem; they make the same attributions, but apes do so earlier than children.

Secondly, apes and children do not solve the problem in the same way but use different mechanisms. They attribute different states, children's attributions being more complex than the ape's. The ape's simpler theory moves directly from seeing to acting, so that the ape has only to distinguish the seeing from the non-seeing onlooker. Moreover, although seeing has subtle determinants (such as attention, which the ape is unlikely to understand), other determinants are less subtle (such as the difference between an obstructed and an unobstructed view, which the ape is likely to understand).

On the child's more complex theory, seeing is necessary but not sufficient for action. Knowing (believing, thinking, etc.) stands between perception and action. Perception, on the child's theory, makes knowledge or belief possible; knowledge or believe *plus* desire leads to action. The ape omits the intermediate epistemic state, that concerned with the representation of what the other one knows. On the ape's theory, perception plus desire leads to action. Thus, when the ape and child choose between onlookers, they do so on the basis of different theories. The ape takes into account whether or not the onlooker has an unobstructed view. Were this the only distinction that children took into account, they too would pass (indeed pass at a still earlier age, for by 3 years of age they can use clear versus obstructed view to distinguish seeing from not seeing, Flavel *et al.* 1981). But the child's theory requires that it take other factors into account, i.e. knowledge of the other one, a factor not yet available to the 3½-year-old child. In brief, the ape not only matures earlier than the child, the theory that matures in the ape is simpler than that which matures in the child. Although apes and children mature at different rates, they cannot be likened simply to vehicles travelling at different speeds. The slower vehicle has a more distant destination.

Our own data support this distinction. For although children did not choose the correct viewer, they invariably took the advice of the onlooker they chose, always approaching the container to which he pointed. By contrast, even though the chimpanzees chose the right viewer, they sometimes ignored his advice. Of the three who chose correctly, only one did so consistently, another followed advice most of the time, while a third rejected it as often as not. This kind of borderline performance makes the chimpanzee an intriguing puzzle. When an animal acts 'intelligently', does it do so for the same reasons as does the human? Indeed, does the animal know why it acts?

Whether or not chimpanzee and child attribute mental states concerning (specifically) knowing and believing was tested more directly (Premack and Woodruff 1978). In one test, the chimpanzee and child were shown a videotape in which an actor confronted four opaque containers; a trainer placed food into one container, and then allowed the actor to make a selection. On some trials, the actor watched the trainer, whereas on others he paid no attention to the trainer. After each such trial, the subject (child or chimpanzee) was given two photographs, one showing the actor selecting the container containing food, another showing him selecting an incorrect container. Both the chimpanzee and 3½-year-old child failed. They did not choose the photograph depicting the actor selecting: (i) the correct container (trials of the actor paying attention), or (ii) the incorrect one (trials of the actor paying no attention).

The chimpanzee's failure was ambiguous. Given the test on parallel videotapes, one in which the actor was a person she liked, the other a person she did not like, the chimpanzee was influenced by the affective factor. If she liked the actor, she chose the photograph depicting the actor selecting the correct container; if she disliked him, the photograph depicting him selecting the incorrect one. This motivational factor, which proved to be surprisingly strong, may have overridden her knowledge (of the actor's knowledge), leading to an underestimation of her ability. However, 3½-year-old children failed this problem as well, and their performance could not be attributed to a motivational factor.

Why did the chimpanzee fail this test while younger animals passed the earlier one in which they correctly chose the trainer with the unobstructed view? Possibly because of the difference between a gross determinant of seeing, such as an unobstructed view, and a more subtle determinant, such as looking in the right place at the proper time.

Additional tests followed — these were aimed at answering whether chimpanzee and child attributed knowledge or belief to others (Premack and Woodruff 1978). They were both shown videotapes in which a 2- and a 6-year-old child confronted the same problems. The problems varied from simple tasks, such as colouring pictures in a book, to more complex ones, such as doing analogies. Following each trial, the subject was shown two photographs, one in which the actor (either a 2- or 6-year-old child) solved the problem accurately or inaccurately. For example, the analogy (and other problems) were answered correctly or incorrectly; the colouring in the book was either tidily within the boundary lines, or outside the lines.

The chimpanzee and 3½-year-old children both failed the tests; they did not choose photographs depicting accurate or tidy solutions for the 6-year-old, inaccurate or untidy ones for the 2-year-old. Both the chimpanzee and children were themselves capable of performing all the problems (except for colouring that stayed within boundary lines). While they selected photographs of both good and bad solutions (the chimpanzee's choices differing somewhat from the children's), the photographs assigned to the 6-year-old did not differ significantly from those assigned to the 2-year-old child.

By 4 years of age, however, children assigned poorer solutions to the 2-year-old actor. Four-year-olds differed from 3½-year-old children and the chimpanzee, not in assigning better solutions to the 6-year-old — but solely by downgrading the performance of the 2-year-old. In fact, the 4-year-old underestimated the abilities of the 2-year-old (such as matching like objects, putting round forms in round holes, etc.). This inaccuracy notwithstanding, the 4-year-old takes the age of the two actors into account, assigning significantly different performances to each. Neither the ape nor the younger child is able to take this step.

Instantiation versus attribution

We have concentrated on the *attribution* of mental states — but how are such states instantiated in species? It seems evident that exceptionally few species attribute mental states. The monkey has intentions and other mental states, for example, but does not attribute these states to others. In attempting to explain the monkey's behaviour, we attribute mental states to the monkey. It is not the monkey who, attempting to explain either our behaviour or its own, attributes mental states; monkeys, so far as we know, do not attempt to explain anything. They do not engage in causal reasoning; even chimpanzees evidently do not.

Why do many species instantiate (have) mental states while exceedingly few attribute them? What is the difference in the demands made by these two conditions? We can answer this question by first establishing what 'having a mental state' requires. Consider, for example, acting *intentionally*. What does this require? How do we distinguish behaviour that is and is not intentional?

The usual criterion is goal-directedness. George Mead (1934), for instance, observed that while the act of sitting on a chair may be quite automatic, if one sweeps the chair of crumbs before sitting down, the act is intentional. However, any act or sequence of acts which, although thoughtful when first planned and executed, can become automatic, sweeping the chair no less than sitting down.

We must determine instead what a person would do if, when sweeping the chair, he/she found that it was not clean. Would he/she decline to sit down? If so, we can conjecture that he/she has a goal — sitting on a clean chair — and that if prevailing conditions do not permit realizing this goal, he/she will not sit down. A minimal test for intentional behaviour requires that we are able to: (i) identify a goal, (ii) stipulate the conditions on which the realization of the goal depends, and (iii) demonstrate that if the conditions are not met, the individual does not carry out the act.

Dickinson and his colleagues (Hayes and Dickinson 1990) have refined this analysis, noting that the conditions on which a goal depends can be under-mined in two different ways. One can remove the conditions on which the goal depends — removing clean chairs will have this effect in the example above — or one can eliminate the underlying motivation — allow the party to sit on a clean chair so often, he/she loses all interest in doing so.

In an interesting series of studies, Dickinson (1989) showed that certain rat and pigeon behaviour was sufficiently 'corrigible' to be called intentional. The animals were deterred from pursuing a goal by either: (i) being satiated on food or (ii) finding that a lever press no longer produced food. Dickinson also cites examples in which animals were not corrigible and thus failed the test for intentionality.

For example, if pigeons are presented food automatically on a schedule unaffected by their behaviour, and food is preceded by the onset of a light, they quickly come to peck the light. Is this behaviour on the part of the pigeon intentional? If so, what is the goal? Perhaps the birds expect, without war-rant, that 'if they peck, food will appear'. To test this possibility, Williams and Williams (1969) changed the delivery of food so that it was no longer unaffected by the birds' behaviour. Now, when the birds pecked the light, food was *not* delivered. If the birds' goal in pecking the light was to produce food, they should stop pecking the light. But they did not. One can therefore argue that the birds were neither goal-directed nor acting intentionally.

If one finds this unacceptable, and wishes to claim that the behaviour was intentional, one must propose an alternate goal, i.e. show that when pecking the light militates against the achievement of this goal, the bird desists. Or else, it seems, one must agree that in this circumstance the bird was not acting intentionally.

One might compound the interpretive complexity of this case by arranging that when the bird pecks the light, while no change is made in the delivery of food, the light is turned off. This change, unlike that of food delivery, may well reduce the bird's pecking. What should we say if it does? The bird's goal is to peck a lighted key (not just any lighted key, of course, but one with a history of having been associated with the presentation of food)? And should we say that the bird is behaving intentionally with respect to this goal?

In describing the conditions on which the apparent instantiation of mental states depend, we must take care not to make these conditions too strong. If they are too strong we will have difficulty explaining why the individual is able to instantiate the state, yet not able to attribute it.

For instance, were we to say that the rat or pigeon *knows* when it is hungry or *wants* food — further, that it *expects* a lever press or key peck will deliver food — would we not expect an animal with all this knowledge concerning its own mental states, to attribute comparable mental states to others?

We can escape this trap in either of three ways. We can argue that self-knowledge is simpler than knowledge of the other one. Hunger, for instance, is more easily detected in self than other. This is a standard position but not, I think, necessarily a good one. Wittgenstein (1953) argues the opposite. All such knowledge, he says, is based on observation, and the self is less easily observed than the other one.

Alternatively, one can distinguish between the representation and meta-representation of a state, holding that the former applies to self, the latter to other. For instance, the state 'Henry wants food' requires a representation, 'Henry thinks John wants food', a representation of a representation. Then it is specifically the difficulty of coping with meta-representations that accounts for the disparity between instantiation and attribution. This point of view, taken earlier by Plyshyn (1978) and Premack and Woodruff (1978), makes the chimpanzee's achievement all the more notable. It credits the animal with meta-representation.

Finally, one can argue that the would-be 'knowledge' states of most species are not based on representation of any kind. Strictly speaking, most species not only do not attribute mental states, they do not even instantiate them. The learning demonstrated in aplysia, for example, is not based on represen-tation of any kind, there is simply no capacity for representation in this species.

But if we adopt the view that the behaviour of most species is caused by neural states that do not entail representations of any kind, where does one draw the line? How does one separate species for which this is the correct view from those for which it is not?

Conclusions

1. Although there is an enormous amount of information 'in the world', not all species are equally able to read it. Humans can use differences in 'motion' to distinguish physical objects from intentional ones. A physical object moves only when acted upon by another object, whereas an intentional object starts and stops its own motion. Physical objects do not pursue goals, whereas intentional objects do.

2. The infant has, we speculate, separate circuits for the two kinds of motion and they compute different parameters. For instance, valence is computed for intentional objects. The action of one intentional object on another is coded positive or negative, in the simplest case, on the basis of the intensity of the movement; in more complex cases on the basis of formulae functionally equivalent to 'helping' and 'hurting'. In addition, reciprocation is computed in the intentional case, i.e. the expectancy that if A acted positively (negatively) on B, B will act positively (negatively) on A.

3. None of the above parameters is computed in the case of non-intentional objects. When a moving object strikes a resting one setting it into motion, the human infant perceives the temporal/spatial contiguity between the two events, and interprets the perception as causality in an evidently adult man-ner. The interpretation of causality and of intentionality are both based upon an analysis of movement, and both are automatic or hard-wired.

4. We distinguish between two kinds of causal relations, *natural* and *arbitrary*. The natural relation is one for which there is an underlying theory that explains the relation, as for example, Newtonian physics explains the launching of one object by the impact of another. In some instances, humans interpret such a relation as causal on the basis of a single perception. By contrast, an arbitrary causal relation is composed of *any* pair of events presented together in time. The relation is not interpreted as causal unless repeated; humans do not believe the relation to be permanent; the relation is not explained by an underlying theory. There are no restrictions on the kind of events that can be paired or on the role they can play; event A can serve as the cause of event B in one pair, or as caused by B in another.

5. 'Learning' and the 'arbitrary causal relation' are identical: arbitrary events are paired in time (no less in instrumental conditioning than in classical). Humans not only interpret instrumental conditioning causally, their causality judgements are affected by the same parameters that affect the conditioning of rats and pigeons (Shanks and Dickinson 1987). Interpretation depends not on language but on the ability to use concepts instrumentally. This would require that the rat or pigeon respond differentially to positive and negative exemplars of causality; there is no evidence for the instrumental use of this or any other concept in these species. Even if rats or pigeons had language, they would not make causality judgements. The 'interpretive level' (in which events are not only perceived but given a conceptual interpretation) are to be associated with several hallmarks of human mentation: causal explanation or reasoning, concepts that can be named and used instrumentally, and consciousness or the recursive use of knowing.

6. Consider a three-step progression from simple learning to reasoning. First, in the simplest learning, two events must be temporally contiguous, a condition to which all creatures are sensitive. Second in observational learning, the individual need not experience the temporal pairing of the two events but need only observe them experienced by another. Third, in reasoning (as in observational learning) the individual need not experience the temporal pairing of the two events; it need not even observe them. One of the two events can have an imaginary or inferred status, and the other can occur at a temporal and spatial remove. Thus, whereas the efficacy of learning is restricted to actual, directly experienced, contiguous events; reasoning escapes all three constraints. Although largely restricted to humans, weak exemplars of reasoning can be found in chimpanzee.

7. Most species act to affect what the other *does*; few to affect what the other *wants* or *believes*. Species acting to affect the mental state of others have what is called a 'theory of mind'. The presence of such a theory can be established only by test, not by field observation. Monkeys, although neuronally sensitive to 'social cues', evidently do not use these cues to make inferences about the mental state of others, for they fail all tests for theory of mind. The chimpanzee, although easily distinguished from a 4-year-old child, passes at least some of the tests.

References

Bennet, J. (1989). Thoughtful brutes. *Am. Phil. Assoc. Proc. Addr.* **62**, 197–210.

Brothers, L. and Ring, R. (1992). A neuroethological framework for the representation of minds. *J. Cog. Neurosc.* **4**, 107–18.

Dickinson, A. (1989). Expectancy theory in animal conditioning. In *Contemporary learning theories: Pavlovian conditioning and the status of traditional learning theory* (ed. S.G. Klein and R.R. Mower). Erlbaum, Hillsdale, NJ.

Flavell, J.H., Everette, B.A., Croft, K., and Flavell, E.R. (1981). Young children's knowledge about visual perception: Further evidence for the Level 1–Level 2 distinction. *Dev. Psychol.* **17**, 99–103.

Golinkoff, R.M. (1975). Semantic development in infants: The concept of agent and recipient. *Merrill-Palmer Quart.* **21**, 181–93.

Heyes, C. and Dickinson, A. (1990). The intentionality of animal action. *Mind Lang.* **5**, 87–104.

Hume, D. (1748). *An enquiry concerning human understanding.*

Leslie, A.M. and Keeble, S. (1987). Do six-month-old infants perceive causality? *Cognition* **25**, 265–88.

Mead, G.H. (1934). *Mind, self and society: From the standpoint of a social behaviorist.* Chicago University Press, Chicago.

Michotte, A. (1963). *The perception of causality.* Methuen, London.

Perrett, D.L., Rolls, E.T., and Caan, W. (1982). Visual neurones responsive to faces in the monkey temporal cortex. *Exp. Brain Res.* **47**, 329–42.

Perrett, D.L., Mistlin, A.J., Harries, M.H., and Chitty, A.J. (1989). Understanding the visual appearance and consequences of hand actions. In *Vision and action: the control of grasping* (ed. M.A. Goodale). Ablex, Norwood, NJ.

Povinelli, D.J., Parks, K.A., and Novak, M.A. (1991). Do rhesus monkeys (*Macaca mulatta*) attribute knowledge and ignorance to others? *J. Comp. Psychol.* **105**(4), 318–25.

Premack, D. (1976). *Intelligence in ape and man.* Erlbaum, Hillsdale, NJ.

Premack, D. (1983). The codes of man and beasts. *Behav. Brain Sci.* **6**, 125–67.

Premack, D. (1986). *Gavagai! or the future history of the animal language controversy.* MIT Press, Cambridge, MA.

Premack, D. (1988*a*). Minds with and without language. In *Thought without language* (ed. L. Weiskrantz). Oxford University Press, Oxford.

Premack, D. (1988*b*). 'Does the chimpanzee have a theory of mind?' revisited. In *Machiavellian intelligence: Social expertise and the evolution of intellect in monkeys, apes and humans* (ed. R.W. Byrne and A. Whiten). Clarendon Press, Oxford.

Premack, D. (1990). The infant's theory of self-propelled objects. *Cognition* **36**, 1–16.

Premack, D. (1993). Prolegomenon to evolution of cognition. In *Exploring brain functions: models in neuroscience* (ed. T. A. Poggio and D. A. Glaser) Wiley, New York.

Premack, D. and Dasser, V. (1991). Perceptual origins and conceptual evidence for theory of mind in apes and children. In *Natural theories of mind: evolution, development and simulation of everyday mindreading* (ed. A. Whiten). Blackwell, Oxford.

Premack, D. and Woodruff, G. (1978). Does the chimpanzee have a theory of mind? *Behav. Brain Sci.* **1**, 512–26.

Premack, D. and Premack, A. J. (1994). Levels of causal understanding in chimpanzes and children. *Cognition* **50**, 347–62.

Pylyshyn, Z.W. (1978). When is attribution of beliefs justified? *Behav. Brain Sci.* **1**, 592–3.

Shanks, D.R. and Dickinson, A. (1987). Associative accounts of causality judgement. In *The psychology of learning and motivation* (ed. G. Bower). Academic Press, New York.

Watson, J.S. (1966). The development and generalization of 'contingency awareness' in early infancy: Some hypotheses. *Merrill-Palmer Quart. Behav. Dev.* **12,** 123–35.

Watson, J.S. (1979). Perception of contingency as a determinant of social responsiveness. In *Origins of the infant's social responsiveness* (ed. E.B. Thoman). Erlbaum, Hillsdale, NJ.

Williams, D. R. and Williams, H. C. (1969). Auto-maintenance in the pigeon: sustained pecking despite contingent non-reinforcement. *J. Exp. Anal. Behav.* **12,** 511–20.

Wittgenstein, L. (1953). *Philosophical investigations*. Macmillan, New York.

DISCUSSION

Participants: B. Latour, G. Lewis, D. Premack

Inappropriate attribution of intentions

Answering a question from **Latour**, **Premack** predicted that an infant, if not a child, would probably attribute no intentions to an anthropomorphic figure whose motion does not show such critical criteria as self-propulsion and valence preservation. Motion information may well override object quality information.

Lewis pointed to the fact that children, as well as adults in other societies, readily attribute intentions to inanimate objects, and asked whether chimpanzees behave in a similar fashion.

Premack replied that if his model is correct, humans should be ready to attribute intentions (inappropriately) to any inanimate object meeting the adequate motion criteria. However, this process operates at a very primitive level and its outputs are not necessarily to be used explicitly. Experimental procedures now exist that enable us to assess implicit as well as explicit competence in the attribution of intentions.

Author index

Subject index